GUIDE TO PAVEMENT MAINTENANCE

GUIDE TO PAVEMENT MAINTENANCE

A Comprehensive Guidebook to Understanding Pavement Maintenance for Parking Lots and Streets

Thomas McDonald and Patrick McDonald

iUniverse, Inc.
New York   Bloomington

Guide to Pavement Maintenance

iUniverse books may be ordered through booksellers or by contacting:

iUniverse
1663 Liberty Drive
Bloomington, IN 47403
www.iuniverse.com
1-800-Authors (1-800-288-4677)

Because of the dynamic nature of the Internet, any Web addresses or links contained in this book may have changed since publication and may no longer be valid.

ISBN: 978-1-4502-0869-7 (sc)
ISBN: 978-1-4502-0868-0 (ebk)

Library of Congress Control Number: 2010912271

Printed in the United States of America

iUniverse rev. date: 8/19/2010

Also by Thomas McDonald

Property Managers' Guide to Pavement Maintenance

Also by Thomas McDonald and Patrick McDonald

"Guide to Bidding Pavement Maintenance"

Contents

Illustrations and Figures

Foreword

This handbook was developed from years of research and experience to help assist in pavement maintenance. Using this guide will teach the reader how to do the following:

- Identify distresses in asphalt pavement
- Determine what it will take to repair those distresses
- Estimate how much it will cost to do the repairs
- Evaluate what benefits will be attained on the investment for the repairs

This book includes several tables and charts designed to help the reader determine the maintenance strategy to repair one distress or a number of distresses. The cost of pavement maintenance keeps escalating upward, especially as refining crude oil technology increases, leaving less liquid asphalt, which is the byproduct from refining. There is also a shortage of raw materials due to shortages of asphalt binder and availability of usable aggregates. Mining permits are becoming harder to obtain for gravel sources, and the reclamation of gravel mine sites is becoming more stringent and costly. As the cost of aggregate bases and asphalt pavements increase, the need for more effective pavement maintenance and more accurate budgeting is needed. As a result, property owners (private and public) and homeowners associations (HOAs) will be spending more funds for pavement maintenance than ever before.

The authors have been in the asphalt pavement construction and maintenance industry for several decades. This handbook was developed from years of research and experience, which includes the following:

- Lead field materials laboratory and testing technician for the Colorado Department of Transportation (DOT)
- Laboratory and quality control supervisor for a heavy highway contractor, major city public works department, and geotechnical consulting engineering firms
- Pavement maintenance specialist for consulting engineering firms
- Construction materials instructor for the Federal Highway Administration
- Development of pavement maintenance plans, budgets, and specifications and project management for homeowners associations, commercial and industrial properties, and multihousing communities
- Owners of a pavement maintenance consulting firm (PMIS, Inc.)
- Authors of several nationally distributed books and publications on pavement maintenance
- Graduates of University of Colorado in business finance and the Del Webb School of Construction at Arizona State University, College of Engineering

This step-by-step book is written and developed for anyone involved in using and maintaining asphalt pavement. This handbook can be used by property owners or managers, maintenance or operations engineers, inspectors, contractors, engineers, and architects. There are instructions on how to identify distresses, establish maintenance budgets, and develop scopes of work and specifications for pavement maintenance projects and a general troubleshooting table for each distress. Formulas are included to aid in calculating areas, and curing characteristics for each maintenance application are included. These curing characteristic documents will aid in explaining how each application cures after being installed and what can be expected after the application material has set up and cured.

A Web site has been developed at www.pavementmaint.com, which presents some articles on various maintenance problems. Should there be problems involving structural damage requiring further design standards or geotechnical failures and construction defects, please contact a geotechnical consultant or qualified pavement consultant (such as PMIS). Should drainage problems exist, please consult a licensed and certified drainage engineer who is a registered professional engineer.

Warning!! Recent studies by the United States Geological Survey (USGS) have shown that coal tar seal coats contain high amounts of polycyclic aromatic hydrocarbons (PAHs). PAHs have been found to be carcinogenic and heavy exposure may cause cancer. As a result, several cities (including Austin, TX and Washington DC) have banned the use of coal tar sealer. Therefore, the contractor and the owner/manager can be subjected to any claims that may arise due to the use of coal tar sealers. As a result, it is the authors' recommendation that coal tar sealers not be used. Because of its ability to resist fuel spills and endure cold weather climates, cold tar is still in demand, therefore, information on coal tar is still included in this book. It is important to check and understand any federal, state, and local rules and regulations prior to using coal tar. Ultimately, it is recommended that asphalt-based seal coats (which can be modified to meet specific environmental needs) be used in lieu of coal tar sealers.

Preface

This handbook is designed as a reference guide to assist property owners and managers, contractors, estimators, inspectors, engineers, architects, and consultants in preparation and optimal utilization of pavement maintenance applications and budgets for pavement maintenance. Using this handbook will enable persons in the listed categories to become familiar with identifying distresses, learn how to repair identified distresses, and develop a maintenance plan and budget to repair those distresses. The information herein is a suggestive methodology of maintenance applications, developing budgets, developing proper bidding, and using contract documents with defined scopes of work and specifications. This handbook is not intended to be the empirical solution to all pavement maintenance problems; however, if used correctly, it can aid in proper selections of correct maintenance applications and development of proper budgets.

In this handbook, readers can identify the types and severity of distresses that exist in their streets and parking lots, learn how to properly repair them, and estimate how much they can expect to budget or spend. Also, the reader can learn how to correctly bid the work to be performed and develop a binding contract to enforce the scope of work and specifications. This handbook contains all the tools needed to develop a proper distress inventory and surface areas with formulas and layout instructions. An appendix is also contained to assist the reader in troubleshooting problems that can arise before, during, and after maintenance applications are applied.

It is not the purpose of this book to provide any professional designations or licensing unless authorized by any organization, association, or licensing agency designated to authorize use of any part of or the whole book as such. For any professional designations or licensing requirements, please contact local chapters of property management associations or governing agencies.

Introduction

Asphalt was first used in ancient times on Persian and Roman roads, including some portions of the Apian Way, where it was used to bind cobbles, known as *suma crusta*. This use continues today in heating asphalt (a natural by-product of the evaporation of petroleum) and blending it with heated aggregates. This end product is referred to as hot mixed asphalt (HMA) pavement (or asphalt concrete [AC]) and is transported to a project site and placed with special machinery. As it cools, it becomes a solid hard surface for use on roadways, streets, or parking lots.

Asphalt pavement is used in place of Portland cement concrete pavement (PCCP) because it costs less to install and is more cost effective to maintain. Because PCCP costs more to install, there are only a few situations where the need for PCCP's strength offsets its initial cost. Another advantage of PCCP is longevity; however, due to the expense and difficulty of repair, once maintenance is required, the cost advantage swings back into asphalt's favor. If there are constant heavy loads or extensive environmental exposure, asphalt pavement will require continuous heavy maintenance; as a result, the high initial costs of PCCP will be offset. If asphalt will be sufficient in design and is able to perform with the usual maintenance, then it will be far more cost effective than PCCP.

When a highway, street, or parking lot is constructed, it is built from a design developed by a licensed professional, usually a licensed geotechnical engineer. This design is based on several factors involving traffic loadings (especially large heavy trucks), number of loading repetitions, climate, soil classification, "R-value," and overall use. Once this data is collected, a pavement section can be developed. A pavement section consists of an aggregate base course (ABC) and asphalt pavement (AC) placed on existing soil (subgrade), with each layer compacted to a laboratory design and control (Proctor). Proper construction is required to support known loadings and the number of times (repetitions) those loads will travel over the pavement section.

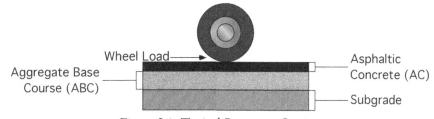

Figure I.1: Typical Pavement Section

Figure I.1 shows a typical pavement section with a wheel loading. The arrows indicate the movement of the loading force. A wheel has one point of loading on the surface. The pavement distributes this loading where the wheel touches the pavement (usually the width of the tire). The ABC is designed and placed to distribute those forces evenly so the subgrade can provide adequate support. If any of these layers are not installed properly, are disturbed by water, or are overloaded or oxidized and aged, pavement fatigue and failure occurs.

Parking lot pavements, roads, and drives are one of the larger capital investments in property management and ownership. After the completion of construction, pavements begin to deteriorate from aging, traffic loadings (especially large trucks), poor compaction and construction, water (natural and irrigation), ultraviolet rays, and other weather conditions. These distresses can be classified as structural or environmental. Structural distresses are those that cause the pavement to fall apart and generally occur from the bottom up. Environ-

xviii ▶ Thomas McDonald and Patrick McDonald

mental distresses are caused by outside influences causing the pavement to wear away and usually occur from the top down. The Asphalt Institute, through extensive research, has shown that restoration of pavements not maintained for fifteen years will cost five times more than pavements that are continuously maintained. Using similar data, and escalating the cost by inflation, shows that pavements maintained on a constant basis over the same number of years yield a substantial savings, as illustrated by the bar graph in Figure I.2. To play on an old cliché, an ounce of pavement maintenance is worth a ton of reconstruction.

Maintenance vs. No Maintnenace

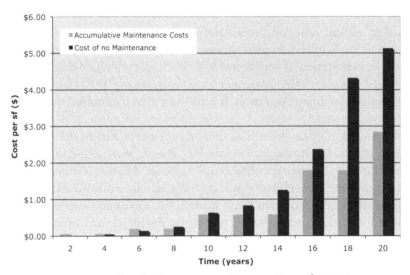

Figure I.2: Accumulated Maintenance Cost vs. Cost of No Maintance

Environmental factors can adversely affect the life of pavement but do not cause major deterioration or rehabilitation costs. However, if the environmental deterioration is not prevented or reduced by maintenance, structural damage can occur. Structural deterioration *is* very costly and, if not repaired, will ultimately require total reconstruction.

Reconstruction and repair of structural failures can be prevented or held to a minimum if the property manager or owner protects against environmental deterioration by establishing an effective maintenance policy. This can be accomplished by inventorying the types and severities of distresses, compiling materials and labor costs, and implementing budgetary planning for a desired period of years. A good rule of thumb or ballpark figure for annual budgeting should be a percentage of the actual cost of the lot. If it is in good condition, 1 to 2 percent per year should be adequate. If the pavement is in fair to good condition, 3 to 5 percent is a good figure to bring the pavement up to a good condition the first year, and then 1 to 2 percent the following years.

A comprehensive inventory of all distresses on a lot (including size and location) is known as a *distress inventory*. Once a distress inventory is completed on the pavement, a maintenance policy can be established. Applying current costs, budgets, and previous maintenance to the inventory, a current and future repair and maintenance schedule can be developed. Distress surveys are discussed further in chapter 4.

A distress inventory and budget planning should be completed no more than six months prior to the work being started. Once the inventory and budget have been established, a repair and maintenance schedule can be implemented. In addition to a timetable, the maintenance plan should include the following:

• A list of preferred contractors
• Materials and specifications
• Request for bids (RFBs) from selected contractors with a specific scope of work to be done
• Specific job specifications

Once bids are submitted, evaluate them carefully to make sure all quantities of materials and time schedules are the same or close to your estimates and other proposals. Also, a plat (or map) of the property with approximate locations of patches and a corresponding table with patch type (skin or R&R) and dimensions should be supplied for reference. This will enable the proposals to be compared accurately and assist in contractor selection.

The purpose of this book is to assist property managers or owners, contractors, and inspectors in methodically establishing distress inventories, maintenance policies, proposal requests, and evaluations; determining materials and labor; establishing specifications; selecting contractors; developing contractual agreements; and finalizing project acceptance. Make this guide a permanent part of any manager's, owner's, contractor's, consultant's, and inspector's library and refer to it often. Please refer to local and federal government and other public printed material for more specifications.

Remember, keep a line of communication open with everyone involved in the project and all material suppliers during and after construction and throughout the warranty period. If in doubt about anything dealing with the project, do not be afraid to ask questions. It is the responsibility of all persons to control the project in the owner's or company's best interests.

Chapter 1

Pavement Distress Inventory

Several exposures affect the longevity of pavements over their design life. If no maintenance is performed, pavements will deteriorate until they convert back to gravel. Exposures that adversely affect pavements include the following:

- Weather conditions (especially snow and ice)
- Deicers for colder climates (chemical and salts)
- Ultraviolet rays
- Water (natural and irrigation)
- Vehicle weight, petroleum (gasoline, oil, hydraulic fluids, antifreeze, brake and power steering fluids, and other petroleum-based products)
- Aging

These exposures affect the pavement from the moment it is constructed to the end of its useful life. There are two categories for distresses:

- Environmental
- Structural

Environmental distresses are outside influences that affect the pavement performance. These influences include snow and ice, chemicals, petroleum products, ultraviolet rays, water, and natural aging. These distresses normally occur from the top down and are usually remedied by surface applications, such as a skin patching, crack sealing, seal coating, type II slurry seal, chip seals, and in some instances, thin hot mixed overlays.

Structural distresses are physical failures in the pavement and sub-bases. They can occur due to overloading, wet subgrade, frost, or substandard design. These distresses usually occur from the bottom up, and the only remedies are overlays, removal and replacement (R&R) of the affected areas, or total removal and repaving. This includes R&R patching, overlays (with or without a fabric inner layer), or some form of removal and repaving (total removal, milling, pulverizing, and paving back).

Once a new pavement surface has been exposed to environmental factors, cracks, raveling, loose rocks, or a light gray appearance develops. If the cracks are not sealed with a crack sealant or by seal coating, water will enter into the aggregate base and eventually into the subgrade, causing future structural problems. If the structural integrity of the pavement has been weakened, major rehabilitation and maintenance will be required. If pavement defects are not repaired and the pavement maintained, reconstruction will become the only maintenance option available. The resulting large capital expenditures of major reconstruction are more costly than routine maintenance and are usually not included in the maintenance budget.

It is the responsibility of a community manager, property manager, or owner to ensure that proper maintenance applications are implemented to keep pavements in good to very good condition and maintain the overall appearance of the property as well as mitigating liabilities associated with poor and deteriorated surfaces. Poor pavement and highly distressed parking lots discourage prospective tenants (commercial and industrial properties) or residents (multihousing properties). Also, poorly or improperly maintained streets in a homeowners association can reduce the value of a property, not to mention the liability issues that are associated with potholes and other tripping and ankle-twisting exposures. **Remember, first impressions are very important, and poorly maintained parking lots and streets may indicate a poorly managed property and can result in loss of prospective tenants or residents.**

The U.S. Army Corps of Engineers and American Public Works Association (APWA) has developed a pavement management program for asphalt and concrete.

This program takes into consideration nineteen types of distresses that affect the design life of pavement, including airports, highways, and streets. Of theses nineteen distresses, the following apply to parking lots and homeowners association (HOA) streets:

- Alligator cracking
- Bleeding
- Corrugation and shoving
- Depressions
- Fuel spills
- Longitudinal and transverse cracking
- Potholes
- Raveling
- Rutting
- Thermal or block cracking

Alligator cracking, depressions, potholes and rutting are structural failures. Transverse cracking is a base failure. Longitudinal cracking is a form of structural failure found in wheel paths that have rutted or a cold joint in the laydown direction of the original construction. All other distresses are environmental. The following are the most common types of distresses found in asphalt streets and parking lots.

STRUCTURAL DISTRESSES

ALLIGATOR CRACKING—A series of interconnection cracks caused by fatigue failure of asphalt concrete surface under repeated traffic loading. Cracking begins at the bottom on the asphalt surface (base), where tensile stress and strain are highest under a wheel load. The cracks propagate to the surface, initially, as a series of parallel longitudinal cracks. After repeated traffic loading, the cracks connect, forming many-sided, sharp-angled pieces that develop a pattern resembling chicken wire or the skin of an alligator. The pieces are less than two feet on the longest side. Alligator cracking occurs only in areas subjected to repeated traffic loading, such as wheel paths. Alligator cracking is considered a major structural distress and is often accompanied by rutting.

CORRUGATION AND SHOVING—A form of plastic movement typified by ripples (corrugation) or an abrupt wave (shoving) across the pavement surface. The distortion is perpendicular to the traffic direction. This usually occurs at points where traffic starts and stops (corrugation) or areas where HMA abuts a rigid object (shoving).

DEPRESSIONS—Localized pavement surface areas with slightly lower elevations than the surrounding pavement. Depressions are very noticeable after a rain, when they fill with water. Depressions are very common in parking lot construction and overlays.

POTHOLES—A type of disruption in the surface of a roadway where a portion of the road material has broken away, leaving a pothole. Sometimes called a kettle and known in parts of the western United States as a chuckhole. Most potholes are formed due to fatigue of the pavement surface. As fatigue cracks develop, they typically interlock in a pattern known as "alligator cracking." The chunks of pavement between fatigue cracks are worked loose and may eventually be picked out of the surface by continued wheel loads, thus forming a pothole. The formation of potholes is exacerbated by cold temperatures, as water expands when it freezes and puts more stress on cracked pavement. Once a pothole forms, it grows through continued removal of broken chunks of pavement. If a pothole fills with water, the growth may be accelerated, as the water washes away loose particles of road surface as vehicles pass. In temperate climates, potholes tend to form most often during spring months when the subgrade is weak due to high moisture content. However, potholes are a frequent occurrence everywhere in the world, including in the tropics. Potholes can grow to several feet in width, though they usually only become a few inches deep, at most. If they become large enough, damage to tires and vehicle suspensions can occur.

RUTTING—Surface depression in the wheel path. Pavement uplift (shearing) may occur along the sides of the rut. Ruts are particularly evident after a rain, when they fill with water. There are two basic types of rutting: pavement rutting and subgrade rutting. Pavement rutting occurs when the pavement surface exhibits wheel path depressions as a result of compaction/mix design problems.

SHOVING—A longitudinal displacement of a localized area of the pavement surface. It is generally caused by braking or accelerating vehicles, and it is usually located on hills or curves, at intersections, and where pavement abuts a rigid object (curb, concrete swale, valley gutter, concrete pad, etc.). Shoving is also present with fabric overlays where traffic has moved the pavement on the fabric inner layer. It also may have associated vertical displacement.

SWELL—A long, gradual wave of more than ten feet long, characterized by an upward bulge in the pavement's surface. Swelling can be accompanied by surface cracking. This distress is usually caused by frost action in the subgrade or by swelling soil.

STRUCTURAL OR ENVIRONMENTAL DISTRESSES

LONGITUDINAL AND TRANSVERSE CRACKING—Longitudinal cracks are parallel to the pavement's centerline or laydown direction. Usually a type of fatigue cracking. Cracking is in the direction of flow of traffic, usually at the edge of wheel paths. Transverse cracks are perpendicular to the pavement's centerline, longitudinal joint. Usually a type of thermal cracking.

ENVIRONMENTAL DISTRESSES

BLEEDING—A film of asphalt binder on the pavement surface. It usually creates a shiny, glass-like reflecting surface, bubbles that appear like blisters, or a buildup of asphalt binder on the surface that can become quite sticky. Bleeding occurs when asphalt binder fills the aggregate voids during hot weather and then expands onto the pavement surface. Since bleeding is not reversible during cold weather, asphalt binder will accumulate on the pavement surface over time. This can be caused by one or a combination of the following: excessive asphalt binder in the HMA (either due to mix design or manufacturing), excessive application of asphalt binder during bituminous surface treatment application, or low HMA air void content (i.e., not enough room for the asphalt binder to expand into during hot weather).

BLOCK (THERMAL) CRACKING—Cracking in the shape of blocks, which are interconnected cracks that divide the pavement into approximately rectangular pieces. The blocks may range in size from approximately one foot by one foot to ten feet by ten feet. Block cracking usually indicates that the asphalt has hardened significantly. Block cracking normally occurs over a large proportion of pavement area, but sometimes will occur only in nontraffic areas. Block cracking is also referred to as "thermal cracking," since it is a result of environmental exposures and aging, which makes it hard and oxidized.

BUMPS AND SAGS—Bumps are small, localized, upward displacements of the pavement surface. They are different from shoves, as they are caused by unstable pavement. Bumps can be caused by several factors, including buckling or bulging of underlying Portland cement concrete slabs. This is where asphalt pavement has been overlaid over concrete pavement. Frost heaves can cause bumps from expansion under the paved surface. Spalling of crack edges is caused by oxidation, especially under chip seals. Roots growing under pavement can cause large bumps and infiltration and buildup of material in a crack in combination with traffic loading (often referred to as tenting). Sags are small, abrupt and localized depression or elevated areas that are a result of settlement or displacement in the pavement surface. Also, pavement shoving and displacement caused by subgrade (native ground) swelling or displacement due to tree roots. Sags cause large and/or long dips in the pavement.

EDGE CRACKING—Edge cracking is generally on the edges of unconfined asphalt pavement. Unconfined (no curbs or other forms of edge barriers) pavement will yield during the compaction process or will develop as the pavement ages, oxidizes, and becomes brittle. Edge cracking is generally in the shape of the letter "C" along the edge of the street, road, or parking lot.

JOINT REFLECTION CRACKING—Cracks in a flexible overlay of a rigid pavement. The cracks occur directly over the underlying rigid pavement joints. Joint reflection cracking does not include reflection cracks, which occur away from an underlying joint or from any other type of base (e.g., stabilized cement or lime).

LANE/SHOULDER DROPOFF—The difference in elevation between the pavement edge and the shoulder. This distress is caused by shoulder erosion, by shoulder settlement, or by building up the roadway or edge of a parking lot without adjusting the shoulder level.

PATCHING AND UTILITY CUT PATCHING—An area of pavement that has been replaced with new material to repair the existing pavement. There are two types of patches: remove and replace, and skin (or surface). A patch is considered a defect, no matter how well it performs.

POLISHED AGGREGATE—Areas of hot mixed asphalt pavement where the portion of aggregate extending above the asphalt binder is very small or there are no rough or angular aggregate particles. This occurs after repeated traffic applications. Generally, as a pavement ages, the protruding rough, angular particles become polished and can reduce skid resistance. This can occur quicker if the aggregate is susceptible to abrasion or subject to excessive studded tire wear.

RAVELING—The progressive disintegration of a hot mixed asphalt layer from the surface downward, as a result of the dislodgement of aggregate particles, which is the loss of bond between aggregate particles and the asphalt binder. This occurs as a result of a dust coating on the aggregate particles that forces the asphalt binder to bond with the dust rather than the aggregate, aggregate segregation, inadequate compaction during construction (poor compaction, cold asphalt pavement, cold weather paving), or exposure to water (from sprinklers or puddles), creating a stripping process.

RAILROAD CROSSING—Railroad crossing defects are depressions or bumps around or between tracks.

SLIPPAGE CRACKING—Crescent or half-moon shaped cracks generally having two ends pointed in the direction of traffic. This can also be referred to as edge cracking.

Other distresses not specifically listed by the Corps of Engineers include cold joint and seal coat checking distresses.

COLD JOINT—Very common in parking lots, inverted crowns (constructed drainways in the center of streets), and low traffic streets. A cold joint is a longitudinal joint that occurs in an asphalt pavement when fresh hot mix asphalt is laid adjacent to an existing pavement. The cold joint is the boundary between the two HMA mats. Most often, differences in the temperature and mat plasticity cause an improper bonding of the fresh HMA with the older asphalt, subsequently causing the longitudinal joint to possess a significantly lower density than the rest of the pavement. Over time, a longitudinal crack usually occurs between the asphalt mats, permitting the intrusion of water, increasing roughness, and potentially limiting the life of the pavement. This type of distress is very common in inverted crowns. Asphalt pavement laid next to a concrete structure (valley gutter or swale, dumpster pad, cast-in-place curb, or combination curb and gutter) will create a cold joint after the asphalt has oxidized and shrunk.

CHECKING IN SEAL COAT—Very common when too much sealer is applied to an existing pavement. This can be the result of too much sealer accumulated over a period of years or too much applied at the time of application. Seal coat, whether asphalt or coal tar emulsion, is not an aesthetic coating. Too much buildup or application will cause the material to shrink and develop a checkered pattern (much the same as a dried lake bed). The edges will eventually curl and allow water to infiltrate between the sealer and the bottom layer, creating a delaminating problem. Once this occurs, no other maintenance application can be applied over or on top of a delaminating seal coat because the sealer will continue to separate, taking any other applications off with it.

The Corps of Engineers has developed a condition rating system for pavement known as the pavement condition index (PCI). With PCI, a pavement is given a rating from 0 to 100, with 100 representing a new pavement in ideal condition. Getting to know and utilizing PCI is advantageous in managing a paved street, parking lot, or other pavement.

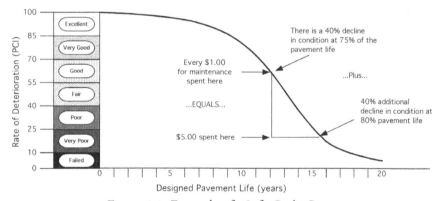

Figure 1.1: Example of a Life Cycle Curve

When a pavement is evaluated for the first time, it must be assumed that the previous condition was a newly constructed pavement, therefore, having a past PCI of 100. Depending on the type and severity (high, medium, or low) of the distresses, values are deducted from 100 to yield a current PCI. The difference of the previous PCI (100 initially) and the current PCI, divided by the age of the pavement, yields a deterioration rate for the period in time. It is important to remember that the PCI graph is a curve and that the deterioration rate will rise or fall depending on the where the PCI is located on the chart. With the current PCI and the deterioration rate, a property manager or owner can prioritize pavement sections, calculate future pavement conditions, and predict condition benefits after maintenance has been performed. All pavements deteriorate at a calculated rate per year similar to Figure 1.1, based on the types and severity of distresses and traffic exposures. As indicated in the deterioration graph, the cost of maintenance increases substantially as time passes. Figure 1.2 depicts a maintenance policy that can reduce the overall cost of pavement maintenance and prolong the life of the pavement.

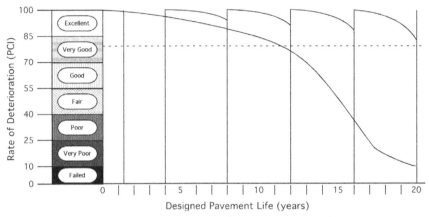

Figure 1.2: Example of a Pavement Life Cycle Curve with Maintenance

The condition of the pavement as shown in Figure 1.2 ranges from good to very good after twenty years versus very poor to failed, as shown in Figure 1.1 for the same period of time. This helps illustrate how a well-planned pavement management program, when adhered to, will reduce the cost of pavement maintenance over the life of the pavement and provide the excellent aesthetics required to show that a property is well maintained.

A simple pavement evaluation is included at the end of this chapter. The only thing that can be calculated is a condition index, not a deterioration rate. If you wish a comprehensive evaluation, with a resulting budgetary planning model and maintenance policy, you can use a computer program, or pavement consultants can provide these services. The software package available for in-house evaluations is MicroPaver, distributed by the American Public Works Association. If the pavement has a large amount of structural failure or a very low PCI coupled with a high deterioration rate, you need the services of a pavement design consulting engineer or a pavement consultant trained and versed in the Paver program.

Simplified Self-Evaluation for Parking Lots

DISTRESS   POINTS DEDUCTED

Alligator Cracking

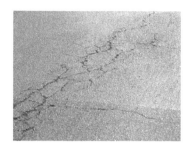

- **Low Severity**   4
  Looks like small cracks in a loading area or wheel tracks
- **Medium Severity**   8
  Looks like an alligator's back, tightly in place, may have a small depression

- **High Severity**   12
  Each piece is separate and loose and may be lying outside the area

Longitudinal/Transverse Cracking

- **Low Severity**   2
  Parallel and perpendicular cracking, less than ⅜-inch opening

- **Medium Severity**   4
  Greater than ⅜-inch, but less than ½-inch opening

- **High Severity**   12
  Greater than ½-inch opening

Potholes

- **Low Severity**   2 each
  1 to 2 inches in diameter

- **Medium Severity**   4 each
  2 to 4 inches in diameter

- **High Severity**   6 each
  Greater than 4 inches in diameter

Thermal/Block Cracking

In square pattern no more than 6 feet by 6 feet (otherwise count as longitudinal/transverse cracking). If in a load-related area, count as alligator cracking.

- **Low Severity**   2
  Less than ¼-inch opening

- **Medium Severity**   4
  Greater than ¼-inch, less than ½-inch opening

- **High Severity**   6
  Greater than ½-inch opening

Ruts and Depressions

- **Low Severity**   4
  Less than ½-inch in depth

- **Medium Severity**   8
  Less than 2 inches in depth

- **High Severity**   12
  Greater than 2 inches in depth

Raveling

- **Low Severity**   2
  Small rocks on the surface

- **Medium Severity**   4
  Some pitting, average amount of small rocks on surface

- **High Severity**   6
  Deep pitting, small or medium rocks on surface

To determine a condition index, add up all the distress value points and subtract the total from 100. The difference will equal the pavement's current PCI.

100 – Total Pavement Points = Condition Index
CONDITION VALUE

Condition Index	Maintenance Plan
85–100	No maintenance required
65–84	Minor crack filling Minor patching Some seal coating
45–64	Crack sealing Patching Seal coating Type I, II, or III slurry seal
25–44	Major crack sealing Major patching Overlay
Less than 25	Reconstruction

Figure 1.9: Maintenance Plan

This is only a quick determination to help the property owner, manager, or HOA board of directors determine the current condition and possible maintenance applications. For a more exact and detailed evaluation, an experienced pavement consultant should be contacted. There are several consultants available, so make sure the consultant you hire has the in-depth experience and education and is not affiliated with a pavement contractor. This will provide the property owner, manager, or HOA board of directors with a correct inventory without the influence of competitive bidding, which will directly affect the overall maintenance budget.

Chapter 2

Causes of Pavement Deterioration

All pavements begin to deteriorate the day construction is completed. This deterioration can be attributed to five factors that adversely affect pavement.

The first factor is ultraviolet rays. Ultraviolet rays (or direct exposure to sunlight) and high temperatures cause pavements to oxidize and become brittle. On a hot day, the temperature of pavement can reach 140°F (the softening point of liquid asphalt) or more, causing the pavement to expand and move; when the temperature is reduced, the pavement contracts. Normal fluctuation of temperatures, along with the expansion and contraction of asphalt pavement, is the cause of block or thermal cracking and can cause the pavement to become hard or brittle. This cracking condition is the most common type of initial cracking.

The second adverse factor, water, enters the base and subgrade below the pavement surface through the cracks, resulting in the start of structural damage.

The third factor of pavement deterioration is fuel spillage from vehicles (e.g., oil, gasoline, antifreeze, and brake fluid), which destroys the pavement integrity by permanently softening the liquid asphalt. Eventually, the pavement surface softens, and any surface applications (seal coats, slurry seals, overlays) become contaminated and prematurely deteriorate. Eventually, potholes develop, requiring total removal and replacement.

The fourth factor involved is aging and raveling due to exposure to ultraviolet rays or water.

The fifth factor is cold climates, which create frost heave and asphalt striping from snow and ice, leading to potholes and surface raveling.

Figure 2.1 shows the effects of these factors on a pavement cross section.

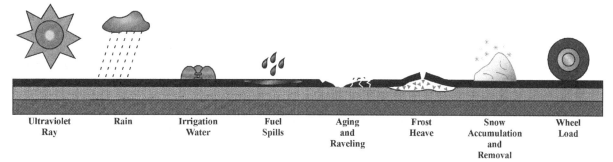

| Ultraviolet Ray | Rain | Irrigation Water | Fuel Spills | Aging and Raveling | Frost Heave | Snow Accumulation and Removal | Wheel Load |

Figure 2.1: Common Pavement Distress Causes

Figure 2.2: Infiltration of Water through Crack

After a crack has formed, water will find its way below the pavement surface and begin to cause instability to the structural integrity of the pavement (see Figure 2.2). When wheel loadings are introduced to the soft area, the pavement moves and alligator cracking and potholes form (see Figure 2.3).

Figure 2.3: Pothole Liability

When water is allowed to remain on a pavement, a natural reaction of the oil separating from the aggregate occurs, causing the asphalt binder to strip off the surface aggregates. This process is known as raveling. Raveling causes the surface to deteriorate by loss of asphalt and rocks. This is evident by rocks and sand on the pavement surface and in gutters (see Figure 2.4). Over a period of time, the thickness of the pavement begins to dwindle away, until the entire pavement has eroded completely. Most raveling can be prevented by proper seal coating or it can be corrected by resurfacing with type II polymer modified slurry seal.

Figure 2.4: Raveling

Ultraviolet rays (exposure to sunlight) cause pavement to turn brittle and oxidize. This oxidation process decreases the pavement's elasticity, causing cracking, and makes the pavement more susceptible to raveling. This exposure is more prominent in Sun Belt locations (primarily the American Southwest), just as frost heaves and freezing are prominent in colder climates. Ultraviolet exposure in colder climates during warmer seasons has an adverse affect on pavement as well, necessitating the appropriate maintenance.

Petroleum spills on asphalt pavements cause irreversible deterioration if not treated immediately and properly. If you ever watch a paving crew clean their equipment after a day's work, you will notice they clean all the equipment with gasoline or diesel fuel, because it completely removes any asphalt buildup and residue. A parked vehicle may leak gasoline, antifreeze, brake fluid, and so on and cause the asphalt to liquefy, separate from the rock, and create soft areas (see Figure 2.5). You cannot ignore the treatment of oil spots on parking lots or other heavy traffic areas.

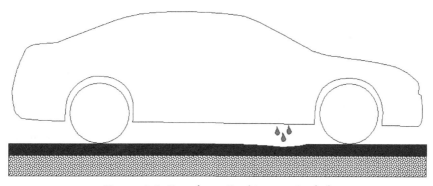

Figure 2.5: Petroleum Leaking on Asphalt

After a period of time, the petroleum product will begin to penetrate the pavement, causing it to become soft and permeable (see Figure 2.6). Once the exposure has reached this stage, the pavement begins to decay.

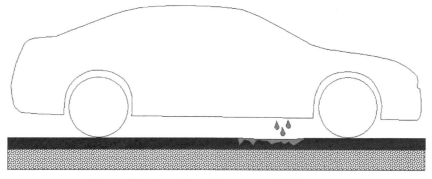

Figure 2.6: Petroleum Penetrates the Pavement

If not repaired, the pavement will deteriorate into a pothole, and the petroleum contamination will enter the base and subgrade (see Figure 2.7). At this point, total removal of the pavement, base, and any contaminated subgrade is necessary. If the distress is patched, sealed, or covered without proper treatment or removal, the petroleum residue will evaporate back up through the new maintenance treatment, causing it to deteriorate as if nothing had been done.

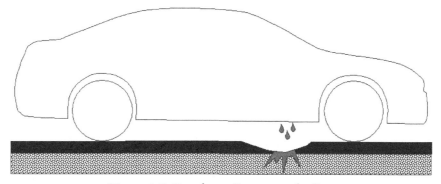

Figure 2.7: Petroleum Penetrates the Base

The proper application of oil spot primer will reduce problems of tracking by promoting adhesion when seal coat material is applied. Tracking happens when tires pick up sealer material and deposit it in other locations on the pavement, curbs, sidewalks and driveways. An acceptable oil spot primer is easy to use and quick to dry. First, scrape off excess oil, dirt, or other chemical or petroleum stains and buildups. Second, using a propane torch, lightly "flash" or burn off the excess petroleum residue on the surface. Next, apply the oil spot primer by using any suitable spray machine, broom, brush, or roller. Fourth, lightly sprinkle clean silica sand over the moist material to provide greater adhesion and to prolong the life of the seal coat being used. Finally, apply an asphalt emulsion sealer or coal tar (where available) using any spray-type application, seal coating machine, push broom, or squeegee. When the sealer has dried, there will be extra protection where it is needed the most (areas exposed to oil, grease, and other petroleum spills and drips).

Aging is the normal life cycle deterioration of pavements and results in accelerated oxidation and cracking. If basic deterioration is not treated properly, water and ultraviolet rays will accelerate the process. With accelerated aging, water can be introduced to the base or subgrade through cracks, causing structural distresses to develop. These distresses will be in the form of block or thermal cracking, longitudinal and transverse cracking, alligator cracking, depressions, ruts, shoving, and potholes, which will require more extensive maintenance by removing, recompacting, and repaving (R&R patch) the structurally damaged areas. If these distresses are not kept in check by proper maintenance, total reconstruction or very expensive major maintenance will be required. When pavements have deteriorated extensively, a pavement consultant or specialist should be retained to advise on maintenance or reconstruction policies, along with establishing budgets and specifications.

In these first two chapters, you have learned about types of distresses, what causes distresses, how to determine pavement conditions, and how to get help in establishing maintenance policies, plans, specifications, and applications. You will learn how to inventory pavement distresses and how to repair them in the following chapters.

Chapter 3

Measuring Distresses and Pavements

An inventory of distresses and the amount of pavement surface area must be determined before a proper maintenance program can be implemented. A solid inventory will be a big help as a budget tool in requesting bids for maintenance work and in selecting a contractor. Figure 3.1 shows a simple parking lot that will be measured.

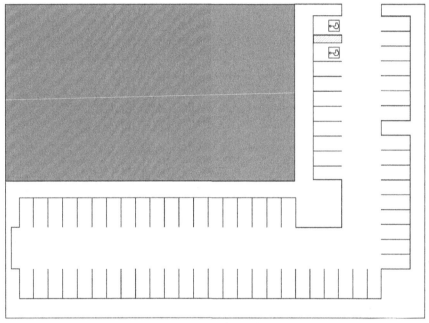

Figure 3.1: Sample Parking Lot

A measuring wheel can be purchased for approximately $100 to $125. Several on-site property managers or maintenance engineers can share this device, because it will probably be used three (3) to four (4) days per property, per year.

The first step is to divide your pavement surface into as many rectangles as possible by parking areas, drive lane areas, and delivery lanes for commercial, industrial, and multifamily properties. Divide HOA streets into rectangles, circles (cul-de-sac), and pull-outs (eyebrow areas), and measure accordingly. Figure 3.2 is an example of properly measuring up a parking lot.

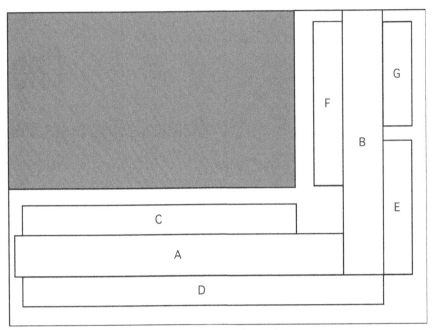

Figure 3.2: Inventory of a Parking Lot

A.   205 ft. × 25 ft.
     Area = 5,125 ft.²

B.   160 ft. × 25 ft.
     Area = 4,000 ft.²

C.   171 ft. × 18 ft.
     Area = 3,078 ft.²
     Stripes = 18
     Stalls = 19

D.   225 ft. × 18 ft.
     Area = 4,050 ft.²
     Stripes = 24
     Stalls = 25

E.   81 ft. × 18 ft.
     Area = 1,458 ft.²
     Stripes = 8
     Stalls = 9

F.   99 ft. × 18 ft.
     Area = 1,782 ft.²
     Stripes = 8
     Stalls = 8
     Handicap = 2
     75 ft. of yellow

G.   63 ft. × 18 ft.
     Area = 1,134 ft.²
     Stripes = 6
     Stalls = 7

Total Surface Area Sq. Ft.   **20,627 ft.²**
Total Surface Area Sq. Yds.   **2,292 yd.²**
Total Stripes (18 ft. each)   **64 stripes = 1,152 ft.**
Total Handicaps   **2**
Total Parking Stalls   **67**

Measuring the drive lanes separately in parking lots is recommended because parking lots that have a seal coat applied after the initial sealing only need one coat in the parking stall areas and two coats in the drive lane areas. The older sealer in the parking stalls was not subjected to traffic to cause any wearing. More than one coat after the initial sealing will cause buildup and lead to checking problems. There isn't a large cost savings; however, it will reduce the effects of sealer checking, which will lead to more expensive maintenance applications. Sealer checking is caused from shrinkage of the new sealer material when placed over a smooth surface. Over time, the checking increases in severity, the edges of the checked area curl or spall, and water infiltrates between the sealer and the original pavement, creating the industry term "sealer rot" or "pavement rot." The extremes of this problem will lead to milling and paving. Once this problem gets to the extreme stages, any applications placed on the top of the checked areas will delaminate and flake off.

The second step is to identify types of distresses that exist and inventory them by type, severity, location, and amount. Structural distresses are measured in square feet, with an additional foot beyond the farthest point of the distress on all sides (for example, if a distress area is 6 ft. × 10 ft., the patch area should be 8 ft. × 12 ft., with an area of 96 ft.[2]). Raveling and skin patching is measured in exact square feet, and cracking is measured in linear feet. When measuring cracking, it is acceptable to measure sample areas of not less than 1,000 ft.[2] and enough sample areas to represent 10 percent of the total area. This is only acceptable for large paved surfaces and more than one mile of streets. Use a map (or plat) of the property to mark locations and dimensions of areas to be patched to aid in locating the patch areas during the bidding process. If possible, mark the patch areas in white paint and label each patch with an "R" for remove and replace (R&R) patch or "S" for skin or surface patch. This will assist in evaluating proposals from contractors and aid contractor estimators when evaluating and submitting the request for bids (RFB).

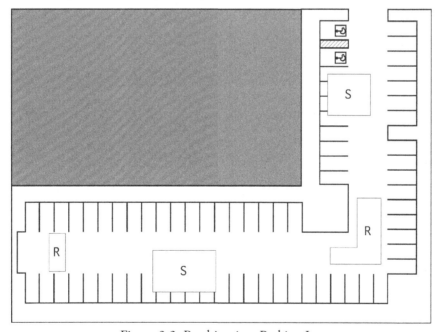

Figure 3.3: Patching in a Parking Lot

Skin Patching (ft.)		R&R Patching (ft.)	
1	27 x 25	3	10 x 23
2	39 x 25	4	32 x 10
			15 x 30

There are two types of patching in pavement maintenance. The first consists of saw cutting the area where structural damage has occurred, removing the area, and patching it back with hot mixed asphalt.

This patch is known as a remove and replace (R&R)patch. The second type of patch consists of applying a thin hot mixed asphalt patch on the surface of the existing pavement where it has been deteriorated or weathered. This patch is known as a skin or surface patch (type II slurry seal can be used in place of the hot mixed asphalt for skin patching). A variation of skin or surface patching is placing a fabric inner layer between the patch and the original pavement to add longevity to the patch. These two types of patches differ greatly in price, with the R&R being considerably more expensive than the skin patch.

With the patch survey, locate areas destroyed by petroleum spills and drips to be patched, along with their dimensions. Also, inventory the amounts and types of cracking using the required sample sections of about 1,000 ft.^{2} per sample. Measure the amount of cracking in these areas and add the totals to get an accumulated total and divide the accumulated total by the number of sample areas to obtain an average linear feet of cracking per square foot. Multiply the total area of the paved surface by the average cracking to obtain the total amount of cracking to be sealed. Make sure the crack opening is a $\frac{1}{4}$ inch or more so material is not wasted on smaller opening cracks.

PARKING LOT CRACK INVENTORY

This is an example of how to determine the crack sealing for the parking area used in Figure 3.1.

The first step is to measure all the cracks with an opening of ¼ inch or more in five sample areas each measuring 1,000 ft.^{2}.

**Sample 1
(only measuring cracks ¼ or wider)**

Sample size = 1,000 ft.^{2}
Amount of cracking in sample area = 119 ft.

**Sample 2
(only measuring cracks ¼ inch or wider)**

Sample size = 1,000 ft.^{2}
Amount of cracking in sample area = 90 ft.

**Sample 3
(only measuring cracks ¼ inch or wider)**

Sample size = 1,000 ft.^{2}
Amount of cracking in sample area = 114 ft.

**Sample 4
(only measuring cracks ¼ inch or wider)**

Sample size = 1,000 ft.^{2}
Amount of cracking in sample area = 120 ft.

**Sample 5
(only measuring cracks ¼ inch or wider)**

Sample size = 1,000 ft.^{2}
Amount of cracking in sample area = 85 ft.

When reinspecting areas, always use the same sample units to ensure consistency in measurements. The next step is to average the two areas together.

First, add the samples together:

```
   119 LF Sample 1
+   90 LF Sample 2
+ 114 LF Sample 3
+ 120 LF Sample 4
+  85 LF Sample 5
   528 LF Total of all sample units
```

Next, divide the sum of the crack samples by the number of samples taken (in this case, there were five samples taken).

528 LF ÷ 5 (sections) = 105.6 LF, rounded to 106 LF per 1,000 ft.^{2}

Next, divide the surface area by 1,000 ft.^{2}

20,627 ÷ 1,000 ft.^{2} = 20.63 (number of 1,000 ft.^{2} sections)

Then, multiply the number of sections by the average of cracking in the sample sections to get the total amount of crack sealing needed, in linear feet.

20.63 x 106 ft.^{2} = 2,187 LF (rounded to the nearest foot)

Finally, round the product up to the nearest ten feet to get the total amount of crack sealing to be used for the bid.

2,187 ft. rounded up to the nearest 10 ft. = 2,190 LF.

There now exists a total inventory of the pavement area, patching, cracking, and any other items (e.g., painting, bumper blocks, and speed bumps).

The total inventory is as follows:

1. **2,190 linear feet** of cracks.
2. **2,292 yd.^{2}** (20,627 ft.^{2}) of pavement area. For seal coats and slurry seals, always use square-yard figures for costs. For overlays and paving, use square-foot figures for costs.
3. **1,650 ft.^{2} of skin patching** of badly damaged, but not structurally inadequate, areas.
4. **1,000 ft.^{2} of R&R patching**.
5. **64 stripes at 18 feet per stripe or 1,152 LF of stripes, 68 parking stalls, 2 handicap stalls, and 1 hash out area.**

SURFACE AREA OF A STREET SYSTEM

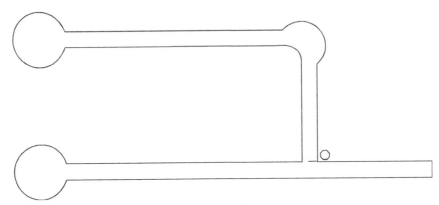

Figure 3.4: Sample Streets

With the patch survey of the streets, locate areas deteriorated by petroleum spills to be patched, along with their dimensions. Also, inventory the amounts and types of cracking using the required sample sections of about 1,000 ft.², as discussed above. Measure the amount of cracking in these areas and add the totals to get an accumulated total, and divide by the number of sample areas to obtain an average linear feet of cracking per square foot. Multiply the total area of the paved surface by the average cracking to obtain the total amount of cracking to be sealed. When measuring cracking on streets, choose sample units to represent each street. Also, when identifying cracking in streets, include the joint between curb line and along concrete structures (valley gutters, swales, pads, etc.). These joints need to be sealed every eight to ten years to prevent water from flowing into the base materials and creating soft pavement and structural damage.

Break the streets into measurable sections, as seen in Figure 3.5. Measure the area of each section and then add the sections together for the total area of pavement. Do not forget to inventory all striping and any reflectors or ceramics.

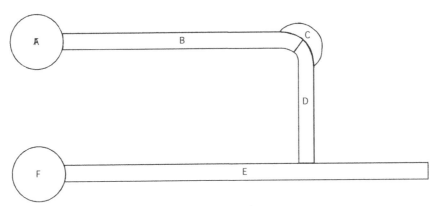

Figure 3.5: Sections for Streets

A.   Radius = 44 ft. (π x r^2
     where π = 3.14)
     Area = 6,079 ft.^{2}

B.   28 ft. x 400 ft.
     Area = 11,200 ft.^{2}

C.   C-Turnout (Eyebrow)
     Run = 135 ft.
     Rise = 15 ft.
     Area = 1,363 ft.^{2}
     (from Table 3.1)

D.   200 ft. x 28 ft.
     Area = 5,600 ft.^{2}
     1 Stop Bar
     1 Blue Reflector

E.   600 ft. x 28 ft.
     Area = 16,800 ft.^{2}
     1 Blue Reflector

F.   Radius = 44 ft. (π x r^2
     where π = 3.14)
     Area = 6,079 ft.^{2}

Total Surface Area	47,129 ft.^{2}
Total Surface Area	5,237 yd.^{2}
Stop Bars	1 each
Blue Reflectors	2 each

Now that the inventory of distresses has been made, you are ready to develop a scope of work and a maintenance policy and define an appropriate rehabilitation strategy.

A quick review of units for the inventory is as follows:

✓ **Cracks** are measured in **linear feet.**
✓ **Surface area** is measured in **square yards (SY) or square feet (SF)** for seal coats and slurry seals.
✓ **Patching, paving, and overlays** are measured in **square feet.**
✓ **Striping and hash outs** are measured in **linear feet,** and **handicap stalls** are counted as **single units (each). Red, yellow, or white curbing is** measured in **linear feet.** If the side and top of a curb is painted, the linear footage must be doubled. **All stenciling (letters and numbers)** is counted as a **single unit (each).**
✓ **All reflectors and ceramic dots (if applicable)** are counted as a **single unit (each).**

All public use parking lots are subject to American with Disabilities Act (ADA) requirements for parking stalls and accessibility. Please refer to all local agency requirements, as some city and county requirements supersede some ADA requirements.

TABLE 3.1
Areas under a Curve or in a Turnout (Eyebrow)

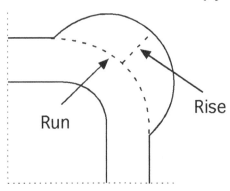

Figure 3.6 Diagram of Turnout

RISE (Feet)

RUN (Feet)	10	15	20	25	30	35	40	45
50	343	534	745	982				
55	376	581	806	1,053	1,064			
60	409	629	867	1,128	1,414	1,167		
65	441	677	929	1,202	1,500	1,511		
70	474	725	992	1,278	1,588	1,924		
75	507	773	1,055	1,355	1,677	2,025	2,036	
80	539	822	1,118	1,510	1,859	2,127	2,513	
85	573	871	1,182	1,510	1,859	2,231	2,628	
90	606	920	1,246	1,589	1,951	2,336	2,745	3,180
95	639	969	1,310	1,668	2,044	2,441	2,863	3,310
100	672	1,018	1,375	1,747	2,137	2,548	2,982	3,441
105	705	1,067	1,440	1,827	2,231	2,656	3,102	3,573
110	738	1,116	1,504	1,907	2,326	2,764	3,224	3,706
115	771	1,165	1,570	1,987	2,421	2,764	3,346	3,842
120	804	1,215	1,635	2,068	2,516	2,982	3,468	3,977
125	838	1,264	1,700	2,149	2,612	3,092	3,592	4,114
130	871	1,314	1,766	2,229	2,708	3,202	3,716	4,059
135	904	1,363	1,831	2,311	2,804	3,313	3,814	4,390
140	937	1,413	1,897	2,392	2,900	3,425	3,967	4,529
145	970	1,462	1,962	2,473	2,997	3,536	4,093	4,669
150	1,004	1,512	2,028	2,555	3,094	3,648	4,219	4,809
155	1,036	1,562	2,094	2,636	3,191	3,760	4,346	4,950
160	1,070	1,611	2,160	2,718	3,288	3,872	4,473	5,091
165	1,103	1,661	2,226	2,800	3,386	3,985	4,600	5,233
170	1,136	1,710	2,292	2,882	3,483	4,098	4,728	5,375

Values are derived from the following calculus formula and expression:

$$\int_{a}^{b} f(x)\, dx = \lim_{n \to \infty} \sum_{i=1}^{n} \left[f(x_i) \cdot \left(\frac{b-a}{n} \right) \right]$$

Chapter 4

Maintenance Applications and Policies

In chapters 1 and 2, you learned to identify various distress types and causes of pavement deterioration. In this chapter, we will discuss the types of rehabilitation applications that are used for pavement maintenance. Once the applications have been chosen, they can be used to develop a comprehensive maintenance policy. A maintenance policy is a combination of rehabilitation strategies and applications that include current projects as well as scheduled projects projected over a period of ten to fifteen years. However, it can be difficult to project future maintenance requirements, as pavement distresses can increase in quantity and severity with weather and traffic exposure changes.

The following is a basic description of major pavement maintenance applications. An experienced and knowledgeable practitioner or consultant should be retained to assist in properly developing an accurate distress inventory and a scope of work using these applications to repair the distresses you have identified in your pavement. Each of these applications requires a certain existing condition to be effective. The distress inventory will aid you in determining the type of maintenance application to choose and the proper strategies to develop and implement. There are five basic types of maintenance applications: crack sealing, patching (mastic, skin or surface, and remove and replace), surface coatings (seal coatings, slurry seals, and chip seals), resurfacing or overlays, and reconstruction. Once the proper application are chosen or decided upon, you will need to become familiar with the characteristics of those application by referring to the proper applications in this chapter. Also, you will need to become familiar with the curing process of each application you choose so you can understand each phase of that application's curing process. Each application cures differently and will have a different aesthetic appearance (some poor) for several weeks (and sometimes months) before it settles and cures. Once the applications have settled and cured, the appearance will change. In some cases (type II slurry sealing and chip sealing), you will have to perform some form of cleanup until the material has completely cured. There will be a great deal of concern from tenants in commercial and industrial properties and homeowners in homeowners associations about the appearance of these applications while they are curing. When you provide these groups with the proper answers, there will be less confusion and fewer confrontations.

In this chapter, a curing characteristic statement is listed after each application to explain how the chosen application cures and settles. Some of these applications take weeks to cure, while others take months. During that time period, many things will appear and develop; however, understanding the curing process, you will know what is happening and what is developing during the curing process of each application. During the life cycle of pavement, several distresses form and develop. A problem and distress development table (Appendix A) has been created to help identify problems and suggest a corrective action that should be applied (troubleshooting table for pavement problems). This chapter will guide you through the various maintenance applications and curing characteristics you can expect.

The first part of a good pavement program starts with a survey of the current pavement condition. The first part of the survey is identifying various pavement distresses. The first distress that will develop in new pavement is cracking. Cracking is generally identified in four forms:

- Longitudinal and transverse
- Block (thermal)

- Crescent (cracking along the edge of the pavement)
- Reflective (in slurry seals and overlays)

Longitudinal, transverse, and block cracking are generally the first cracking distresses to develop in pavement. Longitudinal and transverse cracking develop from paving joints separating, shrinkage of the pavement, and base or subgrade movement. Block cracking is a result of thermal or temperature changes and fluctuations. The pavement will crack from expansion and contraction and oxidation. The cracking is in the form of squares or blocks. Crescent cracking is a result of failure along the edge of an unconfined street, road, or parking lot. This type of crack is caused from movement of the paved surface outward when exposed to a loading. Reflective cracking develops from below a new surface (slurry seal, chip seal, or overlay) from existing cracking. As the older pavement expands and contracts, the cracking will reflect through the new surface. A fabric (stress-absorbing membrane) can be used to retard the rate of reflective cracking to appear; however, the cracking will reflect through.

Once cracking begins, other structural distresses will begin to develop. All cracks allow water to infiltrate (percolate) through the paved surface and saturate the base course and subgrade below the pavement. Once the base or subgrade is saturated, structural failures can begin to develop. That is why it is very important that cracks are sealed whenever they have an opening of ¼ inch or more. This recommended crack opening width was determined by studies conducted by several research agencies; it was found that cracking with an opening of ¼ inch or more went all the way through a 2- to 3-inch paved surface. Also, you need a minimum of ¼ inch opening to get proper penetration of any crack sealant. For less than a ¼ inch opening, you are only wasting maintenance dollars because the crack seal material is not penetrating the crack properly and is only spread on the surface. Seal coating will fill cracks less than ¼ inch wide in opening and provide the protection needed. Smaller or thinner cracks in openings sometimes do not go all the way through the paved surface. As the pavement oxidizes and shrinks, smaller cracks will develop into larger cracks, and new cracks will develop. Also, the pavement next to concrete curbs and valley gutters (concrete swales) will shrink and separate from each other. These openings will need to be crack sealed when these openings reach ¼ inch in opening, since curbs and valley gutters are exposed to more water than the actual paved surface. Separation of the pavement from the curbs and gutters generally occurs after five-plus years from installation. Cracking is the first distress that will develop in asphalt pavements, and constant inspections should be conducted; all cracks with an opening of ¼ inch should be kept sealed.

CRACK SEALING

An elastomeric material should be used as the crack seal material, which will expand and contract with the pavement from exposure to temperature changes. Crack seal material should have a high softening point to eliminate tracking problems on sidewalks, pool areas, and interior carpets. Different geographical areas will require material with different softening points (colder climates vs. desert climates). Several types and grades of crack seal materials are available and may be applied as cold-poured or hot-poured. These material types will be discussed later.

A crack must be properly prepared before the sealer is applied to allow proper adherence of the material on either side of the crack. The normal crack to be sealed is ¼ inch or greater. Less than ¼ inch can be sealed by the seal coat application. If a crack that is less than ¼ inch is scheduled to be crack-sealed, it should be routed to open the crack to ¼ inch. Routing is an expensive task and, therefore, not always recommended for parking lots and HOA streets. When cleaning out the crack, compressed air should be used to blow out debris and any loose pavement particles. Cracks should then be sealed with a crack seal material (hot or cold applied) and banded by squeegeeing after application. Figure 4.1 shows three steps for proper crack seal application. The crack must be thoroughly cleaned by high pressure air. The crack seal material is applied in the crack by a pressure wand and hose from a hot mechanical applicator, then squeegeed with hand squeegees or by a stationary foot on the pour pots to band the crack seal material on either side of the crack. This banding completes the crack sealing process and aids in sealing water from entering the crack.

Figure 4.1: Cleaning Cracks, Sealing Cracks, and Squeegee Banding

Crack seal is a preventive maintenance application to seal cracks with a surface opening of ¼ inch or more. Crack sealing is applied into the crack after it has been cleaned of all sand, dirt, debris, and all non-compressible materials. Also, the crack must be dry and free of moisture prior to applying the sealant. Crack sealant is applied as hot-poured or cold-poured. Hot-poured sealant is heated in an on-site melting pot and applied to the cracks from pour pots or from an on-site hot pot, through a hose and wand applicator under pressure. Cold-poured sealant is applied by pour pot only. Material applied with the wand and under pressure is designed to penetrate into the crack; this is an excellent way to apply the crack sealant material. However, the tip of the wand must be placed at the entry of the crack in order to be applied correctly and get maximum benefit of the application. It is not recommended or acceptable for the material to be "shot" at the crack opening from a distance, because the material will not have the proper, recommended, and specified flow entry. Also, it creates a messy application and increases the cost. Pour pot application is a gravity installation, and the material gets the penetration needed from gravity flow and pressure applied to the pour pot by the operator. Pour pot application is an acceptable method; however, the pour pot must have the proper application foot that is in contact with the surface, forcing material into the crack and banding the material to both sides of the crack.

The purpose of crack sealing is to prevent water from entering into the aggregate base and subgrade material below or under the pavement. The crack sealant must be banded or spread on the pavement next to the crack and cover a width of ¾ to 1 inch on both sides of the crack. Banding of the crack seal product is accomplished by a "V"-shaped hand squeegee when using hot pot application. Banding is accomplished by the application foot of the pour pot. Banding the crack seal on each side of the crack will keep water from entering the cracks.

CRACK SEAL CURING CHARACTERISTICS

Crack sealant will leave a solid black band where it is installed, and this will be permanently in place. When applied prior to seal coating, the crack sealant will be visible within hours after sealing. After the seal coat wears off, the crack sealant will still be present and very visible. Crack sealant will shadow through slurry seals and chip seals, but it will not be as visible as it is with seal coats. There is very little or no apparent visual appearance of crack sealant with overlays, especially when a fabric inner layer is installed.

Seal coat materials will not adhere to crack seal material, because crack sealant becomes soft and pliable or can become liquid (depending on the softening point) from high ground and ambient temperatures. Slurry seals and chip seals can become soft and can move or displace when laid over crack sealant that has reached its softening point. Because of this, crack sealant has to have a specified minimum softening point that is comparable to the geographical area where it is installed, which also relates to the maximum highest daily temperature. Highest daily temperature directly affects the ground temperature, which, in turn, affects the softening of crack seal materials. When a lower grade crack sealant, with a lower softening point, is installed in a climate where the ground and ambient or air temperature can exceed the designed softening point of the material, it becomes liquid and can be tracked on concrete, walks, and pool decks; inside buildings; and so on. There are several grades of crack sealants designed for specific climate areas and conditions. Should ambient and ground temperatures cause the crack seal material to soften and the temperatures stay high consistently, it is recommended to apply a blotter material on the crack sealant as soon as it is applied. This blotter material is usually sand or Portland cement. The blotter material aids in preventing pickup by vehicles or pedestrians until the sealant has ample time to set up or cure. Because crack sealant becomes soft and pliable when hot, striping paint may flake off or disappear when applied over crack sealant. This is caused by the elastic characteristics of the crack sealant as it moves with the expansion and contraction of the asphalt pavement.

When driving on newly applied crack sealant, take care to prevent power steering shear and pulling up of the crack sealant material. When stopping a vehicle, start driving forward or backward, then start turning the steering wheel. When the material is still hot or warm after application, do not park on the sealant. The sealant will take the shape of the tire tread; after it cools and when the vehicle moves, the crack sealant and, in some cases, the pavement will come up on the tire. This is very hard and difficult to remove. When crack sealant is on the tread of tires, it creates a very bad vibration to the wheels and steering. In a short period of time, the material will wear off, and it will not damage the front end alignment. It is just very annoying and bothers the driver until it has worn off. Cleaning the material off is extremely messy and very difficult, but it can be accomplished. Check with the contractor or material manufacturer for the best method and cleaning chemicals to remove the crack seal material.

PATCHING

Patching is the next application to be addressed. There are two major forms of patching: remove and replace (R&R) and skin or surface patching. Following is an explanation of each form of patching.

REMOVE AND REPLACE (R&R) PATCHING

Two forms of distress that develop are fatigue cracking and structural failures. These distresses are caused from repeated traffic loading, overloading pavement with vehicular traffic beyond design standards, and water infiltration, resulting in the saturation of the subgrade. These cracks are visible in the form of alligator cracking, potholes, sags, ruts, depressions, and lateral displacement. In colder climates, frost can cause potholes through an expansive heave under the pavement caused by moisture freezing and expanding, and then thawing, causing contraction of the base reducing its size. This results in pavement collapsing under loading, which causes potholes. With all these structural distresses, the only option is to remove-and-replace patch the failed area. This type of patching is referred to as R&R patching. Fuel and solvent spillage, which includes gasoline, motor oils, antifreeze, hydraulic fluids, brake fluids, and power steering fluids (to name a few petroleum-based products), is another form of distress found in parking lots and parking stalls. Thoroughly cleaning and scraping the area and applying a skin (surface) patch, removing and replacing, or temporarily seal coating are the only three options available for repair of petroleum spillage.

To prepare the area for R&R patching, the distressed area must be removed. This is accomplished by saw cutting and removing the entire deteriorated asphalt pavement. The proper method to install an R&R patch is saw cutting a minimum of 1 foot beyond the outermost portion of the distress and removing the existing pavement. When the area is saw cut, creating right angles at the corners, forming a square (Figure 4.2), the life of the patch is lengthened far more than when removed by gouging out with a loader or picking it out with picks and shovels. Picking out (or removing with equipment without saw cutting) an area leaves too many jagged and irregular corners, which can cause future cracking to develop in the adjacent pavement or even in the new patch. In the case of an overlay, this cracking can increase reflective cracking in the new paved surface.

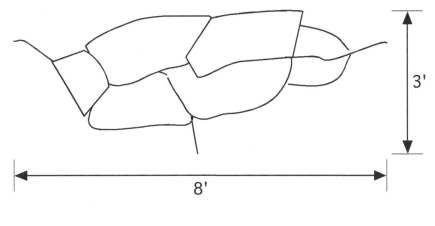

Distress Area

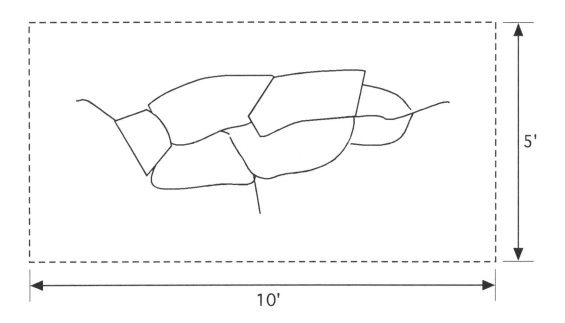

Saw Cut Area
Figure 4.2: Remove and Replace Area

After the deteriorated material is removed, the subgrade should be recompacted to 95 percent Proctor density and the exposed edges of the pavement coated with a tack coat, either by brooming or by spraying to cover the entire edge surface to ensure proper bonding between the older pavement and new hot asphalt patch material. The new patch material should then be installed and compacted. The new hot asphalt material must be at least ½ inch above the existing older pavement before compaction to yield a finished height of ¼ inch. Compacting new hot asphalt with a vibratory plate compactor to a finished compacted surface of ¼ inch above the existing pavement will also compensate for settling of the patch from traffic and natural settlement as the patch ages. The patch must now be compacted using a vibratory plate compactor (minimum pavement thickness of 2 inches in depth) or a minimum of a 3- to 5-ton static or vibratory roller to ensure proper density. Also, it is very important that the temperature of the patch material have temperatures between 260°F to 280°F in order to attain required compaction.

Figure 4.3: Phases of R&R Patching:
Saw Cutting, Removing, Placing Hot Mix Material, and Compacting

SKIN OR SURFACE PATCHING

Another surface distress that will lead to more costly structural repairs is raveling. Raveling from water, petroleum spills, and ultraviolet rays causes the asphalt and aggregate to separate and develop surface-type distresses. If left untreated, raveling will continue to erode away the surface until potholes and other structural damage develop. Also, the surface will become extremely rough and rocky as the severity of raveling increases. This distress is handled with skin (surface) patching and seal coating in the earlier stages. Medium- and high-severity raveling is corrected with more aggressive patching and new wearing surfaces, such as slurry seal, to correct the extreme rough and rocky surface. Potholes and structural damage from advanced high-severity raveling will need to be repaired using more expensive applications (R&R patching or total removal).

Skin or surface patching is applied over a distress that is not structurally compromised (e.g., high-severity raveling or water-eroded areas) and will satisfactorily remedy the problem. As mentioned earlier, alligator cracking is a structural failure and is normally found in areas where there is traffic loading. However, there may be areas that look like medium-severity alligator cracking in areas of little or no traffic loading. These are usually caused by standing and flowing water, resulting in raveling and erosion (see Figure 2.4 in chapter 2). Repairs are made by cleaning the surface and applying an asphalt tack coat, followed by a hot asphalt patch of approximately ¼ to ½ inch in thickness. Make sure the edges are straight and extend beyond the distress. The patch edges should extend beyond the distressed area by at least 1 foot on all sides, and the corners should be square or at 90° angles. Type II polymer modified slurry seal can be used and the edges kept straight with roofing paper at the time of application. When the roofing paper is removed, a straight edge is left. Each edge should be straight and all corners square in order to form a perfectly square patch. This will aid in the longevity of the patch and is aesthetically appealing to the property or streetscape (commonly referred to as good curb appeal). It is recommended to use hot mixed material with smaller size aggregate (3/8-inch hot mixed or hot sand seal), or type II (modified) slurry seal, to keep the height of the patch low. Too high of a patch will result in a traffic barrier and reduce ride quality.

Figure 4.4: Phases of Skin Patching:
Tack Coating, Placing Hot Mix Material, and Compacting Material

In Sun Belt and hot climate areas, it is important not to seal a patch until the new asphalt has had a chance to cure through the hottest part of the year. In all other geographical areas, it is recommended to seal a patch a minimum of four to six weeks after installation.

The delay in sealing a new patch is directly related to the retained volatiles (i.e., gasoline, kerosene, naphtha) in the liquid asphalt used as the binder in the hot asphalt pavement. These volatiles must oxidize (evaporate) out while the asphalt pavement sets up or becomes hard (cures). If sealed too soon, especially in hot climates or in hot temperatures, the asphalt will remain soft, causing future problems with surface scuffing, aesthetic imperfections, tracking, and so on. An exception to this rule is applying a skin patch in cooler temperatures. Should the ground or the ambient temperatures be below the specified minimum, the surface patch should be lightly sealed after compacting to prevent raveling from cooling too rapidly.

FABRIC SKIN PATCHING

Another form of skin or surface patching is a fabric skin patch. Fabric skin patches are regular skin patches placed over a fabric inner layer, which has been tacked to the older distressed paved area. Fabric skin patching is an effective method to correct a badly cracked area or area that will need to be R&R patched in the future.

This type of patching is especially effective in parking lots that need to be scheduled for full removal and repaving within two to four years. This happens due to budget constraints or a possible property or building improvement. A piece of fabric material is tacked to the surface to be patched, and a skin or surface patch is installed over the fabric. The patch length and width will have to extend (overlap) the area to be fabric patched on all edges. This type of patch takes longer to cure, since there is more asphalt material with the tack coat used to glue down the fabric. Reflective cracking is retarded by the fabric, and in the case where this type of patch is used to hold together a badly damaged area, the existing distresses or damage will hold together longer, and the reflective characteristics of the distresses are retarded due to the fabric inner layer. A fabric skin patch is a very good, short-term alternative to more expensive R&R patching on alligatored areas where it would be too expensive to remove and replace patch for a short period of time.

Figure 4.5: Phases For Fabric Skin Patching:
Installing Fabric, Installing Hot Mixed Material, Compacting

PATCHING CURING CHARACTERISTICS

The area where an R&R patch is to be installed is cut with a diamond saw or jackhammer. This task leaves a joint between the original pavement and the patch, which will require crack sealing in future years to prevent water infiltration. When the pavement is saw cut, the patch area has a "+" mark in the corners. These cut marks do not penetrate the total thickness of the pavement since the saw is circular. However, in colder climates (during winter months), these cut marks should be crack sealed to prevent water from infiltrating, freezing, and creating a spalled distress that will later require future maintenance. Another problem when installing R&R patches is creating a surface puddle. R&R patches are going to settle after installation, and they may have an uneven surface, created by workmanship, which will hold water. The larger the R&R patch in a surface area, the higher the probability of puddles forming (referred to as "bird baths"). These low areas that retain water are caused from rolling at the time of compaction, not enough asphalt material, or inadequate base preparation. R&R patches are visible after they are installed and will remain visible until the pavement is resurfaced with type II slurry seal or overlaid with hot mixed asphalt. After a period of time (maximum six months for slurry seal, around twelve months for overlays, depending on the overlay thickness), the cut patch edges will become apparent since the cut joints reflect through. This is a result of the pavement used for the patch material oxidizing, aging, and shrinking, which will create reflective cracking in the new surface.

Skin or surface patches will ravel as they cure and are very visible; in most cases, they are higher than the existing pavement, since they are installed on the top of the paved surface. Skin or surface patches will ravel and spall on the edges when installed in locations where water can flow along the edges (especially curb lines, inverted crowns, etc.). Traffic can cause shear marks (twist marks) in skin or surface patches if vehicles stop on them and turn their wheels. Also, any cracking (sealed or unsealed) under the patch will reflect through even if a fabric inner layer is placed between the older surface and the patch.

Fabric skin patching has the same curing characteristics as regular skin or surface patches and should be used as a temporary repair for badly damaged areas on pavements that will be scheduled for total removal in the near future (two to four years). When installing this type of patch, it is very important to inform the contractor they exist when the pavement is totally removed and repaved so he/she can adjust his/her removal techniques for these areas.

Both skin or surface and R&R patches are very visible when installed and will be visible after seal coating. The appearance of the patch will return in full view after seal coat material wears off from oxidation and aging. Skin and R&R patching cause problems when installed in the flow line of an inverted crown, especially when they are installed incorrectly and block or dam water. Inverted crowns are drainways constructed in the center or flow line of streets or drives. Skin patching creates a barrier for water, since the patch is installed on the surface, which adds height to the area where it is placed. R&R patches can have low spots (bird baths) and high spots (humps, caused from compacting with rollers or a vibratory plate). Both types of patching will create problems with holding back water and developing puddles in inverted crowns or areas designed for water flow.

When driving on newly placed patches, care should be taken to help prevent power steering shear marks. When a vehicle is stopped on the patch, the driver should start driving forward or backward, then start turning the steering wheel. This will help reduce the tire and twist marks in the new patch.

After patching and crack sealing is complete, a surface treatment is applied. There are three basic types of surface treatment for parking lots and homeowner association streets. These are preservative seals (asphalt and coal tar seal coats), wearing surfaces (slurry seals, chip seals, asphalt concrete friction courses [ACFC], and hot-applied precoated chip seals), and structural treatment (conventional overlays and fabric inner layer overlays). A surface treatment is selected based on the current condition of the existing surface, the cost of maintenance, and the benefit derived from a particular application.

TYPES OF SURFACE APPLICATIONS

There are five types of surface applications: fog seals, emulsions with mineral fillers (coal tar emulsions and asphalt emulsions), asphalt acrylic seals, slurry seals, and chip seals. Each application is used for a specific purpose, such as weather exposure or existing surface conditions. In this section, each surface application will be addressed separately. Before we discuss surface applications, we need to address the preparation work before a contractor can begin applying seal coat materials.

Figure 4.6: Cleaning for Seal Coating or Slurry Sealing:
Using Blowers and Sweeper

Before applying any of the surface applications, the contractor must crack seal, patch (skin or surface and R&R), and clean the surface of all debris, dirt, and dust using a high-power air broom (blower), or a rotating broom (street) sweeper, or a combination of the two. Broom sweepers cause a great deal of particulates, so some local laws require or may soon require air brooms to be fitted with dust- and dirt-retaining devices. Any striping or stenciling changes will need to blacked out using a flat black paint. Should these striping and stenciling changes not be blacked out, they will wear back through in a short period of time, causing confusion to drivers. Any petroleum spillage areas (gasoline, motor oil, hydraulic fluid, antifreeze, etc.) must be treated by one of the following: use of a bonding material, scraping and burning, or (in very severe cases) R&R patching.

There are two material types of seal coats: asphalt emulsion and coal tar emulsion. Seal coats are applied as three basic applications: fog seals, mineral-filled seals, and acrylic seals.

SEAL COATS

Seal coats are used to protect the pavement from ultraviolet rays, water, and petroleum products (oil, gasoline, antifreeze, brake fluid, etc.) and to retard the effects of early oxidation. Seal coats will normally wear off in two to three years for a one-coat application and four to six years for a two-coat application. The durability of a seal coat is a function of the dilution rate of the concentrated sealer, the application rate of the material per square yard, the amount and type of traffic exposure, power steering shear, exposure to standing water (overflow from sprinklers, puddles, etc.), and ambient and ground temperatures at the time of application. Ambient and ground temperatures are a very important factor to the longevity of seal coats. Should the high temperatures be less than 55°F, then the material will freeze and degrade. Likewise, should the temperature of the existing pavement exceed 120°F at the coolest part of a twenty-four-hour period, the material will degrade. Where application temperatures are below 55°F for a daily high, then application should be cancelled. Where the surface temperature exceeds 120°F at the coolest part of a twenty-four-hour period, the existing surface should be fogged with a water mist, and the concentrate dilution rate and material application rate should be increased. Check with the manufacturer for application guidelines. In intersections, collector streets, and main drives of parking lots and streets, seal coats will wear faster from the amount and type of traffic exposure. In some instances, it is recommended to apply a third coat to even out the wear pattern where there is a large or heavy traffic count. Crack sealing, patching, and all distresses that were present in the existing pavement will be visible after applying a seal coat. Striping and painted lines, lettering, and logos will shadow through a seal coat in a short period of time, unless they are blacked out with a flat black paint prior to applying a seal coat. All painted markings and directional symbols will reappear as the seal coat material wears off. The black-out paint will be visible after the sealer wears off; however, it will not result in white or yellow stripes, lettering, or colored logos which would confuse the driver.

FOG SEALS

Fog seals are an asphalt emulsion or (occasionally) coal tar, heavily diluted with water (as much as 70 percent), with no mineral filler additive. In some rare cases, a rejuvenating agent (which should be used with caution) is used. This type of seal coat material is sprayed on the surface and provides some remedial protection, but offers little benefit to the longevity of the pavement. Fog seals are mainly used for aesthetic purposes and should be treated as such. If a rejuvenating agent is used as a fog seal, care should be taken, because overapplication can result in bleeding, tracking, a slippery surface, and many other undesirable conditions that could create expensive corrective measures or liability exposures. If you select to use a rejuvenating agent to fog seal your pavement, involve the manufacturer to ensure the correct application rate and specifications, and to assist you with selecting a contractor familiar with that particular application.

EMULSION OR EMULSIFIED SEAL COATS (MINERAL FILLED)

Emulsified seal coats are either asphalt or coal tar based and are mixed with water and an emulsifying agent. These seal coat applications contain a mineral filler and 20 to 30 percent (per the manufacturer's recommendation) water (diluting agent), a mineral filler (usually silica sand in coat tars and a fine-grade slag or sand for asphalts), ash, clay, and a rubber, polymer, or other synthetic additive to enhance the tensile strength characteristics of the material. Since seal coat setting and curing is a function of warm ambient and ground temperatures, applications during the colder months require a coalescing agent to be added to aid and accelerate the curing process. For asphalt emulsions in colder climates, an anti-stripping agent is added to retard the stripping effects of snow and ice. The base type of sealant (coal tar, asphalt, or a combination of the two) is generally determined by the climate and geographical location of the property and the availability of seal coat materials. In colder climates, coal tar emulsions, coal tar and asphalt emulsion blends, and asphalt emulsions with an anti-stripping agent are more commonly used, and in warmer climates, asphalt emulsions are the material of choice.

Asphalt is derived from the "bottom of the barrel" portion of refining. This material contains many volatiles (sulfur, naphtha, diesel, etc.), yet is a good product for asphalt binder material for roads, roofs, crack sealants, and asphalt pavement products. Asphalt emulsions are developed by a unique process of blending water, asphalt, mineral fillers, and an emulsifying agent to produce very good seal coat materials and binders for seal coats and slurry seals. Coal tar emulsions, a derivative of coal, are manufactured much the same way, but are not often used as binders in slurry seals or pavements.

Coal tars are developed from the waste product of coal usage and have been used for many years in the eastern United States, whereas asphalts are derived from asphalt petroleum and have been used extensively in the western United States. There are several differences between them. Coal tar emulsion sealers are more resistant to fuel and petroleum spills (e.g., gasoline, oil, brake fluid, and antifreeze), while asphalt-based sealers are not. Coal tar will hold a blacker appearance longer than asphalt-based sealers for aesthetic purposes. However, asphalt emulsions have better filling characteristics around raveled, rough, or pitted surfaces. Coal tars are a harder-based material and will require an additive to reduce the rigidity of the cured material and more grit or sand to reduce the slippery effects when wet. Because of the hard solid surface, it is difficult to apply an asphalt-based sealer over coal tar in future applications. However, once the coal tar sealer material has worn off, asphalt emulsions will adhere to the surface and can be used. Coal tars will develop a distress identified as checking quicker than asphalt-based materials. Checking is when the material shrinks and leaves an appearance of a drying lakebed, or too much paint on a wall. This distress cannot be corrected, and eventually the surface will have to be removed and overlaid or replaced. Checking is a progressive distress, and when more sealer is applied over checked material, the distress will become worse. After too many applications, the buildup of material will cause the sealer to curl at the edges of the checking patterns and allow water to infiltrate between the sealer and the original pavement. This will cause what is referred to as "sealer rot" or "pavement rot," and the sealer material will flake off the original pavement. Even when type II slurry seal or overlays are applied, the sealer can still separate from the original pavement, causing the type II slurry seal or overlay to delaminate, separate, and detach from the original pavement.

Warning!! Recent studies by the United States Geological Survey (USGS) have shown that coal tar seal coats contain high amounts of polycyclic aromatic hydrocarbons (PAHs). PAHs have been found to be carcinogenic and heavy exposure may cause cancer. As a result, several cities (including Austin, TX and Washington DC) have banned the use of coal tar sealer. Therefore, the contractor and the owner/manager can be subjected to any claims that may arise due to the use of coal tar sealers. As a result, it is the authors' recommendation that coal tar sealers not be used. Because of its ability to resist fuel spills and endure cold weather climates, cold tar is still in demand, therefore, information on coal tar is still included in this book. It is important to check and understand any federal, state, and local rules and regulations prior to using coal tar. Ultimately, it is recommended that asphalt-based seal coats (which can be modified to meet specific environmental needs) be used in lieu of coal tar sealers.

Asphalt emulsions will wear at a quicker rate from exposure to water and traffic; however, they do not check as quickly or as severely as coal tars. Care must be taken not to allow too much seal coat material to be applied to the surface. Also, make sure the surface is ready and due for a seal coat. Seal coats that are applied too frequently on top of each other will also create checking. When using asphalt emulsions in colder climates where there is ice and snow, it is strongly recommended to use an anti-stripping agent, to reduce the effects of the ice and snow, or a coalescing agent to accelerate the setting or curing process due to colder temperatures. Ice and snow will cause asphalt emulsions to strip off the pavement and substantially reduce the longevity of the seal coat; however, the addition of an anti-stripping agent will retard this process and add longevity to the seal coat. In many colder climates, manufacturers will blend coal tars and asphalt emulsions together to provide a material with increased longevity and resistance to ice and snow. A coal tar and asphalt blend is an excellent application in colder climates to reduce the frequency of seal coating. Both materials, when blended to manufacturer specifications and properly applied, should last two to four years before a resealing may be required.

ACRYLIC SEALS

Acrylic seals are a blend of asphalt emulsion and an acrylic (sometimes referred to as a plastic) or other curing (yet pliable) additive. For the most part, there is no mineral filler in acrylic seal coat material, and this particular type of sealer will set and cure more rapidly than seal coats with a mineral filler. As a result, the absence of mineral filler reduces the durability of this type of product. Because of the quicker curing or setup time, this product is used by several government agencies, since streets and highways can be opened to traffic sooner than with sealers containing mineral fillers. There are no smoothing characteristics with this material, which allows it to fill around and between exposed rocks or rough surfaces, but it will provide the same protective benefits as the mineral-filled sealers. Most parking lot owners prefer the mineral-filled sealers as opposed to the acrylic sealers because of aesthetics of the final product application. Most HOAs use the mineral-filled sealers, again for the aesthetics; however, several larger HOAs and community associations use the acrylic sealers for the convenience of reopening the streets to traffic sooner. The cost is similar with each product, so there is very little to no savings from using one material over the other.

In all cases, all the types of seal coat materials mentioned will protect the existing pavement from environmental, chemical, and water exposures and increase the life of the existing pavement. However, seal coating is a preventive and corrective type of maintenance, not a structural resurfacing. All maintenance applications (patching, crack sealing, etc.) will be visible after seal coating, and all imperfections in the original surface will be visible and highlighted. This is particularly a problem with crack sealing, because smaller cracks that were not previously sealed will appear larger and will be questioned by residents or tenants.

SEAL COAT APPLICATIONS

There are two methods of applying seal coats: squeegee or spray. Squeegee-applied sealer is placed on the paved surface and the material is spread over the surface with a large squeegee affixed to the back of a wheeled machine. Spray-applied sealer is applied by hand using a spray wand, applied mechanically with spray bars on a squeegee machine or a distributor truck. Prior to applying a seal coat, the pavement must be trimmed around curbs, islands, pavement edges, or anywhere else the sealer will be difficult to squeegee or spray. This is usually accomplished by hand trimming with soft brooms and hand squeegees or by spraying against a spray shield held directly against the curb, pavement edge, swales (valley gutters), and so on. Brooms and hand squeegees should be used to seal tight areas such as stairway landings, small parking areas, and niches. When soft brooms and hand squeegees are used, it is not necessary to trim for a second coat application. However, if the trimming is applied by spraying against a shield, a second trimming will need to be applied with the second seal coat application. On projects with curbs (usually streets), water is engineered to move along the curb, so it is important to ensure the paved edges are well protected with the proper sealer application. One-coat trim applications will not provide the protection needed over a period longer than two to three years. Water will remove the sealer along curb lines more rapidly than the center of the street or parking lot.

Figure 4.7: Preparation For Seal Coat:
Trimming and Treating Petroleum Spots

The differences between these application methods (squeegee or spraying) are the finished appearance after application, surface appearance while the material is wearing off, the removal or wearing of the material on the tops of the surface aggregates (referred to as "capping"), and the filling characteristics.

When using the squeegee method of application, the material is deposited on the surface in the front portion of the squeegee machine. A weighted or hydraulically locked rubber squeegee bar in the rear of the squeegee machine is pulled over the material, distributing it on the surface. Squeegeed material fills the open or gapped areas (voids) between and around the surface aggregates. This produces a thicker application, applying more material around exposed surface rocks, yet places a thinner layer of material on the surface of the exposed rocks protruding from the pavement. The material covering the tops of the exposed rocks is very thin and will wear off rapidly, exposing the underlying rock color. This is referred to as capping because the sealer material has worn off the caps of the protruding rocks. Capping is caused by the material wearing off the rocks due to water, ultraviolet ray exposure, or tire friction caused by everyday traffic. Other than aesthetics, capping does not have an adverse affect on the performance of the material as long as there is a sufficient application around the surface rocks. The thicker amount of sealer has filled down in and around the surface rocks, sealing the pavement surface from water, ultraviolet rays, and traffic exposures. The squeegee method is a better application for rough or highly raveled pavement.

Figure 4.8:
Squeegee-applied Seal Coat

When using the spray method of application, the material is sprayed on the surface with a handheld spray wand connected to a supply tanker or by spray distribution from a spray bar attached to a distributor (truck, tanker, squeegee machine, or similar piece of equipment). The material is distributed by the person using the spray wand moving from side to side or in straight passes by the use of a spray bar and distributor. Spray-applied materials place an even coating across the entire surface, including the area in between the aggregate, as well as the rock surface itself. As a result, spray-applied material covering the tops of the exposed rocks is thicker than squeegee-applied material. It will wear off, exposing the underlying rock color, but not as rapidly as squeegee-applied material. The wearing characteristics of each application varies as well. Squeegee-applied material leaves an inconsistent drag pattern of thick and thin passes of material, whereas spray-applied material will leave the pattern of the wand movement or direction of the spray application.

Figure 4.9: Spray-applied Seal Coat
by Wand and Distributor Method

As stated before, when a seal coat caps, it normally doesn't indicate a poor sealer product or application. The more traffic and water exposure, the quicker capping will occur. Sealer is meant to seal the voids and gaps in and around the rocks protruding on the paved surface. When surface rocks cap, check the wear of the sealer material below the rock surface and around the rocks. If there is sufficient material around the rocks, the sealer is still in good condition. If the material is wearing in between the rocks and the sealer is less than one year old, contact the contractor to have it corrected.

SEAL COAT CURING CHARACTERISTICS

Seal coats typically take twenty-four (24) hours to cure and become dry enough to drive on. Seal coats can be walked on in four (4) to six (6) hours, depending on ground temperatures. Again, there are two types of seal coat materials: coal tar emulsions and asphalt emulsions. In colder climates, manufacturers will sometimes blend the two materials together using a maximum of 20 to 30 percent asphalt emulsion and 80 to 70 percent coal tar emulsion. The asphalt emulsion will add flexibility and darker coloring to the coal tar. Coal tar emulsions wear better when exposed to water and ice (freeze/thaw), especially surface exposure in colder climates. Water from sprinklers, rain, melting ice, and snow will leave stains on the surface. This is not a defect of the pavement or seal coat material; it is the result of water reacting with the emulsifying agents in the seal coat, fertilizers in the adjacent garden areas, or other debris that can cause color stains with exposure to water.

Seal coats contain a mineral filler, which will have the tendency to roll out or shed. This roll-out is noticeable by loose sand on the surface and in curbs. Roll-out will track into buildings, on carpets, and into vehicles. When sand starts appearing on the surface or in the curbs (or when residents or tenants complain about sand in their homes or suites), the pavement will need to be swept with a broom (street) sweeper. This will pick up the loose sand or roll-out. Where there are covered parking stalls or a large number of parked vehicles, blowers will need to be used to move the sandy material to the drive lane so the sweeper can pick it up. Vehicles will leave power steering twist marks in seal coat material until it has had time to completely set up. In most cases, these marks will knead out or go away. Also, seal coat material will cover stripes as vehicles pick up a small residue on the tires and cover the stripes. This is not a defect of the seal coat material or the striping paint. There is a thin petroleum slick on the surface after the seal coat dries, which is picked up by vehicle tires, or there is seal coat material on tires from power steering twists, which is tracked onto the stripes.

Coal tar-based emulsions, after they have dried, will have a slate dark black appearance; however, they will eventually turn a gray color. It is also important to have some kind of mineral binder added to the material to make the surface abrasive in order to reduce slippery conditions when it gets wet. Coal tar emulsion products and blends are very slippery when wet if a granular mineral is not added. Usually 3 lb. of silica sand per gallon of coal tar emulsion or equivalent should be added to reduce any slippery condition. The cured coal tar seal coat is very hard and rigid, which makes it very susceptible to checking if it is applied too thick (high application rate). It is very fuel resistant, and any petroleum-based material spilled or leaked on it will only set on top and not penetrate. Coal tar sealers are the best for protection against oils, gas, diesel, jet fuel, antifreeze, and so on. Also, coal tar emulsions are very resistant to water. Coal tar emulsion seal coats have a strong offensive odor at the time of application, from the base chemicals that are in the material itself, and can cause skin and eye irritations. Over a period of time after application (usually one week), the odor will subside and go away. Skin and eye irritation will not be a problem after the material has cured (usually two [2] to three [3] days). The mineral binder in the emulsion will roll out as the coal tar cures and may be tracked into homes, offices, and vehicles. Sweeping with a broom (street) sweeper is recommended about sixty (60) days after application and around six months after the initial sweeping.

Asphalt-based emulsion will have a coal black appearance and look soft. All vehicles will leave tire track impressions in the material for several days after application. This is not a defect of the material or application, but a normal characteristic of asphalt emulsion sealers. These areas will knead out and become unnoticeable approximately three to four weeks after application. The amount of scuffing and soft appearance is directly proportionate to ground and air temperatures. The higher the temperatures, the longer it will take to set up. Asphalt emulsions absorb ultraviolet rays and become very thermal. If nighttime temperatures are cool, the asphalt emulsion will lose the thermal heat and cool down, causing it to cure more rapidly. Asphalt emulsions contain a mineral filler from manufacturing and do not require any additional sand or mineral additives for adhesion and antislippage on the existing surfaces. Asphalt emulsions are not a hard surface like coal tars and do not become as slippery when wet. The main benefit to asphalt emulsions is the filling characteristics for filling around surface aggregates. Asphalt emulsions are not fuel or petroleum resistant, but do not develop the distress of checking as rapidly or with the same frequency as coal tars.

A common characteristic of all seal coats is a light oil film that develops on the surface after they have cured. This film attracts normal air dust, and this dust is picked up by tires of vehicles. This will leave a light gray mark on drives and other concrete surfaces. This is normal and will go away over a short period of time. All crack sealing and patching will appear through seal coats, usually within twenty-four (24) hours of application. Also, seal coats will flake off when placed over crack sealant, as the crack seal expands and contracts with the existing pavement. Curing of seal coats can be retarded or slowed by high humidity or dew point. When this occurs, the material will remain sticky and striping will darken or turn darker or white paint will turn yellow or brown. When the humidity and dew point are high, allow three (3) to four (4) days for curing in lieu of twenty-four (24) hours prior to striping. Grocery carts, skateboards, baby strollers, and any type of object with small wheels will track seal coat onto walks and into buildings before fully curing. Cleanup is easy and should be done as soon as possible or floor protection carpets can be placed on floors during the curing period. Also, any crack sealing, patching, older stripes, or other maintenance applications installed prior to seal coating will reappear as the sealer wears off from exposure to ultraviolet rays, water, and friction from vehicle tires. The main thing to remember is seal coats are a preservative application, *not* a new wearing surface application.

When driving on a new seal coat surface, take care and help prevent power steering shear marks. When stopping a vehicle, start driving forward or backward, then start turning the steering wheel. This will help reduce the tire marks in the new surface.

You will notice light gray marks in driveways or other flat concrete surfaces after seal-coated streets or parking lots are put back in use. These marks are caused from normal air dust settling on the surface, adhering to the thin petroleum film, and being picked up by tires of vehicles. These marks are not from sealer material being tracked onto driveways, swales or valley gutters, or other flat concrete surfaces. These marks can be removed by power washing with a garden hose and broom. Tracked asphalt is jet black in color. Should jet black marks in driveway or flat concrete surfaces appear, notify the contractor or consultant to schedule a more aggressive form of cleaning. Should asphalt emulsion material get tracked on carpets or tile or splashed on a vehicle, use a tar or grease remover (auto mechanic hand cleaner without pumice is a good choice) to remove the material. Do not rub, just dab with a soft cloth. Vehicles and tile or colored floors will have to be washed and waxed after using a cleaning material to cover the treated areas, as the cleaner will remove any wax.

SLURRY SEALS AND CHIP SEALS

Slurry seals and chip seals are used to apply a new wearing surface on a badly raveled or rough surface. These two applications are used in lieu of an overlay, where an overlay would be too excessive for the existing conditions. Both these applications are a type of wearing surface installed on an existing paved surface. If the existing pavement is badly cracked or has a high amount of structural damage, these applications are not recommended. Patching and crack seal material can reflect through slurry seals, but not at the rate they will with seal coats. However, cracking will return or reflect through type I, II, and III slurry seals in a few months, if not weeks. Cracking will reflect through chip seals, yet they are not as visible or are concealed with the rough, rocky texture of the chip seals. These applications are only ¼ inch to ⅜ inch thick, and as the pavement expands and contracts with temperature changes, the existing cracks in the underlying pavement will reflect through in the new surface. These applications are also subject to power steering shear and will peel off when exposed to the twisting tires of vehicles. As these two applications cure or age, they will release (shed) small rocks that did not bind to the tack coat (chip seals) or sand that cannot be used to fill voids in the raveled original surface underneath (slurry seals).

Both applications will require sweeping several times with a street or broom sweeper to clean up the sand and rocks. Slurry seals will require a seal coat in twelve to eighteen months after application to retard the effects of raveling and protect the investment in the new wearing surface. Chip seals are not recommended for parking lots or HOA streets due to the amount of rocks that get flung by fast-moving vehicles, which could chip windshields and paint. Also, in tight turning areas such as cul-de-sacs, trash trucks, delivery trucks, and other large vehicles may strip the chips off the pavement and leave a very unsightly appearance and rocks in gutters and yards. A type II slurry seal (specify the use of type II to ensure getting a slurry seal in lieu of a seal coat) with a polymer additive will resist power steering shear more than chip seals or conventional type II slurry seals, and will release less rock and sand than a standard type II slurry seal.

A very good combination that is less expensive than an overlay is a chip seal with a type II polymer slurry seal applied six to twelve months later. This is referred to as a cape seal and is a very good semi-structural layer where a thin overlay would be used. Another application using open-graded rock like chip seals is Nova Chip®, which provides an open-gapped solid surface. Nova Chip® is solidly bonded by hot asphalt and emulsion. This application is a very good wearing surface and does not shed rocks like conventional chip seals. This application is recommended for HOA streets but not parking lots. It sets up very quickly, and where there is hand trimming (i.e., around islands), it cannot be properly applied before it sets up. Therefore, it is not a good application for parking lots unless the parking lot is wide open and has no barriers or obstructions. For HOA streets, it is a fair substitute for rubberized asphalt at a lower expense. Rubberized asphalt will be discussed next when we talk about overlays and new pavements.

SLURRY SEALS

Slurry seal is a combination of blending graded aggregate (slurry seal sand), emulsified asphalt (classified as CQS, CQS-1h, which is a cationic quick set emulsion, or QS, QS-1h, which is an anionic quick set emulsion), a small amount of cement, water, and an emulsifier that is developed by the emulsion manufacturer. These materials are blended in an on-site distributor, with a large sand container and bins for the other materials and water. The blended slurry seal material is laid directly into a holding sled behind the distributor truck on the pavement and leveled or smoothed by a squeegee on the holding sled. Prior to applying the slurry seal material, the surface must be thoroughly cleaned of all loose debris, dust, sand, rocks, and dirt. During high heat months, the surface should be cooled by using water misting or fogging to retard rapid flocculating and coalescing (premature setting). Because slurry seal is not a structural remedy, crack sealing and patching (skin and R&R) must be completed prior to applying the slurry seal.

Improper preparation of the surface prior to applying slurry seal will reduce the life expectancy of the new surface. Dirt, dust, and fine debris left on the surface will impede solid adherence to the original surface. Cracks and failure areas will reflect back through the slurry seal in a short period of time, leaving the cracking condition much the same. This is the reason for thorough crack sealing prior to applying the slurry seal material. When reflective cracking does develop in the new surface, the cracks have been previously sealed from water infiltration. After the surface is prepared, some agencies require the original surface to be primed or tacked with an acceptable tack coat (usually SS-1h, SS-1) before applying the slurry seal. Prime or tack coating will help the slurry seal adhere to the existing pavement; however, care will have to be taken to not apply too much tack coat material, which can lead to a distress known as bleeding. Bleeding is too much liquid asphalt binder, and the slurry seal sand cannot absorb it. The liquid asphalt binder will then surface, causing a tracking problem.

Figure 4.10: Application of Type II Slurry Seal

There are three types of slurry currently in use. Type III is a thick application with a maximum aggregate size of about ⅜ inch, used primarily for roads and highways. Type II is a medium-thickness application with maximum aggregate size of about ¼ inch. This is used on city and HOA streets and parking lots with very high-severity raveling. Type I is a thin application using fine sands. This particular product is used in walk paths, leveling courses, and on some sports courts. Type I is not commonly used in street and parking lot resurfacing. For streets and parking lots, type II polymer modified slurry seal is used. Keep in mind that slurry seal is a new wearing surface installed on very rough and highly raveled surfaces. Where the surface has high-severity raveling, a two-coat application is recommended. The second coating is applied after the first application has cured and set up, usually twenty-four (24) hours later.

SLURRY SEAL CURING CHARACTERISTICS

Slurry seal is an excellent maintenance alternative to overlays where there is low- to medium-density of cracking and the existing surface is rough in texture with low- to medium-severity raveling, yet is structurally sound. However, slurry seal should be seal coated in approximately one to one and a half years to protect it from natural deterioration. Slurry seals are also used as a cape seal over chip seals to create a smooth wearing surface. Slurry seals are a wearing surface *only* and should not be used as a structural pavement application or alternative. When choosing a slurry seal, the curing characteristics must be taken into consideration. Slurry seals are applied in a wet condition. The color is a muddy brown because of the suspension of the asphalt by an emulsifying agent. Once the material begins to dry and set up (the correct term is "break"), the material will turn jet black. This indicates the emulsifying agent has released the asphalt and the emulsifier is now separated from the mixture. The surface is grainy looking, and the larger stone particles can be seen on the surface. The overall appearance of the new slurry seal is very rough and in many cases causes residents and tenants to question the aesthetic appearance.

Slurry seals will then shed sand and rock for the first five to eight months while the material cures and releases excess material not required to fill gaps or voids in the existing surface. The smoother the existing surface, the more shedding of sand and rock that will occur. A broom sweeping (street sweeper) should be scheduled approximately six to eight months after the slurry seal is installed to clean up the loose rocks and sand. Sweeping may be required one or two more times at six-month intervals. Seal coating will aid in preventing or retarding the shedding effects.

Another curing characteristic of slurry seal is peeling. Slurry seal will shear or peel off very smooth areas that exist in the original surface. The original surface will begin to show through the slurry seal, but this does not indicate a poor application; the slurry seal did not have a rough enough area to adhere to, and the slurry seal material sloughed off. The color of the slurry will gray out, and the smooth area will blend and not be very apparent. Remember, slurry seals are a new wearing surface used to fill in and smooth existing rough surfaces. Upon curing properly, slurry seal wearing surfaces are smooth.

Another concern relative to slurry seal is reflective cracking. Cracking that has been slurry sealed over will reflect back through the slurry seal in a short period of time (usually weeks). This is a natural process from the expansion and contraction of the original surface. This is the reason for crack sealing prior to applying the slurry seal. The crack sealing will eventually shadow through the slurry seal and be visible until the surface is seal coated. Slurry seal is not a crack sealing material and should not be used as such. When slurry seal is installed during cold temperatures in the existing surface or minimum ambient temperatures, or in shadowed areas (trees, buildings, etc.), it will take much longer for the slurry seal to break. Also, the color of the slurry seal will not be jet black after it sets and cures. It is not recommended to apply slurry seals if the ambient and ground temperatures are below 50°F to 55°F and rising. The material is still subject to slow curing or breaking and coloration problems at these low temperatures. Also, should the temperatures approach freezing at night before the slurry material has set, it will freeze and peel off. A large portion of slurry seal is water.

The final curing characteristic is staining in the form of a white or yellow residue. This is the emulsifying agent (usually detergent) that is being washed out from exposure to water. This residue will form bubbles and a soapy white residue as the material oxidizes and turns gray. This staining will go away as the slurry seal cures. It is very important that the property owner or homeowners association board of directors understands the curing characteristics of slurry seal prior to installing it so they are able to answer questions of residents or tenants. A slurry seal surface placed over a chip seal (cape seal) is the next best alternative to an overlay. With a wearing surface like slurry seal, patches, crack sealing, and other defined repairs are not readably visible (yet they can be noticed). This is especially true with patching when the existing patches or new patches are installed higher than the original pavement. As the slurry seal cures and hardens, the cracking or saw cuts around the patch will reflect through the new surface. Also, slurry seal material will flake off crack seal material as the pavement and crack seal material heat and expand from daily temperatures fluctuations.

When driving on a new slurry seal surface, care must be taken to prevent power steering shear marks. When stopping a vehicle, start driving forward or backward, then start turning the steering wheel. This will help reduce the tire marks and power steering shear marks in the new surface. You will notice light gray marks in driveways or other concrete flat surfaces after slurry-sealed streets or parking lots are put back in use. These marks are caused from normal air dust settling on the surface and adhering to the thin petroleum film and being picked up on tires of vehicles; they are not material being tracked onto the concrete. These marks can be removed by power washing with a garden hose and broom. Tracked asphalt is jet black in color. Should jet black marks in driveway or flat concrete surfaces appear, notify the contractor or consultant to schedule a more aggressive form of cleaning. Should asphalt emulsion material get tracked on carpets or tile or splashed on a vehicle, use a tar or grease remover (auto mechanic hand cleaner without pumice is a good choice) to remove the material. Do not rub, just dab with a soft cloth. Vehicles and tile or colored floors will have to be washed and waxed after using a cleaning material to cover the treated areas.

How Emulsion in Slurry Seal and Seal Coat Cures

Seal coats and type I, II, and III slurry seals are blended or manufactured with asphalt or coal tar emulsions for the binder material. When seal coats and slurry seals are first installed, they are brown or gray in color due to the amount of water in the mix design. As the material cures (or breaks), the color will turn to dark black, and the material will change from wet (liquid state) to a solid surface. This is referred to as coalescing, which is shown in Figure 4.11. When the surface temperatures of the existing paved surface or the ambient temperatures are approaching the minimum application temperatures, a coalescing agent is recommended to accelerate the coalescing process.

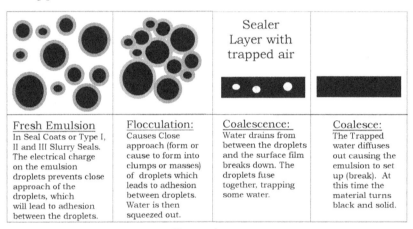

Fresh Emulsion	Flocculation:	Coalescence:	Coalesce:
In Seal Coats or Type I, II and III Slurry Seals. The electrical charge on the emulsion droplets prevents close approach of the droplets, which will lead to adhesion between the droplets.	Causes Close approach (form or cause to form into clumps or masses) of droplets which leads to adhesion between droplets. Water is then squeezed out.	Water drains from between the droplets and the surface film breaks down. The droplets fuse together, trapping some water.	The Trapped water diffuses out causing the emulsion to set up (break). At this time the material turns black and solid.

Figure 4.11:
Curing of Asphalt Emulsion

CHIP SEALS

Chip seals are a combination of clean aggregate, usually ¼-inch or ⅜-inch, applied to a hot liquid asphalt (which is spray applied from a distributor truck) and rolled with a 3- to 5-ton roller and two or more rubber-tired pneumatic rollers This application is used mainly on city or county streets and highways but has very limited use in parking lots or HOA streets. After the chip seal has set up (cured), loose aggregate is swept off, resulting in a sound-wearing surface. In most cases, an asphalt fog seal is placed over the chip seal to firmly set the aggregate and leave a finished black surface. Chip seal applications are not often recommended for parking lots and HOA streets because of power steering shear marks, stripping off of the material in tight turns in cul-de-sacs, and the liability problems generated with loose and flying aggregates. Traffic will propel loose rock chips from tires as the vehicle travels over the surface. This causes problems with broken and cracked windshields and chipped paint. Due to the amount of rock chips applied, there will be plenty of residual rock chips that do not adhere to the hot tack coat and need to be swept off the paved surface. This abundance of excess chips makes this application a messy project with a great deal of cleanup the same day as the application and for several days or weeks after the application. Conventional chip seals are applied with a large chip spreader, distributor truck for the tack coat, rollers, mechanical brooms, and dump trucks. It is a fast-moving application, and traffic can drive on a chip seal shortly after it has been applied.

Another type of chip seal is a stress-absorbing membrane (SAM). This type of application is usually used on streets and highways as a mineral inner layer for overlays to effectively retard reflective cracking and as a leveling course for low areas. In past years, it has become a surface treatment in pavement maintenance. It is applied similarly to a conventional chip seal, using a smaller aggregate. A SAM is comprised of a hot rubber modified asphalt binder (tack coat) and covered with a crushed fine stone. The excess is swept up, and the remaining stone is absorbed into the tack coat. This is a very flexible application and wears very well. It is subject to power steering shear and the same traffic problems as a conventional chip seal.

Figure 4.12:
Application of Rock Chips to Hot Tack Coat and Finished Texture

An alternative to a chip seal is a cape seal. A cape seal is a conventional chip seal with a type II polymer modified slurry seal placed over it after it has had a chance to cure and shed loose rock. This type of application has a higher strength than other type II slurry seals and provides a smooth surface. It is a very good alternative to the hot mixed asphalt concrete (HMAC) wearing surfaces discussed next. The difference is the length of time needed for installation. This process takes several weeks to complete, with two separate scheduled applications.

Chip seal applications should be seal coated eighteen to twenty-four months (with exception of Nova Chip®, which should be sealed twenty-four to thirty-six months) after installation to reduce surface deterioration from ultraviolet rays, water, and oxidation. A seal coat will also aid in securing any rocks that may have come loose in the past few months. Precoated chip seals (the rock chips are precoated with an asphalt or an asphalt and polymer product prior to application) should be sealed only after they have had enough time to oxidize (usually five to six years minimum). A problem that exists when sealing precoated chip seals is bleeding (referred to as cap bleeding). This is caused by too much liquid asphalt product and not enough rock surface area to absorb the liquid asphalt. When bleeding happens from too much asphalt, tracking will occur, taking product into buildings, around pools, and onto other flat surfaces.

CHIP SEAL CURING CHARACTERISTICS

Chip seals are a wearing surface *only* and should not be used as a structural pavement application or alternative. When choosing a chip seal, all curing characteristics must be taken into consideration. Chip seals are applied by spraying a hot tack coat then a coating of rock chips. Once the chips have been placed and rolled, traffic can resume using the pavement, usually in one (1) to three (3) hours, depending on air and surface temperatures. The surface is rocky looking, and stones and rocks will come loose from the surface. Also, there will be a large amount of rock material left loose on the surface because not all of the rocks used in the application are imbedded in the tack coat. After curing, the loose rock is swept up and the surface is sealed with a light coat of an asphalt-based emulsion. The overall appearance of the new chip seal is very rough and, in many cases, this will cause residents to question the aesthetic appearance. Chip seals will shed rock for the first six to eight months while the material cures and releases excess material not required by the existing surface. This is natural and should be expected. The smoother the existing surface, the more loose stones and rocks that will occur. A broom sweeping should be scheduled two months after application and approximately six to eight months after the chip seal is installed to pick up rocks that have loosened during the curing process. Other curing characteristics are peeling or shearing of the chip seal, and in some cases the material will have bleeding of the asphalt tack coat on the surface.

Chip seals will shear or peel off very smooth areas that exist in the original surface. The original surface will begin to show through the chip seal in areas where the original pavement was smooth prior to installing the chip seal, but this does not indicate a poor application. Some areas can have bleeding, which is the asphalt tack coat surfacing through the rock chip. The severity of the bleeding is defined as normal or severe. Normal bleeding is a black tacky-looking surface, yet rock can still be seen on the surface. There can be tire tread marks in the material and, in some cases, a polished-looking appearance. Severe bleeding is the surfacing of the tack coat, and actual puddles of asphalt binder will form on the surface. These areas do not settle out or stiffen over time and, in some cases they will flow off the pavement into curbs and driveways. When bleeding occurs, it can be corrected in two ways. If the bleeding is normal, it will cure with oxidation over six months. Should the bleeding be severe, an application of type II slurry sand, without any asphalt additives, may be spread over it and rolled in with pneumatic rubber-tired rollers. This application can take one or two installations, depending on how severe the bleeding is. Also, bleeding chip seals can be treated with a 1 to 2 percent lime-water wash. This will retard the bleeding; however, the pavement will have a white chalky appearance after the application for several weeks (and in some cases, for months).

Once the tack coat heats up again from thermal heating, the bleeding will begin again. Aging or oxidizing is the only method of curing normal bleeding. Chip seals require one hot season to cure properly, whether applied in spring or fall, and all the bleeding characteristics will occur during that first hot season. However, it is not recommended to apply chip seals in the height of the hot summer months, especially in the high-temperature Sun Belt climates. Chip seals require ample time to cure and for the tack material to oxidize. Thermal heating from direct sunlight and ground temperatures will make the tack coat soft and cause it flow for ten to eighteen months, depending on the amount of tack coat used. In tight-turning radii or cul-de-sacs, the chips will roll in the softened tack coat, causing the rock and stone chips to move off the existing street. When exposed to rapidly moving vehicles and large trucks (trash trucks, delivery trucks, etc.), the rock and stone chips will slide off the surface into the curb or adjacent property and expose the original pavement.

The color of the chip seal will gray out and the smooth areas will blend and not be readily apparent. Remember, chip seal is a new wearing surface with a rough texture. Chip seals are the strongest and hardest of the wearing surface applications. Cracking that has been chip sealed over will reflect back through the chip seal surface in a short period of time. This is a natural process from the expansion and contraction of the original surface. This is the reason for crack sealing all cracks with openings of ¼ inch or more prior to applying the chip seal. The crack sealing will eventually shadow through the chip seal and be visible until the surface is seal coated. With a wearing surface like chip seal, patches, crack sealing, and other defined repairs are not readably visible (yet they can be noticeable). This is especially true with patching when the existing patches or new patches are installed higher than the original pavement.

When driving on a new chip seal surface, care must be taken to prevent power steering shear marks. When stopping a vehicle, start driving forward or backward, then start turning the steering wheel. This will help reduce the tire marks in the new surface. It is recommended to follow this procedure every time vehicles come to a stop. Also, it is imperative to drive 20 MPH or slower to prevent rock chips from flying off the road into any other vehicles, chipping windshields and paint.

Light gray marks can appear in driveways after streets are open. These marks are caused from normal air dust being picked up by tires. These marks are not caused by tracking the asphalt. Tracked asphalt is jet black in color. Should jet black marks appear in driveways or on other flat concrete structures, contact the property owner, manager, or HOA street committee, and they will notify the contractor or consultant.

HOT MIXED ASPHALT CONCRETE WEARING SURFACES

A hot-mixed asphalt concrete wearing surface is applied to the existing pavement to provide a smooth, skid-resistant surface and is usually an open graded (rocky) surface installed to correct medium- and high-severity raveling. This wearing surface is thinner than a hot-mixed asphalt overlay and does not provide some structural benefit of an overlay.

A common wearing surface is an asphalt concrete friction course (ACFC). An ACFC is a combination of small aggregate and asphalt binder, batched in an asphalt hot mix plant and applied as a hot paved surface. It has all the same characteristics of a conventional pavement, but is not as dense or thick. An ACFC does require a tack coat application prior to installing, and if too much tack coat is applied, it will have a tendency to bleed later on. ACFC is applied using the same laydown or paving machine, rollers, and dump trucks as overlays and paving. Traffic is held off the surface until it cools and sets up.

Another application is a process of applying a hot asphalt surface. This combination of a gap graded (rocky appearance) hot mix, using a special modified binder, is applied with a specialized laydown machine. A modified emulsion tack coat is applied ahead of the screed (the portion of the laydown machine where the hot-mixed asphalt material passes under and is spread evenly and leveled) prior to applying the hot-mixed material. The screed sets the grade of the mix and, in this case, also compacts the mix as it is placed. The modified emulsion wicks up into the hot-mixed asphalt and bonds the hot-mixed material to the old surface. This places a very rich asphalt layer at the interface of the older surface and the new one. This layer of modified asphalt helps to mitigate stress cracks from reflecting through the new surface. This is a very effective form of an HMAC wearing surface. A common form of this application is NovaChip®, which is applied to the existing paved surface, compacted, and ready for traffic use in a short period of time. Since this product sets up so quickly, it is a favorite of state highway departments, counties, and municipalities. This product does not shed rocks any more than a hot mixed asphalt overlay and has the same characteristics and wearability of a conventional overlay. This type of application is not equivalent to a hot-mixed asphalt overlay where a structural surface course is desired. This is a thicker form of a wearing surface that is stronger than a slurry seal, does not shear as severely from power steering like slurry seals, ACFC, and chip seals, and does not require the same curing period as slurry seal or ACFC. This form of HMAC does not require the extensive cleanup that conventional chip seals do. This application sets up quickly, but it is not a good application where there is a lot of trimming required around islands, walks, swales or valley gutters, and other concrete surfaces. Because trimming is handwork and this type of application will set up before the trimming is complete, it is not recommended to use this application in parking lots. It is a good application for highways, roads, streets, and HOA streets where there are not a lot of islands or trimming required and the application is in a straight direction. It has many of the desired characteristics of rubber asphalt, at a lower cost. It does not provide the strength of asphalt rubber hot mix pavement, yet this form of HMAC wearing surface is an excellent alternative to rubber asphalt and thin conventional overlays.

These are two of the most commonly used applications for an HMAC wearing surface. There are other applications available, and you can find these by contacting pavement professionals and consultants.

OVERLAYS

Overlays are a structural surface treatment applied to existing asphalt pavements. It is important to remember that crack sealing and R&R patching will still need to be completed prior to installing an overlay. Also, where there are curbs, concrete pads, swales, or valley gutters involved, the pavement should be milled down 1 to 1½ inches and out approximately 3 feet from the edge of curbs, swales or valley gutters, concrete pads, and so on, so the overlay will fit into the concrete or curb line edge. If milling is not be accomplished, then a liability hazard exists for pedestrians and bicyclists, since the edge of the pavement will be 1½ to 2 inches higher than the curb line, pad, or swale (valley gutter). There are four common overlay applications:

- One and one-half inches to two inches of hot-mixed asphalt pavement laid over the existing pavement (conventional overlay), or other thickness as specified.
- Minimum of 1½ inches of hot-mixed asphalt pavement laid on a fabric or mineral inner layer, which is between existing pavement and the overlay.
- One inch of asphalt rubber hot-mixed concrete overlaid on the existing pavement or new pavement.
- Milling off the existing pavement 1 to 2 inches and replacing with one of the above overlays. There must be at least 1½ to 2 inches of existing pavement left after milling to make this a viable option to support construction equipment.

A conventional overlay, the least expensive of overlays, is one where the hot-mixed asphalt is laid over the existing pavement. The new overlay will crack within a period of time (depending on climate temperature changes) as a result of reflective cracking from the underlying surface. These cracks will allow water to infiltrate into the subgrade and reduce the life expectancy of the pavement with structural failures. A fabric overlay is one where there is a stress-absorbing membrane placed on the existing pavement (usually a fabric inner layer), followed by the hot-mixed asphalt overlay. The stress-absorbing membrane retards reflective cracking and reduces water infiltration through cracks once the reflective cracking occurs. The last type of overlay is where the surface is completely milled 1 to 2 inches in depth, crack sealed, and patched (skin and R&R), and an overlay placed over the surface with or without the stress-absorbing membrane (fabric). For this application, there must be at least 2 inches of existing asphalt left after milling in order to provide the proper base and support for the milling and overlay equipment. If there is not sufficient depth, then removal and replacement (discussed in next section) should be considered. To get the best return on your overlay investment, it is recommend that a stress-absorbing membrane be used on all overlays to extend the longevity of the overlay. Also, it is recommended to seal coat the overlaid pavement eighteen to twenty-four months after installation to reduce the effects of raveling and oxidation.

Rubberized asphalt is being used across the country on freeways, highways, county roads, and municipal streets. Rubberized asphalt provides a more flexible surface and reduces road noise for vehicles traveling over 35 MPH. Because of its flexibility, initial maintenance applications (primarily seal coating) do not need to be done as soon as conventional asphalt (seal coat, twenty-four to thirty months after installation) and are required less frequently (every five to seven years) over the life of the overlay pavement. Regular maintenance (crack sealing and patching) will still have to be performed on an ongoing basis because the pavement will still deteriorate from ultraviolet rays, water, deicers, and oxidation. Initially, high-speed traffic (35 MPH or higher) on the rubberized asphalt will seem quieter to adjacent buildings and communities; however, once dust and dirt accumulate on the surface and the pavement oxidizes, the reduction of the traffic noise will diminish. The major noise factor of rubberized asphalt is the difference in traffic noise to the occupants of the vehicle as opposed to traveling on a concrete surface or rough and oxidized asphalt pavement. Rubberized asphalt is very expensive and usually is not cost effective for parking lots and HOAs. Rubberized asphalt is recommended (and is an excellent application) for airports (aprons, runways, and taxiways) since it oxidizes at a slower rate than other asphalt overlays.

Overlays should be the next-to-last rehabilitation choice due to cost and the effects on drainage. Parking lots are designed to drain by sheet draining across the surface to a retention area or to drop inlets positioned throughout the lot. Also, most commercial and industrial buildings have roof drains that empty through the curbs directly onto the pavement. When an overlay is installed, the elevation of the pavement is raised by the new thickness of the overlay. This has the tendency to slow the drainage across the parking lot and blocks or obstructs the roof drains at the curb openings. This obstruction can build up debris in the drains and back up water, since there is a smaller opening. With HOAs, overlays can alter the drainage designed for the streets. This is especially true where there is an inverted crown (drainage in the center of the street, constructed in a "V" configuration). Overlays will raise the street elevation by the thickness of the overlay and, in most cases, impede the inverted crown drainage if it exists. When installing an overlay, it is imperative to mill out the pavement next to the curbing, valley gutters (concrete swales and drains), and other concrete surfaces or pads to make the new paved surface level with the curbs, valley gutters, and all other concrete surfaces. If this is not done at the time of the overlay, the association or owner could be liable for injuries by pedestrians and bike riders. Also, where there are dumpster pads or other flat surface structures, water will accumulate by the depth of the overlay.

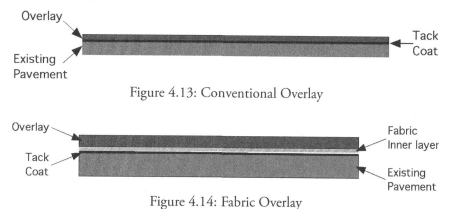

Figure 4.13: Conventional Overlay

Figure 4.14: Fabric Overlay

Preparation of the existing pavement prior to any of the overlay applications consists of sealing cracks (¼ -inch opening or greater) and repairing all structural failures with R&R patching. After surface preparation, the entire area must be thoroughly cleaned and a tack coat applied with a hot-applied asphalt emulsion (CRS2) or hot-applied asphalt tack coat such as MC70 (medium cure asphalt liquid) or an approved equivalent to ensure bonding of the new overlay to the existing surface. After the overlay has been placed, it is rolled and compacted with static or vibratory rollers or rubber-tired (pneumatic) rollers. The ambient and ground temperatures are critical to ensure a proper installation and compaction. Both temperatures should be 50°F and rising in order to obtain a good application, dense compacted mix, and proper compaction. If the ambient or ground temperatures do not meet this criteria, the hot-mixed asphalt will cool too quickly, compaction with the rollers will be difficult, and the end results can be devastating. Also, wind is a factor when overlaying, as it will cool the hot-mixed asphalt quickly and compaction will be difficult, which can also have devastating results.

Figure 4.15: Hot-Mixed Asphalt Overlay

There are three basic types of overlay:

1. One-and-one-half- or two-inch overlay on existing pavement
    a. Placement of overlay
    b. Rolling and compaction
2. One-and-a-half-inch overlay on fabric or mineral inner layer
    a. Placing fabric over tacked existing surface
    b. Placement of overlay
    c. Rolling and compaction
3. Milling of existing pavement and replacing with overlay
    a. Milling 1 inch off existing pavement and hauling off the spoils
    b. Placement of overlay on tacked surface
    c. Rolling and compaction

Figure 4.16: Phases Of Fabric Overlay:
Installing Fabric, Installing Hot-Mixed Material, and Compaction

Overlays should be seal coated one to two years after application. The asphalt will need to be sealed when the pavement turns gray or when the small aggregate (sand) begins to ravel or roll out. This will leave a sandy look, deposit black-colored sand in the curbs, and leave a sandy or gritty feeling on the surface. When seal coating an overlaid surface, refer back to the section in this chapter on seal coating. When installing a new paved surface, puddles may occur or develop during the placing process. Puddles in an overlay are likely caused by one of the following:

- Deposits of more fine aggregates than large ones, causing the pavement to compact tighter, resulting in a thinner area than the rest of the surface. This is a problem of segregation of the larger aggregate out of surface material, leaving the smaller, finer aggregate (sandy mix).
- The roller used for compaction remained in one spot for a long period of time (especially if the roller is vibrating and the vibrator is on).
- There could be a low area in the original pavement. Since an overlay is a 1- to 2-inch layer placed by a machine that is calibrated to lay an exact thickness, the existing low area will remain in the new surface.
- Workmanship while installing the overlay.

When the overlay is completed, the surface should be flooded and left to stand for twelve (12) hours and then all remaining puddles should be measured for depth. Puddles with a depth of more than ¼ inch should be skin or surface patched, using a string line to keep the surface level.

A second problem that may occur with overlays is ridges along doors, which prevents them from opening. As soon as this is discovered, heating the ridge and raking it down will correct the problem. Another problem that may occur is marks caused by the roller (referred to as roller marks and in the form of ridges) in the pavement. This is caused by the weight of the roller and the soft surface during the compaction process. Heating and compacting with a roller or plate compactor can eliminate these marks in most cases.

A third problem that may occur is segregation of the larger aggregates on the surface. This problem is caused by handling of the hot mix from the time it leaves the hot plant to placement on the pavement. Because of segregation, pockets of the larger aggregates will appear on the surface. These areas will need to be skin or surface patched with a hot sand seal mix or they can be treated with a mastic patch (asphalt emulsion sealer with 5 lb. of sand).

CURING CHARACTERISTICS OF OVERLAYS

Caution must to be taken when installing overlays on an existing pavement because the existing elevation will be increased by the thickness of the overlay. The drainage design can be altered when an overlay is installed. On a commercial property where roof drains are constructed to drain onto parking lots, overlays will obstruct the drainage exits. Another problem associated with overlays is the reduction in the height of vertical (cast in place and combination curb and gutter), extruded, or border curbs by the thickness of the overlay (a 2-inch overlay next to a 4-inch curb will reduce the height of the curb to 2 inches). Conventional overlays (overlays placed directly on the existing pavement) will increase the height of the crown on streets (the edge of the overlay next to curbs or other concrete structures has to be milled down to fit the curb line elevation), which will affect the entrances to driveways and can cause problems with vehicles sliding or slipping toward the curbs when wet or iced over (most city and county governments have a height restriction on crowns). Drainage also will be affected when an overlay is installed due to the increased pavement height. This can be particularly problematic on flat elevation parking lots, streets, and streets with inverted crowns. Milling the entire pavement will leave the same elevation, but with a complete new surface. However, soft spots in the base below the pavement can become an issue during milling. These areas will need to be addressed at the time of construction. Without a special type of testing, it is very difficult to locate soft spot areas before pulverizing or milling.

These soft spots will increase the cost projections and overall cost of the project. A contingency budget should be established in order to properly deal with these areas; however, it is very difficult to set an exact reserve for soft spot repairs and treatment. When conventional and fabric inner layer overlays are selected, existing drainage problems or puddles will not go away. The pavement overlay will not allow water to penetrate the surface or drain through to the base since all existing cracking (which helped percolate the water) has been covered. After the existing pavement is overlaid with a conventional overlay or fabric overlay, the prior puddle areas are sealed, and water will puddle and last longer. This is especially true with a fabric overlay.

When a conventional or fabric overlay is installed, the existing drainage flow will remain the same, and any puddles that existed prior to installing these two types of overlays will remain.

Surface patching can be done to reduce existing puddles, but in extreme cases, the asphalt should be removed, regraded, and repaved or patched. However, the original drainage design and curb and gutter, as well as other permanent concrete structures, dictate the rate of flow. Should the existing and established grade for drainage be less than 1 inch in 10 feet for asphalt pavement (½ inch in 10 feet for concrete), drainage will be very difficult to achieve. You should expect puddles to occur after the pavement or concrete is installed when the existing drainage grades are less than previously stated. When an overlay is installed after milling or pulverizing, the elevation next to the curb will vary because the concrete curbing has moved and displaced over the years. The asphalt pavement is laid with a mechanical machine that installs the pavement in a level and straight direction. Because the concrete curb is now wavy, there will be some areas where the pavement is flush with the curb and areas where it will be above or below the curb. There are no alternatives to this problem short of replacing all the concrete curbing or creating a wavy overlay, which will create puddles. Water from sprinklers, rain, melting ice, and snow will leave stains on the surface. This is not a defect of the pavement or the asphalt material; it is the result of water reacting with the fresh asphalt binder in the new pavement, fertilizers in the adjacent garden areas, or other debris that can cause color stains with water. There is a residue left on the surface that will accumulate with water and will cause a light brown, orange, or yellow color and will stain the pavement, stripes, and concrete. This is not a permanent stain and will fade away as the pavement cures.

Reflective cracking will develop in the new overlay surface from expansion and contraction of the older pavement underneath. All cracking that existed in the older pavement (whether sealed or not) will reflect through the new overlay in a short period of time (three weeks to six months). The severity of the reflective cracking is proportional to the severity of the cracking in the older pavement. The larger the width and amount of cracking in the older pavement, the larger the width and amount of cracking that will reflect into the overlay. A fabric inner layer (more commonly referred to as a stress-absorbing membrane) will reduce the severity and amount of the reflective cracking and will also aid in reducing water percolation into the base. During colder months, water will migrate to the cracks; however, the fabric inner layer will stop the percolation of the water through the overlay. This will leave the water to set in the crack and freeze with colder temperatures. This in turn will cause expansion and spalling or widening of the crack. Crack sealing will prevent this process when the cracks reach an opening width of ¼ inch or more.

When driving on a new overlay surface, care must be taken to prevent power steering shear marks. When stopping a vehicle, please start driving forward or backward and then start turning the steering wheel. This will help reduce the tire marks in the new surface, as well as reduce stress on the power steering mechanism in your vehicle.

Light gray marks in driveways may occur after the streets are open. These are caused from normal air dust being picked up by tires. These marks are not caused by tracking asphalt, but by tracking dust and dirt off the new pavement. Tracked asphalt is jet black in color.

REMOVAL AND REPAVING

If the condition of the pavement is not suitable for a surface treatment or overlay, then removal of the existing pavement and repaving is the next step. Because of the totality of removing and repaving, no patching or crack sealing during this application is needed. Examples of reconstruction include total removal (by stripping off the pavement) by loaders and heavy equipment, removal by milling the existing pavement to the aggregate base or pulverizing the existing pavement and base and replacing the surface. These applications, of course, are dramatically more expensive and are only needed if the pavement is totally deteriorated. A breakeven analysis can help determine when the cost of remove and replace (R&R) patching surpasses the cost of reconstruction or replacement of the total paved surface. R&R patching is the most expensive maintenance treatment for a paved surface, at a high price per square foot (depending on geographical location) compared to a medium price per square foot (2 inches thick) for complete removal and repaving. Should R&R patching for structural failure be at or close to the cost of replacing the entire pavement, then reconstruction is the more feasible choice. In addition to the cost, a rebuilt parking lot will have a more aesthetic appearance compared to an older lot with extensive patching.

Removing the existing pavement and repaving provides a brand-new paved surface and will increase the value and appearance of a property. There are three methods of removing the pavement and repaving.

The first method is removing the existing pavement by lifting it off with loaders, loading it into trucks, and hauling the old pavement off to the dump or to a recycling processing plant. This process does not leave any material to use in filling low areas or replacing soft spots that exist or may develop during the grading operation. Should there be any low areas or any locations where the base elevations must be increased or filled, or if soft spots need to be excavated and removed, aggregate base needed to correct these areas must be imported, creating additional costs. A benefit to this process is being able to use smaller equipment in more restrictive areas, causing very little disturbance to the existing grade, while maintaining the original drainage design.

Figure 4.17: Removing Asphalt Pavement

The second method is milling the older pavement off to the aggregate base and hauling off the spoils to a recycle plant. The main advantage to this method is it is quick and efficient, and it maintains the original grade. Another benefit is some of the millings can be stockpiled to adjust grades, fill in low areas, and fill in excavated areas for soft spots. Stockpiling millings can reduce costs of importing ABC at a later time if soft spots or additional grading is needed.

Figure 4.18: Milling Asphalt Pavement

The third method of removing and repaving is pulverizing the existing pavement and aggregate base into a new base material. This method breaks down and blends the existing pavement and base into new aggregate base. Since no asphalt is initially removed, there will be an excess of material that will need to be hauled off. However, this material can be used as base for other projects, and most contractors and local agencies will want to use it for trench backfill, base for pavement, and other uses where aggregate base can be used. A portion of the material can be left to fill in excavated areas for soft spots, and should the elevation of a parking lot or street need to be changed or adjusted, there is plenty of pulverized base to accomplish that. The areas where the material is pulverized will have to be graded to the existing drainage prior to pulverizing or to a new drainage design, as the existing drainage design has been changed or destroyed. The benefit to this process is being able to adjust for new or better drainage configurations and leaving plenty of material to treat soft spots.

Soft Spots (Soft Subgrade): A common problem that occurs when removing older pavements is uncovering soft spots or soft areas in the base or subgrade material. This is caused from water migrating over the years through existing cracks and from behind the curbs from overwatering or surface puddles. Prior to paving, these areas will have to be dealt with by either one of four methods. The first method is uncovering, scarifying, and allowing the area to air dry. This will only work if there will be high enough ambient temperatures for approximately 2 to 3 weeks. Air drying is the least expensive of the four, but the most time consuming. The subgrade material will have to be mechanically turned a couple of times to aid in the drying effort. The second is lime or chemical additive used to quickly dehydrate the areas and allow for work to resume in 2 to 3 days. The problem with this method is premature surface cracking in the new asphalt from release of moisture after paving. The third method is installing a geogrid system between the aggregate base and subgrade materials, which will aide in support over the soft areas. Finally, the fourth is penetrating rip-rap into the soft subgrade with a sheep's foot roller. This method can cause damage to utilities that are not installed deep enough prior to the penetrating process.

Figure 4.19: Pulverizing Asphalt Pavement

When installing a new paved surface, puddles will occur or develop after the new pavement has been placed. There are several reasons puddles will develop after a new surface (overlay or new paving) is installed.

Figure 4.20: Asphalt Paving

Puddles in newly paved surfaces can be caused by the following:

• Segregation of the rock material, where there are more fine aggregates than large ones and the pavement will compact more, resulting in a thinner pavement section than the rest of the surface.
• The roller used for compaction remained in one spot for a long period of time (especially if the roller is vibrating and the vibrator is on).

- Low areas or spots in the aggregate base after final grading was performed. If the contractor did not check the final grade of the aggregate base with a string line or other method (string lining is the best method), low spots will be left behind. Since a new pavement is put down in a 1- to 3-inch layer placed by a machine that is calibrated to lay an exact thickness, existing low areas will remain in the new surface.
- Workmanship while installing the new surface.

When the new paving is completed, the pavement should be flooded and left untouched for twelve (12) hours, and then all remaining puddles should be measured for depth. Puddles with a depth of more than ¼ inch should be skin or surface patched, using a string line to keep the surface level to surrounding surface elevations.

A second problem that may occur with new pavement is ridges along doors, which can prevent the doors from opening. As soon as this is discovered, heating the ridge and raking it down will correct the problem. Another problem that may occur is marks caused by the roller (referred to as roller marks, in the form of ridges) in the pavement. This is caused by the weight of the roller and the soft surface during the compaction process. Heating and compacting with a roller or plate compactor can eliminate these marks in most cases; this problem should be addressed and corrected as soon as possible.

REMOVAL AND REPAVING CURING CHARACTERISTICS

Removing and repaving a parking lot or street pavement may result in several curing problems and deficiencies. The pavement placed over the newly prepared base may have several problems that can occur once the new paved surface is completed. There may be several twist marks from power steering shear, small rocks may come out of the surface, areas may have different shades of black and dark brown, and roller marks may occur. These are common early curing characteristics of new pavement and should not be of major concern in the short term (one to two weeks). It is recommended to wait four to six weeks to see whether any major problems develop. At this point, vehicles may cause the pavement to develop deep twist marks and, in some cases, alligator cracking will develop in areas where the base is soft and wet under the pavement. Also, puddles may develop after rain or watering. These are caused from low areas in the paved surface that were developed by material segregation, the roller, or grading the base problems. These areas can be identified by flooding the paved surface and waiting twelve (12) hours and then measuring to identify which puddles are ¼ inch deep. There will be several puddles in the new surface; however, only those puddles ¼ inch or greater will need to be repaired.

Another curing problem is shedding rocks from the surface or along edges next to curbs. This is not a problem with the pavement or its installation; it is a common characteristic of releasing rocks from high edges and areas. This will continue until the edge or surface next to curbs and concrete flat work adjacent to the pavement is flush with the pavement. Rocks will be released from segregated areas (areas where the hot-mixed asphalt had a large amount of rock), which will need to be skin or surface patched. Also, areas where there are more rocks exposed, creating a rocky appearance, will have to be skin or surface patched (this is caused by segregation of fine and course aggregates during the laydown operation). The majority of the scuff mark areas will knead away in six to twelve months; deeper marks will have to be skin or surface patched during the one-year warranty period. The pavement color will lighten, rocks and sand will develop on the surface, and seal coating will be required to retard this oxidizing phase, about twelve to eighteen months after installing. The pavement stripes will fade away as the pavement cures and oxidizes and will need to be restriped in twelve months or when the surface is seal coated. All the previously mentioned curing problems and characteristics for paving will occur for pulverizing and paving. Water from irrigation sprinklers, rain, melting ice, and snow will leave stains on the surface. This is not a defect of the pavement or the asphalt material; it is the result of water reacting with the fresh asphalt binder in the new pavement, fertilizers in the adjacent garden or grass areas, or other debris that can cause color stains with water. There is a residue left on the surface that will accumulate with water and will cause a light brown, orange, or yellow color and will stain the pavement, stripes, and concrete. This is not a permanent stain and will fade away as the pavement cures.

SPEED BUMPS, HUMPS, AND TABLES

Traffic safety and speeding vehicles are major concerns of homeowners association boards of directors and managers, city and county engineers, law enforcement officials, and commercial and industrial property owners and managers. The most common speed deterrent is the speed bump. For homeowners associations, this causes the biggest political upheaval amongst the residents and has been known to provoke heated discussions and meetings. With commercial and industrial properties, speed bumps are accepted, but not well liked. With city and county streets, they are not liked, but most drivers ignore them unless they have an abrupt strike. Everyone is concerned about the safety of pedestrians, especially children, and the need to slow down speeding vehicles. Therefore, speed bumps, humps, and tables are introduced.

There are the three types of speed deterrents: speed bumps, speed humps, and speed tables. Each type is constructed to slow traffic, and each is used in a specific situation and exposure. In the next few paragraphs, we will discuss each type and the pros and cons of each.

SPEED BUMPS

Speed bumps are the narrowest of the three deterrents, usually 2- to 4-feet wide and span the entire width of the street or parking lot drive lane. Speed bumps are most commonly used in parking lots and are constructed to create the hardest (most aggressive) hit or strike. Since parking lots have a large pedestrian exposure (especially along the fronts of stores), owners and managers want speed bumps to have an immediate impact on vehicles that do not slow down. The speed bump is constructed with a high or thick center, usually 2 to 3 inches in the center. The strike area (part of the speed bump where a vehicle comes in contact with the bump) is the distance and slope from the center of the speed bump to pavement. An example is a 4-foot speed bump with a center height of 2 inches. Each side of the bump is 2 feet from the center on each side. That will make the slope approximately 8 percent, or 1 inch of height in the strike area per 1 foot of distance from the center of the bump. This particular speed bump will slow traffic to about 25 to 30 MPH, maintaining a smooth, easy roll.

The magnitude of the discomfort to the driver will determine how fast a vehicle will travel over a bump. The type of vehicle also plays a role into the effectiveness of the speed bump. An American luxury vehicle or mini-van-type vehicle has a large suspension, and the easy roll does not cause a hard strike or interrupt the ride quality of the passenger as much as for a European luxury vehicle with a tighter suspension. Therefore, should owners or managers want to slow traffic further, they can construct a speed bump with a more aggressive strike area and a harder hit. This is accomplished by narrowing the speed bump from 4 feet in width to 3 feet (or increasing the height of the bump from 2 to 3 inches), increasing the slope of the strike area, and creating a harder hit or vehicle strike. The more the width is shortened or the height is increased, the more uncomfortable the strike and the slower the traffic.

The reverse holds true if a softer strike area is desired. Also, the length of the street or drive lane will figure into how many speed bumps are needed to effectively control speeding traffic. Speed bumps should be placed no less than 100 feet apart, to really slow traffic in a parking lot drive, or 200 feet for a street. Also, the speed bump should be placed where there is minimum or no pedestrian exposure, usually from an island to the curb (never behind a parking stall or in a curb ramp opening). Speed bumps should be painted solid yellow to warn the driver and call attention to them. There are several striping designs for speed bumps other than solid yellow; however, solid yellow reduces liability claims from drivers who insist on driving over the bump faster than it is designed and constructed for. Other common striping configurations are chevrons (or sergeant stripes) or vertical or horizontal stripes with a box around the bump. Never stripe just the box around the bump because it cannot be seen at night, increasing liability.

SPEED HUMPS

Speed humps are a larger variation of the speed bump, which are 8 to 12 feet in width and have a 4- to 6-inch center or height. Speed humps are commonly used for homeowners associations and public streets. Speed humps are very successful at slowing traffic and deterring speeders because of the ramp-type shape. Vehicles traveling more than 25 to 30 MPH will nosedive into the pavement on the other side of the hump as the vehicle crosses the center. A speed hump's slope can be adjusted to create an abrupt start, then a gradual incline to the center on each side. This type of adjustment will determine the maximum speed limit allowed to clear the hump. Unlike speed bumps, a speed hump cannot become a learned speed barrier. Because of the width and center height, drivers will have to travel over the hump rather than learning how to navigate over it at higher speeds. The strike area is calculated the same as a speed bump (8 percent or 1-inch rise for every 1 foot of width from the center of the hump), which makes the strike area an inclining rise. The strike area (part of the speed hump where a vehicle comes in contact with the hump) is the distance and slope from the center of the speed hump to the pavement.

An example is an 8-foot speed hump with a center height of 4 inches. Each side of the hump is 4 feet from the center on each side. That will make the slope approximately 8 percent or 1 inch of rise in the strike area per 1 foot of distance from the center of the hump. This particular speed hump will slow traffic, but the ride quality and roll of this hump is graceful and will slow traffic to about 25 to 30 MPH. The magnitude of the discomfort to the driver after clearing the center of the hump and descending will determine how fast a vehicle will travel over the hump. The type and style of a vehicle will have no effect on the strike area or incline, but rather will determine if the decline has an adverse effect on the vehicle. The more compact or smaller a vehicle, the greater the chance that the vehicle will scrape the street with the bumper or underguard. Therefore, should owners or managers want to slow traffic further, they may construct a speed hump with a more aggressive strike area and a harder hit on the front edge of the hump. This is accomplished by adjusting the initial edge to the speed hump to cause a harder hit before starting the incline or reduce the rise per 1 foot (i.e., using our 8-foot hump, each side of the hump would be 3 to 3½ feet).

The length of the street or drive lane figures into how many speed humps are needed to effectively control speeding traffic as well as the type of street (collector, residential, or private drive). With shopping centers, the amount of speed humps is determined by the owner and how much to slow traffic for safety and liability. Speed humps should be placed no less than 200 to 300 feet apart to really slow traffic on a street. Also, the speed hump should be placed where there is minimum or no pedestrian exposure, usually from an island to the curb (never in a curb ramp opening). The most common striping configuration for speed humps is chevrons (or sergeant stripes). Never stripe just the box around the hump, because it cannot be seen at night and increases liability.

SPEED TABLES

Speed tables are a modified version of a speed hump, creating a level top or landing. The purpose of the speed table is to cause the vehicle to rise and land on a table-top or level landing. This type of barrier has a sharp incline on either side of the level landing, and the landing is created to cause the vehicle to incline at a rate that automatically slows the driver and vehicle. The incline on each side is designed or built with a standard 1-inch rise for each 1 foot in length. Where the design of height or thickness of the table is 4 inches, the incline should be 4 feet on each side of the flat area. A common speed table has a 4-foot incline, 4-foot flat top, and 4-foot decline. With this particular speed table, the height of the table from the original street is 4 inches. Increasing the length of the flat top, height, or thickness of the table, or reducing the length of the incline, will dictate the speed of vehicles. The best deterrent is a shortened incline length.

Using our example again, we build a speed table with a height or thickness of 4 inches with a 3½-foot incline, 5-foot flat or table-top, and 3½-foot decline. This creates a slope on the inclines on both sides of the speed table of 9.5 percent. This will create a sharper hit or strike area and a longer table-top length. A maximum slope would be 11 percent, and when we apply that to our example, the incline and decline would have a length of 3 feet and a table-top of 6 feet. This will create an abrupt hit or strike area and may also create some opposition from homeowners. The latter (3 × 6 × 3 feet) would control traffic at 10 to 15 MPH. Speed tables do not have a ramp incline like speed humps and do not provide an easy or graceful roll and ride quality. The type of vehicle will determine the effects of the speed table, and the tighter the suspension, the harder the strike or hit and the slower the vehicle will travel. Speed tables should be striped with a chevron configuration, and to make the speed table more visible, the incline and decline ramps should be painted solid yellow. The distance between speed tables should be a minimum of 200 to 300 feet.

Cost is usually a big factor in deciding whether to construct speed bumps, speed humps, or speed tables. Our speed bump example (4 feet wide with a 2- to 2½-inch center or thickness) would cost $1,200 to $1,800 a bump. That includes installation and painting with glass beads in the paint. The speed hump installed and painted with chevrons would cost $3,500 to $4,000 each, and speed tables usually cost $4,000 to $5,000 each with solid yellow inclines and a chevron pattern painted on the top. This is using our examples above (12-foot-wide speed hump with a 4-inch-high center and a speed table with an incline of 4 feet, a table-top of 4 feet, and a decline of 4 feet). The larger the speed bump, hump, or table, the more the cost. These are example costs, and all costs should be verified for your area. (Note, these values are hypothetical values based on 2010 costs.)

Speed tables are used more by cities and would be a preferred choice of HOAs where a major (collector) street is two miles long or more. Controlling speeding vehicles is a major problem, so it is important to involve the input of a consultant or other traffic professional to help design the proper application. Also, signage should be placed on both sides of the speed bumps, humps, or tables to warn drivers of the barrier ahead. Rectangle or triangle signs can be used as warning devices stating "Speed Bump Ahead," "Caution: Speed Bumps Ahead," and so on. In gated HOA communities, a warning can be posted on each gate—"Traffic Controlled by Speed Bumps"—to reduce the amount of signage within the community. However, it is recommended to install warning signs at each bump, hump, or table for liability purposes.

There are some variations to the bumps, humps, and tables. Ceramic speed bumps are more decorative than practical and will break often and need replacement. There are virtual speed bumps, which are painted in a fashion to resemble an actual speed bump, but after a very short period of time, drivers realize they are only painted on the pavement and speeding problems resume. Another speed bump application is rubber or plastic speed bumps. These bumps are installed by bolting the bumps to pavement or concrete. These bumps provide a narrow width and a hard hit or strike area. Rubber and plastic speed bumps are an excellent alternative to asphalt speed bumps. They are not used or recommended as alternatives to speed humps or tables. There are several types and kinds, fabricated by many manufacturers, and provide many different designs, reflectors, and ride quality. Considerable research is recommended to make sure the proper speed bump is used. Rubber and plastic speed bumps are not permanent like asphalt speed bumps and can be moved to new locations or to allow for pavement maintenance. They do not require painting and come with color, stripes, and reflectors manufactured into the product. It should be noted that some plastic products will oxidize and deteriorate rapidly in Sun Belt areas due to ultraviolet ray exposure. In these climates, rubber products are recommended.

Most airports and city and county governments install speed humps and tables to ensure maximum speed control and safety. Some airports use very large speed tables with a 12-inch height, which controls traffic to the maximum in passenger pickup and dropoff drives. Most shopping centers, parking lot owners, and HOAs install speed humps and wide speed humps and, in some cases a speed table, to control their traffic, while strip shopping centers, grocery stores, and some parking lot owners install a speed bump that is narrow in width and has a high center for an abrupt hit and strike area or a speed table that has a high flat top and abrupt ramps up and down to the table-top to provide maximum speed control. These particular speed bumps and tables will cause front end alignment damage and tire damage if struck at speeds higher than 5 to 10 MPH. These bumps and tables are used to protect pedestrians from speeders as they exit stores and cross to the parking areas. They are very effective but are not recommended for HOA streets. Where the streets or drives have an inverted crown (V shape in the center for water flow), a speed bump is the best option, with an opening in the center to allow for water flow. The speed bump is actually two speed bumps in one location on either side of the inverted crown. Speed humps and tables are too large to construct a waterway through. However, a separate hump or table could be built in each lane.

Figure 4.21:
Speed Bump, Speed Hump, and Speed Table

PAINTING, STRIPING, AND DIRECTIONAL MARKINGS

After resurfacing maintenance has been performed, directional and informative message markings have to be replaced or redesigned, particularly striping and blue reflectors marking fire hydrants.

Figure 4.22: Striping

The following material addressing parking area planning and layout is intended solely as a general introduction and basic information on the subject. Each parking lot requires its own design and parking configuration for stalls and traffic flow. There are zoning and building code requirements for the number of parking stalls, fire lanes, and handicap stalls (with ADA guidelines) that need to be adhered to when a parking lot is resurfaced or serviced. Please refer to local city and county regulations for these requirements prior to reconfiguring your parking area.

There are three basic types of parking programs utilized today, each of which requires its own flow or layout plan. These are the park-and-lock or self-park programs, the attendant assist plan, and the upscale valet parking service. While a valet service may appear to be the optimum program, wherein visitors are provided with service to and from a specified point at a shopping center, theater, casino, or restaurant, this plan is increasingly structured for low-volume, high-end users. The reason for this is increased exposure to liability brought on by the creation of bailment, when a valet accepts a customer's keys, thereby assuming care, custody, and control of the property. Different from valet service is park-and-lock or self-park, which greatly reduces liability predicated on the fact that the customer has retained the key to the vehicle and, therefore, custody and control. Attendant assist programs were borne out of necessity for additional parking space. Not a valet service, customers enter the parking facility and are directed to the most available space by an attendant, who would park the vehicle only when no spaces were available, usually in the aisles. When a space opens up, the vehicle is moved from the aisle by the attendant. Another form of park-and-lock is total self-park, where the driver locates an empty stall, parks the vehicle, locks it, and keeps possession of the keys (there is no attendant available).

These three methods play an important role in planning your parking lot layout. Space efficiency could vary from as little as 175 to 200 ft.^{2} per space for a stack park valet plan to more than 400 ft^2 for a self-park design. Parking stall sizes are controlled by local zoning and building codes. In most cases and most local building codes, the number of stalls is determined by the square footage of office or retail space. Local building codes also have input into the design and size of a parking area. The ratios for parking include space utilized for driveways, landscape islands, and surfaces rendered unusable by the angle of parking stall selected for the project. Everything depends on the property or floor size and any column spacing as to the angle selected for a specific parking bay. Optimum efficiency of a parking lot is realized through a 90° angle of a stall on a 60-foot design (24-foot drive lane and 18-foot stalls on either side of the drive lane).

Figure 4.23: 90° Parking Stall Design

A parking bay consists of two rows of vehicle parking, separated by an aisle or drive lane, whereas a half-bay is considered one row of vehicle parking and one drive lane. Ninety degrees is optimum, since the general rule is this angle will produce spaces at 162 ft.^{2} each (18 × 9 feet). It is generally more difficult for drivers to negotiate vehicles into or out of a 90° space than those with a parking stall on an angle. As an example, shopping centers, grocery stores, and other high-turnover users utilize angle parking. The added angle adds to the ease of pulling into and leaving the space and exiting, which is accomplished in one simple maneuver. Considering that the customer cannot begin to turn the wheels when backing from a 90° space until the vehicles on either side have been cleared, it takes up to three maneuvers for some drivers to clear this type of space. This is time lost and an inconvenience to the driver and people waiting to park, not to mention additional wear and tear on the parking lot due to power steering shear. The use of 90° parking is excellent for long-term parking areas, which fill early with vehicles that sit throughout the day.

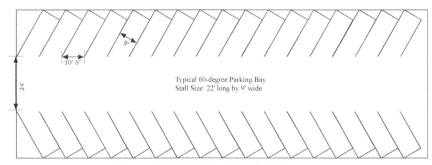

Figure 4.24: 60° Parking Stall Design

One problem with making a decision on 90° versus angle parking and driver convenience versus maximum space utilization is that angle parking will require less aisle space, but more square footage per actual space (refer to Figure 4.23). When looking at the drawing, the first thing that comes to mind is that in each scheme, the property is 189 feet in length; however, at 60°, 2 feet in width is saved. Also, it is very noticeable that eight spaces have been lost by using a 60° layout.

When a decision is made to institute angle parking, consideration should be given to the 70° angle configuration. With a width of the stall at the aisle at 9 feet for an 8½-foot space versus 12 feet on a 45° angle, you have nearly the efficiency of 90° with the driver convenience of lesser angles to the drive lane. There are literally hundreds of variables to be considered when designing what may be considered the smallest of parking areas. It would be wise to consult a parking operator or consultants in the field to assist you in your parking design and configurations.

When reviewing a current parking facility or proposed drawings, some of the items to look for include the following:

- Proper floor and ramp markings
- Location of ingress and egress points, with particular attention to one-way street access

  Make sure your plans allow for the following:

- Enough stacking room for exiting or entering vehicles not to block internal flow patterns or streets and sidewalks.
- Control has been considered to bring a maximum return from parking as an investment.
- Traffic flow, cross-traffic turning radius areas, lighting levels, and proper markings and local code requirements. Some municipalities require a specified amount and type of parking lot lighting for safety purposes. These requirements are part of the agency code and should be investigated prior to constructing or expanding a parking lot.
- The parking plan maximizes the amount of parking needed or desired and meet local city and county zoning and building codes.

ADDITIONAL REQUIREMENTS FOR CONSTRUCTING A PARKING LOT

When a parking lot is removed and repaved, there are some ADA requirements that will need to be adjusted for ramps to buildings and stores, slope of ADA parking spaces, and any other requirements listed by the local county, city, or town building or public works departments. Also, when removing and repaving parking lots, check with local public works departments for permitting requirements. Some entries may be on the public right of way, and many governing agencies require permitting for work performed in these areas or entries.

CURING CHARACTERISTICS AND TROUBLESHOOTING

Appendix A at the end of this book presents problems that can develop after application, during the curing processes, and also during the life of the pavement. This table can be used as a general identification of a problem and possible correction strategies.

Chapter 5

Scopes of Work and Specifications

This chapter is designed to discuss developing a scope of work and specifications. Many times the expression "comparing apples to apples" is used when referring to bids. There isn't anything common to comparing "apples to apples" when requesting bids to do pavement maintenance for parking lots or homeowners association streets. The problem is, What type of apples are you comparing? There are inexpensive apples and gourmet apples. The only way to communicate with contractors is to define a scope of work and establish specifications. A defined scope of work will tell the contractor what type of pavement maintenance is desired, and the specifications tell the contractor how the work will be performed. This is the only way to communicate with the contractors, and it helps them bid properly and accurately. Using a defined scope of work and specifications is a win-win situation for the property owner and the contractor. The property owner (commercial, industrial, or residential) and the HOA board of directors will obtain the correct maintenance application at a fair price, and the contractor can provide the highest quality of workmanship at a fair profit. The bid documents—Request for Bids (RFB), Request for Quotations (RFQ), or Request for Proposals (RFP)—will contain the Scope of Work and Specifications, as well as other project-specific instructions, which will be discussed in a later chapter.

Which Apple Is Being Referred To?

The scope of work is a part of the request for bids and tells the contractor what is desired or needed to repair or replace your pavement. The scope of work is listed on the first or opening page of an RFB or RFQ. The following is an example of a scope of work for seal coating.

ALL PAVEMENTS TO BE SEAL COATED SHALL BE THOROUGHLY CLEANED OF ALL DEBRIS, DIRT, AND DUST TO PROVIDE A COMPLETELY CLEAN SURFACE FOR ADHERENCE. ALL PETROLEUM SPOTS AND DIRT BUILDUP AREAS IN PARKING STALLS OR ON CURB LINES SHALL BE COMPLETELY SCRAPED AND ALL MATERIAL REMOVED. THE AREAS SHALL BE HEATED WITH AN OPEN FLAME (PROPANE WEED BURNER, ETC.) IN A MANNER TO NOT BURN THE PAVEMENT AND ALL PACKED MUD AND CONCRETE SPILL SPOTS SHALL BE REMOVED PRIOR TO INSTALLING THE SEAL COAT.

SEAL COATING WITH ASPHALT EMULSION

Two Coats Spray Applied
Seal coat all surface areas with asphalt emulsion. First coat applied at an application rate of 0.14 to 0.16 gal./yd.^{2}, and the second coat applied at a rate of 0.14 to 0.16 gal./yd.^{2}; both coats spray applied and diluted not more than 20 percent of concentrate or the manufacturer's specified dilution rate.

The scope of work helps you control the bidding by narrowing down the choices of surface treatments to the one that is needed. This scope of work is generally listed on the first page of the request for bids or request for quotes.

This scope of work tells the contractor what type of work is expected and what type of product is to be used. Also, it tells the contractor what application rate is expected, what dilution rate is required, and what project preparation is needed. Once the scope of work is developed, the specifications are attached to instruct the contractor how to properly prepare the work site for the requested application and how to apply the requested material.

Scopes of work are listed at the beginning of the Request for Bids to inform the contractors of the type of work you are requesting. A scope of work is a brief and simple description of what is to be accomplished. The specifications are detailed instructions of how the work is to be accomplished to enforce the scope of work.

EXAMPLE SCOPES OF WORK

The Contractor shall supply all required insurance certificates as shown with the appropriate named "additional insureds."

All work outlined within the Scope of Work shall be applied with the material, application rates, construction standards, and specifications as outlined in the Specifications section of this Request for Bids. All Bidders will be responsible for reading and understanding each and every specification for each maintenance application, verifying all quantities, and visiting the work site.

CRACK SEALING

Fill all cracks with an opening of ¼ inch or larger with hot pour elastomeric crack sealant. All cracks shall be thoroughly cleaned by blowing out all debris, dirt, dust, noncompressibles, and loose pavement with compressed air at a minimum of 200 PSI; should any vegetation be present, that vegetation shall be removed, and the crack blown clean. All cracks shall be thoroughly dry of all moisture prior to applying sealant. Should the ambient and ground temperatures cause the crack sealant to remain tacky and viscous, the Contractor shall apply a blotter material consisting of fine sand, silica sand, or other acceptable material that will prevent traffic from picking up the crack sealant and prevent tracking.

CRACK SEALING WITH ROUTING

Rout all existing cracks with an opening of less than ¼ inch to an opening of ¼ inch wide and 1 inch in depth. Fill all cracks with an opening of ¼ inch or larger with hot pour elastomeric crack sealant. All cracks shall be thoroughly cleaned by blowing out all debris, dirt, dust, noncompressibles, and loose pavement with compressed air at a minimum of 200 PSI; should any vegetation be present, that vegetation shall be removed, and the crack blown clean. All cracks shall be thoroughly dry of all moisture prior to applying sealant. Should the ambient and ground temperatures cause the crack sealant to remain tacky and viscous, the Contractor shall apply a blotter material consisting of fine sand, silica sand, or other acceptable material that will prevent traffic from picking up the crack sealant and prevent tracking.

SKIN PATCHING

Skin patch all areas marked on the pavement outlining the area to be patched and marked with the letter "S" or as plotted on the attached property plat, leasing plan, or map.

SKIN PATCHING WITH FABRIC INNER LAYER

The surface shall be cleaned of all loose debris, sand, dirt, and any foreign substances prior to applying tack coat and fabric. The surface shall be dry of all standing water prior to applying tack coat and fabric. The surface will have a tack coat applied (asphalt emulsion CRS2 or CRS2P or AC-10 asphalt tack coat) prior to installing the fabric. A fabric inner layer shall be installed over the entire surface to be skin patched, free of any crinkles, folds, etc. Install R-9.5 modified (⅜-inch maximum size aggregate) hot-mixed asphalt pavement and compact with 3- to 5-ton steel wheel roller. All manhole covers and all utility services (water valves, etc.) shall be elevated to the new paved elevation. Skin patch all areas marked on the pavement outlining the area to be patched and marked with the letter "S" or as plotted on the attached property plat, leasing plan, or map, if attached.

R&R PATCHING

Remove and replace patch all areas marked on the pavement outlining the area to be patched and marked with the letter "R" or as plotted on the attached property plat, leasing plan, or map. All patching shall conform to PMIS specifications or local specifications.

INTRODUCTION TO SEAL COATING

ALL PAVEMENTS TO BE SEAL COATED SHALL BE THOROUGHLY CLEANED OF ALL DEBRIS, DIRT, AND DUST TO PROVIDE A COMPLETELY CLEAN SURFACE FOR ADHERENCE. ALL PETROLEUM SPOTS AND DIRT BUILDUP AREAS IN PARKING STALLS OR ON CURB LINES SHALL BE COMPLETELY SCRAPED AND ALL MATERIAL REMOVED. THE AREAS SHALL BE HEATED WITH AN OPEN FLAME (PROPANE WEED BURNER, ETC.) IN A MANNER TO NOT BURN THE PAVEMENT, AND ALL PACKED MUD AND CONCRETE SPILL SPOTS SHALL BE REMOVED PRIOR TO INSTALLING THE SEAL COAT.

SEAL COATING WITH COAL TAR

<u>One Coat, Spray Applied</u>

Seal coat all surface areas with coal tar. One coat applied at an application rate of 0.16 gal./yd.², spray applied and diluted not more than manufacturer's recommendations. The diluted coal tar product must meet or exceed Federal Specification RP355e.

<u>Two Coats, Spray Applied</u>

Seal coat all surface areas with coal tar emulsion. First coat applied at an application rate of 0.14 gal./yd.², and the second coat applied at a rate of 0.14 gal./yd.², both coats spray applied and diluted not more than manufacturer's recommendations. The diluted coal tar product must meet or exceed Federal Specification RP355e.

<u>One Coat in Parking Stalls and Delivery Drives, and Two Coats in Drive Lanes, Spray Applied</u>

Seal coat all surface areas with coal tar emulsion. One coat should be applied in the parking stall and rear delivery areas at an application rate of 0.12 to 0.14 gal./yd.²; two coats should be applied in the drive lanes at an application rate of 0.12 to 0.14 gal./yd.² for first coat and an application rate of 0.10 to 0.12 gal./yd.² for the second coat. Each application should be spray applied and diluted not more than manufacturer's recommendations. The diluted coal tar product must meet or exceed Federal Specification RP355e.

<u>One Coat, Squeegee Applied</u>

Seal coat all surface areas with coal tar. One coat applied at an application rate of 0.18 to 0.22 gal./yd.², squeegee applied and diluted not more than manufacturer's recommendations. The diluted coal tar product must meet or exceed Federal Specification RP355e.

<u>Two Coats, Squeegee Applied</u>

Seal coat all surface areas with coal tar. First coat applied at an application rate of 0.18 gal./yd.², and the second coat applied at a rate of 0.10 to 0.14 gal./yd.², both coats squeegee applied and diluted not more than manufacturer's recommendations. The diluted coal tar product must meet or exceed Federal Specification RP355e.

<u>One Coat in Parking Stalls and Delivery Drives, and Two Coats in Front Drive Lanes, Squeegee Applied</u>

Seal coat all surface areas with coal tar. One coat applied in the parking stall areas at an application rate of 0.18 to 0.22 gal./yd.², two coats applied in the drive lanes at an application rate of 0.18 to 0.22 gal./yd.² first coat, 0.10 to 0.14 gal./yd.² second coat; each application squeegee applied and diluted not more than manufacturer's recommendations. The diluted coal tar product must meet or exceed Federal Specification RP355e.

SEAL COATING WITH ASPHALT EMULSION

One Coat, Squeegee Applied

Seal coat all surface areas with asphalt emulsion. One coat applied at an application rate of 0.16 gal./yd.², squeegee applied and diluted not more than 20 percent of concentrate or the manufacturer's specified dilution rate.

One Coat, Spray Applied

Seal coat all surface areas with asphalt emulsion. One coat applied at an application rate of 0.16 gal./yd.², spray applied and diluted not more than 20 percent of concentrate or the manufacturer's specified dilution rate.

Two Coats, Squeegee Applied

Seal coat all surface areas with asphalt emulsion. First coat applied at an application rate of 0.18 to 0.22 gal./yd.², and the second coat applied at a rate of 0.10 to 0.14 gal./yd.², both coats squeegee applied and diluted not more than 20 percent of concentrate or the manufacturer's specified dilution rate.

Two Coats, Spray Applied

Seal coat all surface areas with asphalt emulsion. First coat applied at an application rate of 0.14 to 0.16 gal./yd.², and the second coat applied at a rate of 0.14 to 0.16 gal./yd.², both coats spray applied and diluted not more than 20 percent of concentrate or the manufacturer's specified dilution rate.

One Coat in Parking Stalls and Delivery Drives, and Two Coats in Front Drive Lanes, Spray Applied

Seal coat all surface areas with asphalt emulsion. One coat applied in the parking stall and rear delivery areas at an application rate of 0.14 to 0.16 gal./yd.², two coats applied in the drive lanes at an application rate of 0.14 to 0.16 gal./yd.², each application spray applied and diluted not more than 20 percent of concentrate or the manufacturer's specified dilution rate.

One Coat in Parking Stalls and Delivery Drives, and Two Coats in Front Drive Lanes, Squeegee Applied

Seal coat all surface areas with asphalt emulsion. One coat applied in the parking stall and rear delivery areas at an application rate of 0.14 to 0.16 gal./yd.², two coats applied in the drive lanes at an application rate of 0.14 to 0.16 gal./yd.², each application squeegee applied and diluted not more than 20 percent of concentrate or the manufacturer's specified dilution rate.

SEAL COAT WITH ASPHALT AND COAL TAR BLEND
ONE COAT, SPRAY APPLIED

Seal coat all surface areas with asphalt emulsion and coal tar blend (20 percent / 80 percent) (20 percent asphalt emulsion with 80 percent coal tar emulsion). The material shall be applied at an application rate of 0.14 gal./yd.^{2}, spray applied and diluted not more than 50 percent of concentrate or as specified by the manufacturer.

SEAL COAT WITH ASPHALT AND COAL TAR BLEND
ONE COAT, SQUEEGEE APPLIED

Seal coat all surface areas with asphalt emulsion and coal tar blend (20 percent / 80 percent) (20 percent asphalt emulsion with 80 percent coal tar emulsion). The material shall be applied at an application rate of 0.16 to 0.22 gal./yd.^{2}, squeegee applied and diluted not more than 50 percent of concentrate or as specified by the manufacturer.

SEAL COAT WITH ASPHALT AND COAL TAR BLEND
TWO COATS, SPRAY APPLIED

Seal coat all surface areas with asphalt emulsion and coal tar blend (20 percent / 80 percent) (20 percent asphalt emulsion with 80 percent coal tar emulsion). First coat applied at an application rate of 0.14 gal./yd.^{2}, and the second coat applied at a rate of 0.14 gal./yd.^{2}, both coats spray applied and diluted not more than 50 percent of concentrate or as specified by the manufacturer.

SEAL COAT WITH ASPHALT AND COAL TAR BLEND ONE COAT IN PARKING STALL AND REAR DELIVERY AREAS AND TWO COATS IN THE DRIVE LANES, SQUEEGEE APPLIED

Seal coat all surface areas with asphalt emulsion and coal tar blend (20 percent / 80 percent) (20 percent asphalt emulsion with 80 percent coal tar emulsion). One coat applied in the parking stall and rear delivery areas at an application rate of 0.14 to 0.16 gal./yd.^{2}, two coats applied in the drive lanes at an application rate of 0.14 to 0.16 gal./yd.^{2}, squeegee applied and diluted not more than 50 percent of concentrate or as specified by the manufacturer.

SEAL COAT WITH ASPHALT AND COAL TAR BLEND ONE COAT IN PARKING STALL AND REAR DELIVERY AREAS AND TWO COATS IN THE DRIVE LANES, SPRAY APPLIED

Seal coat all surface areas with asphalt emulsion and coal tar blend (20 percent / 80 percent) (20 percent asphalt emulsion with 80 percent coal tar emulsion). One coat applied in the parking stall and rear delivery areas at an application rate of 0.14 to 0.16 gal./yd.^{2}, two coats applied in the drive lanes at an application rate of 0.14 to 0.16 gal./yd.^{2}, both coats spray applied and diluted not more than 50 percent of concentrate or as specified by the manufacturer.

SEAL COAT WITH ASPHALT AND COAL TAR BLEND
TWO COATS, SQUEEGEE APPLIED

Seal coat all surface areas with asphalt emulsion and coal tar blend (20 percent / 80 percent) (20 percent asphalt emulsion with 80 percent coal tar emulsion). First coat applied at an application rate of 0.18 to 0.22 gal./yd.^{2}, and the second coat applied at a rate of 0.14 to 0.16 gal./yd.^{2}, both coats squeegee applied and diluted not more than 50 percent of concentrate or as specified by the manufacturer.

NOTE: The higher the percentage of asphalt emulsion in the asphalt and coal tar blends, the lower the dilution rate will be. Make sure you have reviewed the manufacturer's material spec sheet prior to stating the dilution rate in the scope of work .

TYPE II SLURRY SEAL

Apply type II quick set slurry seal to all surfaces after proper cleaning and surface preparation have been accomplished. The slurry seal shall have an aggregate in compliance with local agency specifications for type II slurry seal. The slurry seal shall be rolled and compacted as soon as the compacting equipment can move on the material without picking up or scuffing the new slurry seal, or no later than twenty-four (24) hours.

TYPE II POLYMER MODIFIED SLURRY SEAL

Apply type II quick set slurry seal with a polymer modified additive to all surfaces after proper cleaning and surface preparation have been accomplished. The slurry seal shall have an aggregate in compliance with local agency specifications for type II slurry seal. The slurry seal shall be rolled and compacted as soon as the compacting equipment can move on the material without picking up or scuffing the new slurry seal, or no later than twenty-four (24) hours.

CHIP SEAL

Clean all debris from the existing paved streets by use of power "kick" brooms. The surface will have a hot-applied asphalt emulsion CRS2 or CRS2P. A coating of ⅜-inch chip seal will be installed at an application rate of 25 to 30 lb./yd.^{2}. The chip seal shall be rolled and compacted by the use of pneumatic rollers immediately following the cover application. At least three self-propelled pneumatic rollers should be used on this project. Surplus aggregate or chips shall be removed at the completion of the application by broom sweeping, kick broom sweeping, or a combination of both. The excess chips shall be hauled away and disposed of according to local agency requirements. The chip seal shall have a coat of SS-1h spray applied six months after the loose stones have been cleaned off the surface.

RUBBER/POLYMER COATED CHIP SEAL
STRESS-ABSORBING MEMBRANE (SAM)

Clean all debris from the existing paved streets by use of power "kick" brooms. The surface will have a hot-applied asphalt rubber modified emulsion CRS2 or CRS2P. A coating of ⅜-inch polymer or rubber precoated chip seal will be installed at an application rate of 25 to 30 lb./yd.^{2}. The chip seal (SAM) shall be rolled and compacted by the use of pneumatic rollers immediately following the cover application. The minimum number of rollers to be used on this project is three self-propelled pneumatic rollers. Surplus aggregate or chips shall be removed at the completion of the application by broom sweeping, kick broom sweeping, or a combination of both. The excess chips shall be hauled away and disposed of according to local agency requirements.

EARTHWORK

All areas designated for new pavement structure shall have the existing soils prepared for construction of a new pavement section. The existing topsoil, all vegetation, and undesirable roots, organic materials, and so on shall be removed and disposed of. The existing suitable and acceptable soils shall be scarified to a depth of 8 inches or as designated in the job site geotechnical report. The existing soils shall be watered, processed, and compacted to the specified density as set forth in the project geotechnical report or as stated by the engineer. Where there is no geotechnical report, the soils shall be scarified 8 inches in depth, watered, and compacted to 95 percent standard Proctor density (ASTM D-698) for clays and clayey materials, or 95 percent modified Proctor density (ASTM D-1557) for granular materials.

AGGREGATE BASE COURSE

Aggregate base course material with a maximum size aggregate of $3/4$ inch (unless otherwise specified) shall be installed after the subgrade material has been compacted and accepted. The aggregate base material shall have finished grades established and set for matching existing or establishing optimal drainage. The aggregate base course material shall be compacted to 95 percent modified Proctor density (ASTM D-1557).

INSTALL NEW PAVEMENT

Remove all overburden, topsoil, vegetation, and any concrete curbing and sidewalk and dispose of according to local agency regulations. Relocate all utilities (water pipes, PVC, valve boxes, etc.) to a new location out of the work site, yet in a location to properly service the area (as required). Excavate the existing subgrade material 6 inches, scarify the finished subgrade material, and compact to 95 percent standard Proctor density for clayey material and 95 percent modified Proctor density for granular materials. Install new curb and gutter and sidewalk as required. Install 4 inches of Class A aggregate base course and compact to 95 percent modified Proctor density. Install 2 inches hot-mixed asphalt pavement designated R12.5 and compact to 95 percent Marshall density. Stripe as required according to local striping regulations.

REMOVE ALL PAVEMENT AND REPAVE

Inspect all bumper blocks in the parking lot, remove all bumpers, and stack off work site. Remove and replace all the pavement as shown on the attached property plat, leasing plan, or map; haul off and dispose of removed material in an acceptable disposal area. Grade the existing aggregate base course to obtain optimal drainage. Compact the aggregate base to 95 percent modified Proctor density (ASTM D-1557). Install 2½ inches of hot-mixed asphalt pavement designated as R12.5 modified (more commonly known as: ½-inch maximum sized aggregate) and compact to 95 percent Marshall density (ASTM D-1559). Replace all bumper blocks and restripe to the existing configuration. All manhole covers and all utility services (water valves, etc.) shall be elevated to the new paved elevation. Refer to "Paving Specifications" for rolling and compaction requirements to be met.

PULVERIZING AND PAVING

Inspect all bumper blocks in the parking lot, remove all bumpers, and stack off work site. Pulverize the existing pavement and ABC to the equivalent of a ¾ -inch aggregate base course and grade to obtain optimal drainage. Compact the pulverized base to 95 percent modified Proctor density (ASTM D-1557). Install 2 inches of hot-mixed asphalt pavement designated as R19 modified (more commonly known as: ¾ inch maximum size aggregate) or R12.5 modified (more commonly known as: ½ inch maximum size aggregate) and compact to 95 percent Marshall density (ASTM D-1559). Replace all bumper blocks and restripe to the existing configuration, except the handicap stalls. Install universal handicap stalls with ADA signs and poles. Upon completion of the pulverizing, all irrigation systems will be turned on and any leaks in the irrigation system that are detected will be repaired prior to grading the base and paving. Included with the Scope of Work is repair to all irrigation lines damaged by pulverizing. All manhole covers and all utility services (water valves, etc.) shall be elevated to the new paved elevation. Refer to "Paving Specifications" for rolling and compaction requirements to be met.

MILL AND PAVE

Inspect all bumper blocks in the parking lot if applicable, remove all bumpers, and stack off work site. Mill off the existing asphalt pavement to the existing aggregate base course; haul off and dispose of milled material in an acceptable disposal area. Grade the existing aggregate base course to obtain optimal drainage. Compact the aggregate base to 95 percent modified Proctor density (ASTM D-1557). Install 2 inches of hot-mixed asphalt pavement designated as R19 modified (more commonly known as: ¾ inch maximum size aggregate) or R12.5 modified (more commonly known as: ½ inch maximum size aggregate) and compact to 95 percent Marshall density (ASTM D-1559). Replace all bumper blocks and restripe to the existing configuration, except the handicap stalls. Install universal handicap stalls with ADA signs and poles. Upon completion of the pulverizing, all sprinkler systems will be turned on and any leaks in the sprinkling system that are detected will be repaired prior to grading the base and paving. Upon completion of the milling, all irrigation systems will be turned on and any leaks in the irrigation system that are detected will be repaired prior to grading the base and paving. Included with the Scope of Work is repair to all irrigation lines damaged by milling. All manhole covers and all utility services (water valves, etc.) shall be elevated to the new paved elevation. Refer to "Paving Specifications" for rolling and compaction requirements to be met.

STANDARD CONVENTIONAL OVERLAY

All edges along curbs, gutters, sidewalks, block walls, garage floors, and so on shall be milled 6 feet from the edge toward the centerline of the drive lane on both sides of the drive or away from the building fronts. Where there are drainways (weep holes) in block walls, these drainways shall be kept free of all overlay material and shall be left in a manner to allow for all water to exit the adjacent property. The surface shall be cleaned of all loose debris, sand, dirt, and any foreign substances prior to applying tack coat. The surface shall be dry of all standing water prior to applying tack coat and fabric. The surface will have a hot-applied asphalt emulsion CRS2 or CRS2P or AC-10 asphalt tack coat. Install 2 inches of hot-mixed asphalt pavement designated as R19 modified (more commonly known as: ¾ inch maximum size aggregate) or R12.5 modified (more commonly known as: ½ inch maximum size aggregate) and compact to 95 percent Marshall density (ASTM D-1559). Replace all bumper blocks. All manhole covers and all utility services (water valves, etc.) shall be elevated to the new paved elevation. All blue reflectors and any other ceramic markers or plastic reflectors shall be removed prior to installing the fabric and shall be replaced as originally located. Refer to "Paving Specifications" for rolling and compaction requirements to be met.

FABRIC INNER LAYER OVERLAY
BIKE PATHS AND WALKWAYS

When listed in the Scope of Work, all edges along curbs, gutters, sidewalks, block walls, garages, and so on shall be milled 2 feet from the edge toward the centerline of the bike lane or walkway on both sides of the drive or away from the building fronts. The milled material shall be hauled off and disposed of according to local requirements. Where there are drainways (weep holes) in block walls, these drainways shall be kept free of all overlay material and shall be left in a manner to allow for all water to exit the adjacent property. The surface shall be cleaned of all loose debris, sand, dirt, and any foreign substances prior to applying tack coat and fabric. The surface shall be dry of all standing water prior to applying tack coat and fabric. The surface will have a hot-applied asphalt emulsion CRS2 or CRS2P or AC-10 asphalt tack coat. A fabric inner layer shall be installed over the entire surface (1 foot from the edges of the pathway), overlaid free of any crinkles, folds, and so on. Install 1½ inches of R12 modified (more commonly known as: ½ inch maximum size aggregate) hot-mixed asphalt pavement and compact to 95 percent Marshall density (ASTM D-1559). All manhole covers and all utility services (water valves, etc.), if present, shall be elevated to the new paved elevation (no depressions or overelevations shall exist after completion). Refer to "Paving Specifications" for rolling and compaction requirements to be met.

FABRIC INNER LAYER OVERLAY PARKING LOTS

All edges along curbs, gutters, sidewalks, block walls, garages, and so on shall be milled 6 feet from the edge toward the centerline of the drive lane on both sides of the drive or away from the building fronts. The milled material shall be hauled off and disposed of according to local requirements. Where there are drainways (weep holes) in block walls, these drainways shall be kept free of all overlay material and shall be left in a manner to allow for all water to exit the adjacent property. The surface shall be cleaned of all loose debris, sand, dirt, and any foreign substances prior to applying tack coat and fabric. The surface shall be dry of all standing water prior to applying tack coat and fabric. The surface will have a hot-applied asphalt emulsion CRS2 or CRS2P or AC-10 asphalt tack coat. A fabric inner layer shall be installed over the entire surface to be overlaid free of any crinkles, folds, and so on. Install 1½ inches of R12 modified (½-inch maximum size aggregate) hot-mixed asphalt pavement and compact to 95 percent Marshall density (ASTM D-1559). Replace all bumper blocks. All manhole covers and all utility services (water valves, etc.) shall be elevated to the new paved elevation. All blue reflectors and any other ceramics or plastic reflectors shall be removed prior to installing the fabric and shall be replaced as originally located. Refer to "Paving Specifications" for rolling and compaction requirements to be met.

FABRIC INNER LAYER OVERLAY STREETS

All edges along curbs, gutters, and valley gutters shall be milled 3 feet from the edge toward the centerline of the street on both sides of the streets. The milled material shall be hauled off and disposed of according to local requirements. Where there are drainways (weep holes), these drainways shall be kept free of all overlay material and shall be left in a manner to allow all water to exit the property. The surface shall be cleaned of all loose debris, sand, dirt, and any foreign substances prior to applying tack coat and fabric. The surface shall be dry of all standing water prior to applying tack coat and fabric. The surface will have a hot-applied asphalt emulsion CRS2 or CRS2P or AC-10 asphalt tack coat. A fabric inner layer shall be installed over the entire surface to be overlaid free of any crinkles, folds, and so on. Install 1½ inches of R12 modified (½-inch maximum size aggregate) hot-mixed asphalt pavement and compact to 95 percent Marshall density (ASTM D-1559). All manhole covers and all utility services (water valves, etc.) shall be elevated to the new paved elevation. All blue reflectors and any other ceramics or plastic reflectors shall be removed prior to installing the fabric and shall be replaced as originally located. Refer to "Paving Specifications" for rolling and compaction requirements to be met.

DRY WELL

Install an acceptable dry well for the existing drainage conditions and to the designed depth of the existing soil and bedrock conditions. The dry well shall be installed in the approximate location indicated on the property map and will be located in the most suitable location. This will require some grade evaluation and survey by the Contractor prior to installing the well to ensure proper location.

DUMPSTER PAD

Remove the existing asphalt pavement, aggregate base course, and any subsoil to a depth of 8 inches. Compact the subgrade to 95 percent Proctor density, form with suitable and stable forms, and add reinforcing mesh to the entire pad area. The wire mesh shall be placed on mortar pedestals prior to pouring the concrete or shall be lifted during the pour to ensure the mesh will be 1 to 1½ inches off the subgrade upon completion. Install Portland type II concrete with a minimum strength of 2,500 lb./in.^{2}, finish, and trowel, with proper contraction and tool joints as specified in local agency specifications.

NEW VALLEY GUTTER

Remove the existing asphalt pavement, aggregate base course, and any subsoil to a depth of 8 inches. Compact the subgrade to 95 percent Proctor density and form with suitable and stable forms. Install Portland type II concrete with a minimum strength of 2,500 lb./in.^{2}, finish, and trowel, with proper contraction joints at intervals of 100 feet and tool joints every 12 feet as specified in local agency specifications and details.

REPAIRING EXISTING VALLEY GUTTER

Remove specified existing broken valley gutter stones and dispose of concrete in an acceptable manner. Excavate to 8 inches in depth, compact the sub-base to 95 percent Proctor density, and form with suitable and stable forms. Install Portland type II concrete with a minimum strength of 2,500 lb./in.^{2}, finish, and trowel, with proper expansion joints at intervals of 100 feet and tool or contraction joints every 12 feet or as specified in local agency specifications and details.

REPAIRING EXISTING CONCRETE CURB AND GUTTER

Saw cut the existing pavement 1 foot from the broken curb and gutter and remove the existing pavement, remove specified existing broken curb and gutter stones, and dispose of concrete in an acceptable manner; compact the sub-base to 90 percent modified Proctor density, and form with suitable and stable forms. Install Portland type II concrete with a minimum strength of 2,500 lb./in.^{2}, finish, and trowel, with proper expansion joints at intervals of 100 feet and tool or contraction joints every 10 to 12 feet as specified in local agency specifications. Patch back the area in front of the curb in the existing pavement as outlined in the Patching Specifications.

SIDEWALK REPAIR

Remove specified existing sidewalk stones and dispose of concrete in an acceptable manner. Excavate to 4 inches in depth, compact the sub-base to 90 percent modified Proctor density, and form with suitable and stable forms. Install Portland type II concrete with a minimum strength of 2,500 lb./in.^{2}, finish, and trowel, with proper expansion joints at intervals of 100 feet and tool or contraction joints every 5 feet or as specified in local government specifications and details.

CURB REPAIR AND REPLACEMENT

Saw cut and remove existing asphalt pavement 1 foot from the edge of combination curb and gutter or vertical curb, saw cut the curbing at a location beyond the broken or distressed curbing or furthest joint, and prepare to remove. Saw cut in place extruded curbing at a location beyond the broken or distressed curbing or furthest joint and prepare to remove. Remove specified existing extruded, combination curb and gutter, vertical curb, or ribbon curb stones and dispose of concrete in an acceptable manner. For combination curb and gutter, vertical curb, and ribbon curb, excavate to 4 inches in depth, grade even in elevation, and compact the sub-base to 95 percent standard Proctor density, and form with suitable and stable forms. Install Portland type II concrete with a minimum strength of 3,500 lb./in.^{2}, finish, and trowel, with proper expansion joints at intervals of 100 feet and tool or contraction joints every 10 to 15 feet or as specified in local government specifications and details. Prior to installing new extruded curbing, an acceptable adhesive shall be applied to the pavement. Once the new combination curb and gutter, vertical curb, or ribbon curb is installed and properly cured, the adjacent pavement shall be patched with hot-mixed asphalt and compacted as outlined in the Specifications for remove and replace patching.

ADA RAMP CONSTRUCTION

Saw cut the existing concrete curb and sidewalk at handicap stall parking and hash mark area to the dimensions established by the attached ADA ramp details. Install a concrete ramp that will provide a 3-foot landing and dispose of the concrete in an acceptable manner, compact the sub-base to 90 percent modified Proctor density, and form with suitable and stable forms. Install Portland type II concrete with a minimum strength of 2,500 lb./in.^{2}, finish, and trowel, with proper expansion joints at intervals of 100 feet and tool or contraction joints.

CONCRETE SPLASH PADS

Saw cut and remove the existing asphalt pavement under the downspouts of roof drains and remove asphalt and base to a depth of 4 inches, and dispose of the spoils in a manner acceptable to local government agency requirements. Install Portland type II concrete with a minimum of 2,500 PSI, float and trowel to a smooth surface, and allow to cure.

RESTRIPING AND PAINTING TO EXISTING CONFIGURATION

Restripe all paved surfaces to the existing configuration, including all handicap and special markings. All handicap markings shall be installed according to current ADA standards and all local agency requirements.

RESTRIPING AND PAINTING TO NEW LAYOUT AND CONFIGURATION

Stripe all paved surfaces to the attached striping plan, including all handicap and special markings. All handicap markings shall be installed according to current ADA standards and all local agency requirements.

Once the scopes of work have been selected, specifications are attached to the RFB or RFQ to instruct the contractor. Following are specifications to atatch to the RFB or RFQ.

GENERAL SPECIFICATIONS

SUBMITTALS FOR MATERIALS OTHER THAN SPECIFIED

Prepare specification sheets for all materials and submit to the Owner, authorized Agent of the Owner, or Consultant.

Before supplying any material to the job site, the Contractor and the materialman's area representative shall prepare a complete schedule of delivery, storage instructions, and specifications of the materials proposed to be used for each maintenance application specified. Submit same to the Owner, the Agent, or the Consultant.

Product Delivery, Storage, and Handling

Deliver all material to the job site in the materialman's original wrapper with label and seals intact. Labels shall give manufacturer's name, brand, type, batch number, and instructions for reducing, diluting, or cutting.

Store all material used on the project as recommended by the materialman in a single designated space, clear of the ground and of moisture. All materials will be protected from rain and weather at all times. Such storage place shall be kept clean. Make good any damage to storage area or to its surroundings. Remove any oily rags, waste, etc. from the building every night and take every precaution to avoid any danger of fire. In no case shall the amount of material stored exceed that permitted by local ordinances, state laws, or fire underwriter regulations and agencies.

Weather and Surface Conditions

Do not apply any materials in damp rainy weather or until the surface has dried thoroughly from the effects of any wet weather or irrigation. Do not apply any asphalt material when temperatures are below 55°F, or as listed in the particular material specification and/or material manufacturer's recommended application.

Materials shall be applied according to the Specifications. All existing surfaces shall be clean of all debris, dust, dirt, and water before application can proceed.

Work and Project Hours

The hours allowed to work on the project will be established to the requirements of the residents, tenants, and customers. These hours shall be established to achieve all the work tasks at the least inconvenience to any residents, tenants, or customers. The hours of operation will be spelled out in the Request for Bids, and the Contractor shall abide by these work hours. The end of the workday shall be considered the finish time, and all equipment and workers shall leave the project at this time. It shall be the responsibility of the Contractor to clean up the project and equipment, and secure the project before the finish time outlined in the Request for Bids. Should the work hours not be specified in the Request for Bids, the work hours shall then be considered to be between 7:00 AM and 4:30 PM

Protection

Before starting work, remove or otherwise cover and/or protect all landscaping, concrete structures, equipment, accessories, signs, lighting fixtures, resident/tenant and customer personal property, and similar items or provide ample protection of such items. Upon completion of each work area, replace any of the above items if damaged during the maintenance or construction phase. Protect adjacent surfaces as required or directed. Any damage incurred as a result of the Contractor, his/her subcontractors, or Sub-subcontractors shall be repaired by the Contractor of record, at his/her expense.

The Contractor shall observe all safety precautions as recommended by OSHA and the governing agency in which the project resides.

All debris from the operation shall be removed daily from the job site and disposed of at an approved, suitable disposal site. **The Contractor will not dispose of any debris or materials in the dumpsters. If the Contractor disposes of any material in the dumpsters, the Contractor will pay all charges and fees incurred to have the material hauled off and disposed of.**

The Contractor shall clean all sidewalks, driveways, store entrances, or any other surfaces adjacent to the work area of all dust, dirt, rocks, or any other debris by sweeping, or air blowing immediately after the maintenance material has been applied. The debris shall be cleaned away from all building entrances.

Prior to beginning the work, the Contractor shall have all preparatory tasks completed to ensure the area is ready to receive the materials for that day as outlined in the material specifications. Should the pavement be replaced or overlaid, or any drainage structures (curbs, valley gutters, or slabs) be replaced, the Contractor will maintain optimal drainage at the completion of the project. Prior to and at the completion of construction, the Contractor and the Owner's/Manager's representative or consultant will inspect the surfaces and report to the Owner any deficiencies or necessary corrections which will require repair or attention. Application of materials will not proceed until all unsatisfactory conditions noted in the site inspection have been corrected.

The Contractor assumes responsibility for acceptance of all materials used by his/her application of those materials. The Contractor and his/her employees shall not track or walk back over the new work or track materials onto adjacent areas, such as walks, patios, curbs, outdoor carpeting, etc.

FINAL CLEANUP

Replace any materials disturbed as the result of erecting or removing the barricades and protection.

Remove all debris, scraps, and containers from the job site.

Repair or replace defaced or disfigured finishes caused by work performed, including curbs where asphalt has been splashed on.

The Contractor shall be responsible for cleaning all walks, floors, carpets, etc., should any material used be tracked onto any of these surfaces. The Contractor shall be responsible for cleaning all property (autos, walls, windows, or any personal property or building fixtures) of material or residue as a result of the construction on all legitimate claims. These claims will be determined legitimate by the Owner, the Agent, or the Consultant.

The Contractor shall NOT dump or put anything into dry wells, in dumpsters, or on landscaped areas at any time. The Contractor shall properly dispose of any and all materials, either removed from the job site or leftover product, in a proper and environmentally safe manner.

Perform all work using only experienced, competent personnel in accordance with the best standards or practices in the trade. When completed, the repairs shall represent a first-class workmanlike appearance.

All work shall be completed in a manner to improve and/or maintain the aesthetic appearance of the property. Also, any work completed that will requ-ire any corrections due to punch lists, etc. must also be completed in a manner to improve and/or maintain the aesthetic appearance of the property.

ASPHALT MATERIAL SUPPLIER'S INSPECTION VERIFICATION

Upon the request of the Owner/Manager or a designated Representative/Consultant, the asphalt manufacturer/supplier representative will make at least one interim inspection early in the project. This is to ensure and verify that the materials being applied are in proper compliance and have been applied in the proper manner, and that the Contractor is following the manufacturer's recommended procedures.

The asphalt or concrete product supplier or manufacturer's representative, upon request, will forward to the Owner a letter, on his/her letterhead, stating the materials applied are in compliance and applied in the proper manner. Final payment will not be released until the Owner has received this information from the asphalt or concrete supplier/manufacturer, or the Owner's representative/Consultant.

ASPHALT MATERIAL SUPPLIER'S MATERIAL SAFETY DATA SHEET (MSDS) HAZARDOUS COMMUNICATION

The Contractor shall supply, with the Contract Documents, all MSDS forms required for compliance to OSHA Safety and Health Standards as outlined in the local government safety standard. This includes any products containing asphaltines, coal tar pitch, and all paints used for striping and marking, whether oil or acrylic based. This information is supplied by the materials manufacturer and is available upon request.

CRACK SEALING SPECIFICATIONS

A. Materials

A-1 The crack sealant shall be a hot elastomeric asphalt sealant that is formulated specifically to be a stiff, non-tracking, yet flexible sealant, which is specifically suited for areas subject to pedestrians and slow-moving vehicle traffic and for the local temperature conditions.

A-2 The crack sealant shall meet the standards as listed in ASTM D-1190 for hot pour elastomeric crack sealant. The material applied shall meet or exceed ASTM D-1190 and shall have a softening point of not less than 210°F and a viscosity of greater than 2500 cps. **A manufacturer's product specification sheet must be submitted for approval ten (10) working days prior to the bid opening.**

A-3 The Owner, Manager, or Consultant may take samples of materials to an independent laboratory for testing or require on-site inspection for application.

B. Preparation

B-1 Prior to crack sealing, ALL qualifying cracks shall be cleaned by blowing out with compressed air (at least 80 CFM), until clean. In any areas where weeds are present, all weeds shall be removed, the crack cleaned, and then an herbicide (sterilant) will be applied into the crack.

C. Application

C-1 The sealant shall be heated to at least 350°F (or manufacturer's recommendations) when applied but not to exceed 400°F (or manufacturer's recommendations). Do not apply when the ambient temperature is less than 40°F. The material shall be mechanically applied with a continuous heating, self-contained hot pot with a pressure pump, hose, and wand. The wand shall have a pointed tip to fit into the crack and pressure-deliver the material. The mechanical applicator shall be manufactured by Crafco or Bearcat, or equivalent, and meet the specifications of the Crafco EZ Pour 200.

C-2 If pavement is wet or if rain or water has been on the asphalt within the last forty-eight (48) hours, the project should be postponed until the cracks have ample time to dry.

C-3 Cracks less than ¼ inch in opening should be left undisturbed. Cracks with an opening of ¼ inch or more shall be thoroughly cleaned as listed above, and filled using hot pour crack sealant.

C-4 If routing is required, the areas will be highlighted on the attached plot plan. Cracks to be routed will be routed out to a width of ¼ inch wide and at least ½-inch deep, then filled with the hot pour crack sealant.

C-5 All cracks will be slightly overfilled and then leveled with a squeegee, leaving a 1-inch band on both sides of the sealed crack. Should the local temperatures stay high enough to leave the crack sealant tacky and increase the possibility of pickup by foot or tire traffic, the crack sealant shall have a blotter material applied over it. This blotter material shall be a fine mesh, sand, or ash.

C-6 Allow crack sealer to cure at least two (2) to four (4) hours before applying seal coating, slurry seals, or overlay.

C-7 Crack sealing shall be performed before any seal coating, slurry seals, or overlays are applied.

C-8   Should the ground temperature remain at a level that will not allow cooling of the material in a reasonable time period, and/or the material will be soft and tacky when traffic is allowed to drive on the crack seal material, a blotter material of silica sand or other small-type grade sand shall be used to prevent pickup of crack sealing material by vehicle and pedestrian traffic.

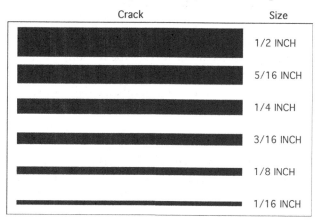

Figure 5.2: Crack Sealing Scale

SKIN PATCHING SPECIFICATIONS

A. Materials

A-1 Asphaltic material shall be ⅜-inch maximum size aggregate and 5.5-6 percent asphalt binder by total weight. The material shall meet all local government construction specifications for ⅜-inch maximum size aggregate. Depending on the severity of the raveling and the existing surface condition, a hot mixed sand seal can be used in place of the ⅜-inch hot-mixed material. When a surface has low density of raveling, it is recommended to use hot-mixed sand seal. The amount of asphalt binder shall be at least 6 percent by weight of the sand seal mix design. The gradation and asphalt contents are based on project mix designs and local materials. Where the skin patch is to be placed in an existing waterway (inverted street crown, asphalt valley gutter, etc.), only the sand seal shall be used.

A-2 The Owner, Manager, or Consultant may take samples of materials to an independent laboratory for testing of proper mix design or require on-site inspection for proper application of materials.

B. Preparation

B-1 Prior to application, ALL areas will be power swept and cleared of all loose material. Oil spots will be manually scraped and cleaned with mild detergent if necessary to remove any petroleum residue. Manholes, valve boxes, drop inlets, and other service entries or markers will be protected from the application of the patch material and elevated where required after the patch is installed.

C. Application

C-1 Treat the entire area to be patched with a coat of SS-1, CRS, or QRS primer tack extending at least 1 foot beyond the distressed area.

C-2 The location edges of the patch area shall be marked with a chalk line to ensure square even edges. The Contractor shall be responsible for installing patches with straight even edges. Any uneven, raveled, ragged edges will be corrected at the Contractor's expense.

C-3 Apply asphaltic material to a depth of not more than ½ inch for ⅜-inch maximum size aggregate material or ¼ inch for sand seal. Make sure the edges are square and extend at least 1 foot beyond the distressed area, and the patch should be as square or rectangular as possible. Care shall be exercised in raking the asphalt to provide a good-feathered edge and for blending of the new materials with the existing pavement and concrete curbs, openings, and walks. The pavement shall be installed in a manner to prevent segregation of the aggregate. The material shall be raked to feather the edges smoothly into the existing pavement to prevent an abrupt edge. This may require raking the edges to remove large stones from the mix approximately 6 to 8 inches from the edge and making an even, smooth transition from the patch to the existing pavement.

C-4 Roll and compact with a 3- to 5-ton static roller, where the roller can move without interruption or sharp turning or where the patch is in line with the flow of water. Where the patch is in the flow of water, especially in inverted crowns or asphalt valley gutters, or where the roller will require sharp turns, a vibratory plate will be used. In the case of an inverted crown or asphalt valley gutter, the patch shall be installed to fit the existing grade and not block or impede the flow of water or cause damming of water behind it.

C-5 Skin patches will only be installed when the high ambient temperature is 40°F and rising, and the ground temperature is 40°F and rising.

C-6   If a patch is installed when the high ambient temperature is below 50°F and above freezing, the patch shall be coated with an asphalt-based emulsion to retard raveling.

C-7   Patching around curbs, walls, walkways, sidewalks, and/or landscaped areas will be done by hand with necessary protection to prevent damage to property.

C-8   The materials shall be applied in such a manner as to restore areas to original elevation and grading as much as possible allowing for good drainage.

C-9   Patched areas shall be allowed to thoroughly cure, based on local specifications, manufacturer's recommendations, and local weather conditions, before seal coat is applied (approximately thirty [30] days). In desert climates, it is recommended to allow the full hot summer season for curing when the patch material is placed in extremely hot weather conditions.

C-10   The combination of particle sizes or sieve analysis of the asphalt material shall be according to local specifications and available materials.

REMOVAL AND REPLACEMENT REPAIRS SPECIFICATIONS

A. Materials

A-1   Asphaltic material will contain ½- or ¾-inch maximum size aggregate and a minimum of 5.5 percent asphalt by weight. Material shall meet all local government construction specifications for ½-inch and ¾-inch hot-mixed asphaltic concrete.

A-2   The Owner, Manager, or Consultant may take samples of materials to an independent laboratory for testing of proper mix design or require on-site inspection for proper application of materials. The gradation and asphalt contents are based on project mix designs.

B. Preparation

B-1   Saw cut or jackhammer the distressed area at least 1 foot in all directions beyond distressed area and remove the existing pavement. All materials removed shall be properly disposed of in an authorized disposal site. Be sure that all cuts are at right (90°) angles and as square as possible.

B-2   The subgrade and ABC shall be compacted to ASTM D-698 (standard Proctor) for clay soils and ASTM D-1557 (modified Proctor) for aggregate bases, sandy soils, and sands.

B-3   Should the aggregate base be water saturated or inadequate, the Contractor shall notify the consultant or manager of the situation and provide remedies for correcting the problem. Should the subgrade be saturated and unstable, the Contractor shall halt all operations and contact the consultant or manager immediately. At no time shall the aggregate base course or subgrade be patched until a suitable and agreeable solution is attained.

C. Application

C-1   Apply an even coat of tack coating of SS-1 or acceptable tack coat to all edges of the existing pavement that surround the patch area. The tack coat shall cover the entire vertical surface of the pavement edge.

C-2   Apply asphaltic material to a depth of at least 2¼ to 2½ inches where the depth of the existing pavement is 2 inches. Where the depth of the existing pavement is greater than 2 inches, the asphaltic material shall be placed in lifts of not more than 2¼ to 2½ inches or a smaller fraction thereof to ensure proper finished thickness. At no time shall the asphaltic material lift be greater than 2 inches after compaction. Each lift shall be compacted before the next lift is applied. Care shall be exercised in raking the asphalt to prevent segregation of the aggregate.

C-3   New asphalt material must be ¼ to ⅜ inch above the existing pavement material before compaction on the final surface lift.

C-4   Roll and compact each lift with a 3- to 5-ton roller to 95 percent compaction (Marshall stability ASTM D-1559). The new asphalt shall be compacted to ⅛ to ¼ inch above existing asphalt. Where prior approval of the property manager/owner or consultant has been given, a vibratory plate compactor can be used.

C-5   The materials shall be applied in order to restore the area(s) to original grading and prevent pounding.

C-6   Patched areas shall be allowed to thoroughly cure, based on local governing agency recommendations and weather conditions, before any seal coat is applied (approximately thirty [30] days). If seal coating is NOT scheduled, and the high ambient temperature is less than 50°F, then apply one coat of asphalt emulsion seal coat over the patched area after twelve [12] hours of curing depending upon local weather conditions. In desert climates, it is recommended to allow the full hot summer season for curing when the patch material is placed in extremely hot weather conditions.

C-7   Any and all areas that are excavated shall have sufficient barricades and warning (flagged) ropes to help protect and prevent pedestrians and vehicles from entering.

SEAL COATING SPECIFICATIONS

A. Materials (There are two base materials as listed in sections A-1 and A-2.)

Warning!! Recent studies by the United States Geological Survey (USGS) have shown that coal tar seal coats contain high amounts of polycyclic aromatic hydrocarbons (PAHs). PAHs have been found to be carcinogenic and heavy exposure may cause cancer. As a result, several cities (including Austin, TX and Washington DC) have banned the use of coal tar sealer. Therefore, the contractor and the owner/manager can be subjected to any claims that may arise due to the use of coal tar sealers. As a result, it is the authors' recommendation that coal tar sealers not be used. Because of its ability to resist fuel spills and endure cold weather climates, cold tar is still in demand, therefore, information on coal tar is still included in this book. It is important to check and understand any federal, state, and local rules and regulations prior to using coal tar. Ultimately, it is recommended that asphalt-based seal coats (which can be modified to meet specific environmental needs) be used in lieu of coal tar sealers.

A-1. <u>Emulsified Coaltar sealer</u> with a minimum of 1-4 percent Latex or Co-polymer added, when specified in the scope of work or by the manufacturer. Mineral filler shall be added in the quantity of 1 to 3 pounds of grade #60 silica sand (or approved equal mineral filler with same sieve size) added per gallon and mechanically agitated (or prior approved mix design). All applications of coal tar products shall have sand added. Where the material has silica or other mineral additives added by the manufacturer, no additional silica will be required. The Contractor must supply the Manufacturer's guaranteed material specification sheet indicating such silica sand and polymer or co-polymer has been added. Dilution of products shall not exceed 40 percent of manufacturers concentrate of solids for coal tar-based products. Coal tar sealants must meet Federal Specification RP-355e, coal tar emulsion (coating for bituminous pavements) and conform to ASTM requirements D490, D3320-74T, D5727 and FAA P627, P628, and P631 for RT12. Proper covering or protection of exposed skin shall be strictly adhered to prevent skin damage. Other standards for testing coal tar material is ASTM C136, D160, D490, D2939, and D4866. All coal tar emulsions must meet the specifications as shown in sub-section C-14.

A-2. <u>Emulsified Asphalt sealer</u> with a minimum of 2-3 percent Latex or Co-polymer added when specified in the scope of work or by the manufacturer. Mineral filler shall be added in the quantity of 1 to 3 pounds of grade #60 silica sand (or approved equal mineral filler with same sieve size) added per gallon and mechanically agitated (or prior approved mix design). Where the material has a polymer or co-polymer added by the manufacturer, no additional polymer or co-polymer shall be required. Where the material has silica or other mineral additives added by the manufacturer, no additional silica or other mineral additives will be required. The Contractor must supply the Manufacturer's guaranteed material specification sheet indicating such silica sand and polymer or co-polymer has been added. The dilution of products shall not exceed 20-30 percent of manufacturers concentrate of solids for asphalt based products as specified in the manufacturers material product specification data sheet. Should the Manufacturer's asphalt emulsion recommend a higher minimum for dilution, the Contractor shall submit the Manufacturer's original specification for approval and amendment to the dilution requirement. The dilution rate can be adjusted to accommodate the application, and will be noted in the Scope of Work. Should there be no change to the application in the Scope of Work, this specification shall be adhered to. ALL ASPHALT EMULSION SEAL COATS SHALL BE TESTED ACCORDING TO ASTM D2939.08, ASTM D2939.07, ISSA (TB-100), ASTM D3810, and ASTM G154. All asphalt emulsions must meet the specifications as shown in sub-section C-13.

An application for equal material must be submitted for approval ten (10) working days prior to the bid opening.

A-3. Where ambient temperature is 55°F at the time of application and the paved surface temperature is 50°F at the time of application and rising, an approved coalescing agent shall be added to the seal coat blend according to manufacturers recommendations

A-4. Where the sealed paved surface will be exposed to colder and freezing temperatures and/or snow and ice, an approved anti-strip agent (ZeTac™ by SealMaster or equivalent) shall be added to retard the striping action of the seal coat and increase adhesion, wearability, and the longevity of the material from exposure to cold climate elements.

A-5. The Owner, Manager, or Consultant may/will take samples of materials to an independent laboratory for testing or require on-site inspection for proper application of materials.

B. Preparation

B-1. Prior to application, ALL areas will be power swept, vacuumed, and cleared of all loose material. All adjacent concrete structures (driveways, walks, gutters, garages, etc.) shall be cleaned of all dirt, dust, and debris prior to applying any materials to the areas designated for seal coating. Any standing or puddled water shall be spread over the surface and dried before any sealer material is applied. AT NO TIME SHALL ANY SEALER MATERIAL BE APPLIED OVER AN AREA THAT IS WET OR PONDING WITH WATER. SHOULD THE CONTRACTOR APPLY SEALER MATERIALS OVER WET OR PONDED AREAS, THESE AREAS WILL BE NOTED AND CORRECTED AS PUNCH LIST ITEMS. Oil spots will be manually scraped to remove any buildup of petroleum products, dust, dirt, etc. and heated with an open flame (propane weed burner) prior to applying any primer or seal coat. A primer coat shall be applied over high saturated petroleum spill areas (Bond Seal or approved equal) prior to applying the seal coat material. Should the area be soft and pliable from exposure to petroleum products to a point after treatment, the area shall be saw cut, removed, and patched prior to installing any seal coat product. (The contractor shall note these areas and report them to the Consultant or Owner's representative for consideration of any compensation). All manholes, adjacent concrete structures (valley gutters, spandrels [aprons], walks, etc.) valve boxes, drop inlets, and other service entrances as well as ALL survey markers will be protected from the application of the product by a suitable method of covering and/or protection. Any material covering, spilled, placed, or running onto any of the above mentioned items will be cleaned by high pressure hot water, or wet sand blasting, and/or other methods acceptable to local agencies. All drywell lids and entrances will be protected by covering with a non-permeable material prior to applying any sealer. Any sealer material placed in the drywell shall be cleaned and removed prior to payment. All existing concrete spills, clay and soil deposits, and any other materials hardened to the pavement shall be removed prior to applying the seal coat materials

B-2. Before seal coating, any new skin patches and/or any new asphalt (R&R repairs) must be allowed a curing time of at least fifteen (15) days BEFORE seal coating is applied. In desert climates, it is recommended to allow the full hot summer season for curing when the patch material is placed in extremely hot weather conditions.

B-3. Prior to application of the sealer, all existing handicap stalls, crosswalks, arrows, and any line stripes not required will be blacked out with flat black paint. The final product will be even square areas with right angles. NO UNEVEN PAINT SPOTS.

B-4. Prior to application of the sealer, all ceramics and reflectors (blue, yellow, etc.) shall be removed and shall be replaced after sealing.

B-5. When the existing paved surface and ambient temperatures are 110°F at the time of application and rising, the surface shall be fogged with a water mist to allow for sealer materials to flow evenly and reduce premature drying causing a splotchy application. This fogging with a water mist shall be applied prior a single coat (one coat) application. On two coat applications, the second coat will provide proper coverage over the first coat.

B-6. When the ambient temperature is a high of 50°F at the time of application and the pavement surface temperature is 50°F at the time of application and rising, a coalescing agent shall be added to the seal coat mix at the manufactures recommendation rate or 2 percent of volume of concentrate. The material shall be blended thoroughly prior to dilution.

B-7. Where the paved surface to be seal coated will be exposed to cold and/or freezing temperatures, snow, stacked snow, and ice during the life of the seal coat, an approved anti-stripping agent shall be added to the asphalt emulsion sealing material. With the agitator turning, add proper amount of water to the sealer. After the water dilution is complete, add 1-2 percent of the anti-stripping agent based upon the pavement sealer concentrate (1-2 gallons of anti-stripping agent per 100 gallons of pavement sealer concentrate or manufacturer's recommendations). As the sealer materials thicken, add sand and agitate thoroughly before and slowly during application.

C. Application

C-1. The ambient and surface temperatures shall be a minimum of 55°F at any time during application and for a twenty-four (24) hours after application when the seal coat is applied without any additives or add mixtures for colder temperatures. When using coal tar sealers, the sealer shall not be applied to hot surfaces under summer sun (over 90°F ambient) without first cooling the surface with clean water. The water should dampen the surface without leaving puddles.

C-2. Seal coating or trimming around curbs, walls, walkways, sidewalks, and/or landscaped areas will be done by hand with necessary protection to prevent over-spray and slopping. Should the sealer material for trimming be applied by broom and hand squeegee, the areas being sealed will only have to be trimmed once during the first application. Should the sealer material for trimming be applied by spraying (with or without a shield), the areas being sealed will have to be trimmed for each application.

C-3. Apply first coat of "SEALER" as listed in A-1 above, at the minimum rate of 0.125 to 0.185 of a gallon per square yard spray applied (or as indicated in the Scope of Work), or the minimum rate of 0.20 of a gallon per square yard squeegee applied (or as indicated in the Scope of Work).

C-4. Allow manufacturer's recommended drying time before applying second coat (This is usually four [4] hours depending on geographical locations and climate conditions)

C-5. Apply the second coat of "SEALER" as listed in A-1 or A-1 above, at the minimum rate of 0.100 to 0.185 of a gallon per square yard spray applied (or as indicated in the Scope of Work), or the minimum rate 0.140 of a gallon per square yard squeegee applied (or as indicated in the Scope of Work).
   *   Squeegeed material will have a higher application rate than spray applied due to the material filling characteristics around surface aggregates. The rougher the surface, the more material is used to fill around surface aggregates

C-6. Where two coats are specified, the first coat shall dry thoroughly before the second coat is applied. The first coat shall not have any wet spots, and shall not pick up on vehicle tires to be considered dry enough to apply the second coat. This timing will vary with ambient and existing ground temperatures at the time of application.

C-7.   Upon completion of the application (one or two coats), the sealer shall be allowed to dry based on manufacturer's recommendations and weather condition before traffic is allowed, (at least twenty-four [24] hours based on local weather conditions). To accomplish this, barricades shall be left in place until the next day before they are removed. Barricades cannot be removed before the application has cured to a point where the sealcoat is not tacky and can not be picked up by foot or tire traffic.

C-8. The Contractor shall clean all concrete structures (driveways, sidewalks, garage floors, curb cuts, valley gutters, spandrels [aprons], curb and gutter, planters, etc.) that have any sealer material tracked or over-sprayed by high pressure steam/water or sand blasting.

C-9.   Where the weather conditions permit, striping will be accomplished with a minimum of twenty-four-(24-) hour cure time from the last moment of application. Should the dew point or humidity be high, then curing time should be forty-eight (48) to seventy-two (72) hours before striping.

C-10. Any material covering, spilled, placed, or running onto any of the above mentioned items will be cleaned by high pressure hot water, or sand blasting, and/or methods acceptable to local agencies. All existing concrete spills, clay and soil deposits, and any other materials hardened to the pavement shall be removed prior to applying the seal coat materials.

C-11. All speed bumps in a resurfaced sections, which are covered with material and not visible, shall be temporarily striped with white or yellow marking paint before any sealed sections are opened to any traffic (vehicle or pedestrian).

C-12. All ceramics and reflectors shall be installed at the original locations prior to removal. All ceramics and reflectors added to the project shall be installed.

C-13. Asphalt emulsified sealers shall meet or exceed the following specifications for concentrate.

TEST	SPECIFICATION	TEST METHOD
PENETRATION AT 77°F 100g, 5sec, dmm	15-55	ASTM D5
Recycled Tire Rubber %	10% of Asphalt Base Stock	Terminal Certificate
Softening Point °F	130 Min	ASTM D36
Solubility %	97.5	ASTM D2042
Nonvolatile Component % by Weight, Min	50	ASTM D2939
Asphalt Content, % by weight of Nonvolatile component soluble in Trichloroethylene	25-35	ASTM D2042
Mineral Aggregate Component, % by weight passing No. 20 sieve Min	100	ASTM C136
Wet Track Abrasion	Less than 5% loss	ISSA (TB-100)
Weight Per Gallon 25°F	10 Lbs Min	ASTM D2939.07
Residue by Evaporation %	50% Min	ASTM D2939.08
Accelerated weathering test	Pass (1,000 hours) Certificate	ASTM G154
Color	Black	

All asphalt emulsified materials shall be prequalified by a certified laboratory prior to submitting proposals. Field samples will be taken and submitted to a certified laboratory for comparison tests and specification compliance. All material will be accompanied with a manufacturers certificate of compliance (COC) for compliance purposes

C-14. Coal tar emulsified sealers shall meet or exceed the following specifications for concentrate.

TEST	SPECIFICATION
Chemical & Physical Analysis -Nonvolatiles % -Ash Nonvolatiles % -Solubility of Nonvolatiles in CS2% - Specific Gravity 25°C	47-53% 30-40 20 Min 1.20 Min
Drying Time	8 hours Max
Adhesion & Resistance to Water	No Penetration or Loss of Adhesion
Resistance to Heat	No Blistering or Sagging
Flexibility	No Cracking or Flaking
Resistance to Impact	No Chipping, Flaking, or Cracking
Resistance to Volatilization	10% Loss in Weight Max
Wet Film Continuity	Smooth, Nongranular Free-form Coarse Particles
Resistance to Kerosene	No Loss of adhesion or penetration
P-625 Fuel Resistance Test	Evaluation of Fuel Resistance
Application Rate	0.11 to 0.13 gallons per square yard
Polymer Additives	1–2 gallons/100 gallons concentrate
Sand (40-70 mesh AFS rating)	300–500 lbs/100 gallons concentrate
Dilution Rate Water	30–40 gallons per 100 gallons concentrate
Color	Grayish Black

The above additive specification (Application Rate, Polymer Additives, Sand, and Dilution Rate) is a typical recommendation. Alternative mix designs may be substituted to account for local pavement conditions and use of other pavement sealer additives.

SAND SHALL BE USED IN ANY MIX DESIGN IN ALL CASES.

All coal tar emulsified materials shall be prequalified by a certified laboratory prior to submitting proposals. Field samples will be taken and submitted to a certified laboratory for comparison tests and specification compliance. All material will be accompanied with a manufacturer's certificate of compliance (COC) for compliance purposes.

SLURRY SEAL SPECIFICATIONS

A. Materials

A-1   The slurry seal shall consist of a mixture of an approved emulsified asphalt, mineral aggregate, water, and specified additives, proportioned, mixed, and uniformly spread over a properly prepared surface as directed by the Owner, Representative, or Consultant. The completed slurry seal shall leave a homogeneous mat, adhere firmly to the prepared surface, and have a skid-resistant surface texture.

A-2   The slurry seal shall be type I, II, or III as outlined by ASTM (American Society for Testing Materials), AASHTO (American Association of State Highway and Transportation Officials), ISSA (International Slurry Seal Surfacing Association), or local governing agencies or the Scope of Work, and meet *all* applicable specifications for aggregate and mineral filler, emulsified asphalt, and application.

A-3   The Owner, Manager, or Consultant may take samples of materials to an independent lab for testing or on-site inspection for application compliance.

B. Preparation

B-1   Prior to application, ALL areas will be power swept, vacuumed, and cleared of all loose material. All adjacent concrete structures (driveways, walks, gutters, garage floors, etc.) shall be cleaned of all dirt, dust, and debris, prior to applying any materials to the designated sealing area. Oil spots will be manually scraped and cleaned with a mild detergent, or heated with an open flame weed burner, and if necessary, a primer applied over high-saturated petroleum spill areas. All manholes, adjacent concrete structures (valley gutters, spandrels or aprons, walks, etc.), valve boxes, drop inlets, and other service entrances, as well as ALL survey markers, will be protected from the application of the product by a suitable method of covering and protection. Any material covering or spilled, placed, or running onto any of the above mentioned items will be cleaned by high-pressure hot water, or sandblasting, and/or methods acceptable to local agencies. All drywell lids and entrances will be protected by covering with a non-permeable material prior to applying any sealer. All existing concrete spills, clay and soil deposits, and any other materials hardened to the pavement shall be removed prior to applying the slurry seal materials. Any slurry sealer material placed in the drywell shall be cleaned and removed prior to payment.

B-2   All cracks greater than ¼ inch shall be properly prepared before the slurry seal is applied by sealing with hot-poured elastomeric crack sealant.

B-3   Before applying slurry seal, any new skin patches and/or any new asphalt (R&R repairs) must be allowed a curing time of at least ten (10) days BEFORE the slurry seal is applied.

B-4   Prior to slurry sealing the existing pavement, the Contractor shall inventory all manholes, valve boxes, sewer clean-outs, survey markers, striping, and markers. The Contractor shall plot the locations of each of the manholes, valve boxes, sewer clean-outs, and survey markers on a site plan, leasing plan, or blueprints, as well as establish a marking on the curbing or edge of the street or parking lot. The Contractor shall protect against the disruption of any of these items during the slurry sealing operation. Should a manhole, valve box, or survey marker be damaged, removed, or covered over, the Contractor shall repair, replace, or uncover that item in as good (or better) condition. All striping, ceramic markings, reflective markings, etc. shall be inventoried prior to slurry sealing and plotted on a site plan. All raised markings, ceramics, reflectors, and any other markers shall be removed and replaced identically and in the same location prior to construction unless specified or directed otherwise by the Owner, Agent, or Consultant. All striping and regulatory markings shall be installed to current local codes.

B-5   The existing bumper blocks shall be inventoried for existing breaks, cracks, etc. Any bumper blocks that are broken and/or have excessive cracking shall be removed and replaced with new blocks and pinned in place. All bumper blocks that are out of alignment with all other blocks or are not pinned shall be placed in alignment with the other blocks and pinned in place. All bumper blocks as well as any required new blocks shall be reinstalled in a straight configuration with each other and pinned to prevent movement. All broken bumpers shall be hauled off the site and disposed of.

B-6   All signs and poles shall be inventoried for condition and code requirements. All signs will be inventoried for damage, and any signs damaged afterwards shall be replaced at the Contractor's expense. All proper signage shall be installed according to current local codes and located at the direction of the Owner and the Agent or designated Consultant.

C. Application

C-1   Trimming around curbs, walls, walkways, sidewalks, and/or landscaped areas will be done by hand with necessary protection to prevent overapplication on curbs, sidewalks, walkways, etc. and slopping.

C-2   Application of the material shall be with slurry seal mixing equipment that has a continuous flow mixing unit, either an individual unit that returns to the stockpile for reloading or a continuous run unit that is resupplied on the job. All units must have suitable means of accurately metering each individual material being fed into the mixer. All feeding mechanisms must be continuous feed, and proportioning must remain constant at all times. The units shall be equipped with approved devices so the machine can be accurately calibrated, and the quantities of material used during any one period can be estimated. In the event these metering devices stop working, the slurry unit will stop the application process until they are fixed. The mixer shall thoroughly blend all materials to form a homogeneous mixture before leaving the mixer.

C-3   The spreader box shall be equipped to prevent loss of slurry seal from all sides and with a flexible rear strike-off. It shall be capable of producing a uniform surface its full width. It shall have suitable means for side tracking to compensate for deviations in pavement geometry. Any type drag used shall be approved by the Owner, Representative, or Consultant and kept in complete flexible condition at all times. The box shall be kept clean, and buildup of asphalt and aggregate shall not be permitted.

C-4   The slurry seal mixture shall be applied at a proper consistency at all times in order to provide the amount of mixture required by the surface condition. The average application rate for type II slurry seal as measured by the Owner, Representative, or Consultant shall be approximately 18 lb./yd.^{2}, $\pm$15 percent based on dry aggregate weight.

C-5   The existing surface should be pre-wet by fogging ahead of the slurry box when required by local conditions, usually when high air and pavement temperatures will cause the mix to break-set prematurely. Water used in pre-wetting the surface shall be applied such that the entire surface is damp with no apparent flowing water in front of the slurry box. The rate of application of the fog spray shall be adjusted during the day to suit temperatures, surface texture, humidity, and dryness of the pavement surface.

C-6 The slurry mixture shall be of the desired consistency upon leaving the mixer, and no additional materials shall be added. A sufficient amount of slurry shall be carried in all parts of the spreader at all times so that a complete coverage is obtained. Overloading of the spreader shall be avoided. No lumping, balling, or unmixed aggregates shall be permitted.

C-7 No streaks, such as those caused by oversized aggregate, will be left on the finished surface. If excess oversize develops, the job will be stopped until the Contractor proves to the Owner, Representative, or Consultant that the situation has been corrected.

C-8 No excessive buildup, uncovered areas, or unsightly appearance shall be permitted on longitudinal or transverse joints. An excessive overlap will not be permitted on longitudinal joints. The Contractor shall provide suitable width-spreading equipment to produce a minimum number of longitudinal joints throughout the project. When possible, longitudinal joints shall be placed on lane lines. Half passes and odd-width passes will be used only in minimum amounts. If half passes are used, they shall not be the last pass of any paved area.

C-9 The slurry mixture shall possess sufficient stability so that premature breaking of the slurry seal in the spreader box does not occur. The mixture shall be homogeneous during and following mixing and spreading. It shall be free of excess water or emulsion and free of segregation of the emulsion and aggregate.

C-10 Areas that cannot be reached with the slurry seal machine shall be surfaced using hand squeegees to provide complete and uniform slurry seal coverage. The area to be handworked shall be lightly dampened prior to mix placement and the slurry worked immediately. Care shall be required. Handwork shall be completed during the machine-applying process.

C-11 Care shall be taken to ensure straight lines along curbs and shoulders. No runoff on these areas will be permitted. Lines at intersections will be kept straight to provide a good appearance.

C-12 All slurry sealed areas shall be rolled by a self-propelled 10-ton pneumatic roller with a tire pressure of 50 PSI and equipped with a water spray system, or a 3- to 5-ton static steel wheel roller. The surfaced area shall be subjected to a minimum of five full passes by the roller. Rolling should not commence until the slurry seal has cured enough to prevent pickup on the tires of the roller.

C-13 The slurry seal shall not be applied if either the pavement or air temperature is below 55°F (15°C) and falling, but may be applied when both pavement and air temperature are above 45°F (7°C) and rising. No slurry seal shall be applied when there is danger that the finished product will freeze before twenty-four (24) hours. The mixture shall not be applied when weather conditions prolong opening to traffic beyond a reasonable time.

C-14 Allow recommended drying and curing time (at least twenty-four [24] hours depending upon local weather conditions, and air and pavement temperatures) before opening to local traffic.

C-15 When the surface temperature will remain above 100°F after the slurry seal is applied, a blotter material consisting of the slurry seal sand, without any emulsion, or silica sand shall be applied to reduce power steering shear and pickup of slurry material by vehicles.

C-16 All areas, such as main ways, gutters, and intersections, shall have the slurry seal removed as specified by the Owner, Representative, or Consultant. The Contractor shall remove any and all debris associated with the performance of the work. This shall be done on a daily basis and done fully at the completion of the project.

C-17 The Contractor shall clean all concrete structures (driveways, sidewalks, garage floors, curb cuts, valley gutters, spandrels or aprons, curbs and gutters, planters, etc.), by high-pressure steam/water or sandblasting.

C-18 Any material covering or spilled, placed, or running onto any of the above mentioned items will be cleaned by high-pressure hot water, or sandblasting, and/or methods acceptable to local agencies. All existing concrete spills, clay and soil deposits, and any other materials hardened to the pavement shall be removed prior to applying the slurry seal materials.

C-19 Rippling, also referred to as washboarding, shall be corrected by reapplication of the slurry material or by hot sand seal overlay over the entire area where the distress exists. The surface rippling condition shall be improved to provide an acceptable ride quality and no steering wheel vibration.

C-20 All speed bumps in resurfaced sections that are covered with material and not visible shall be temporarily striped with white or yellow marking paint before any sealed sections are opened to traffic.

TYPE II POLYMER MODIFIED SLURRY SEAL SPECIFICATIONS

A. Materials

A-1   The slurry seal shall consist of a mixture of an approved emulsified asphalt, polymer additive (solid or liquid form), mineral aggregate, water, and any other specified additives, proportioned, mixed, and uniformly spread over a properly prepared surface as directed by the Owner, Representative, or Consultant. The completed slurry seal shall leave a homogeneous mat, adhere firmly to the prepared surface, and have a skid-resistant surface texture.

A-2   The slurry seal shall be type II as outlined by ASTM (American Society for Testing Materials), AASHTO (American Association of State Highway and Transportation Officials), ISSA (International Slurry Seal Surfacing Association), or local governing agencies or the attached Scope of Work, and meet all applicable specifications for aggregate and mineral filler, emulsified asphalt, polymer additive, and application.

A-3   The Owner, Manager, or Consultant may take samples of materials to an independent lab for testing or on-site inspection for application compliance.

B. Preparation

B-1   Prior to application, ALL areas will be power swept, vacuumed, and cleared of all loose material. All adjacent concrete structures (driveways, walks, gutters, garage floors, etc.) shall be cleaned of all dirt, dust, and debris, prior to applying any materials to the designated sealing area. Oil spots will be manually scraped and cleaned with a mild detergent or heated with an open flame weed burner, and, if necessary, a primer applied over high-saturated petroleum spill areas. All manholes, adjacent concrete structures (valley gutters, spandrels or aprons, walks, etc.), valve boxes, drop inlets, and other service entrances, as well as ALL survey markers, will be protected from the application of the product by a suitable method of covering and protection. Any material covering or spilled, placed, or running onto any of the above mentioned items will be cleaned by high-pressure hot water, or sandblasting, and/or methods acceptable to local agencies. All drywell lids and entrances will be protected by covering with a non-permeable material prior to applying any sealer. All existing concrete spills, clay and soil deposits, and any other materials hardened to the pavement shall be removed prior to applying the slurry seal materials. Any slurry sealer material placed in the drywell shall be cleaned and removed prior to payment.

B-2   All cracks greater than ¼ inch shall be properly prepared before the slurry seal is applied by sealing with hot-poured elastomeric crack sealant.

B-3   Before applying slurry seal, any new skin patches and/or any new asphalt (R&R repairs) must be allowed a curing time of at least ten (10) days BEFORE the slurry seal is applied.

B-4   Prior to slurry sealing the existing pavement, the Contractor shall inventory all manholes, valve boxes, sewer clean-outs, survey markers, striping, and markers. The Contractor shall plot the locations of each of the manholes, valve boxes, sewer clean-outs, and survey markers on a site plan, leasing plan, or blueprints, as well as establish a marking on the curbing or edge of the street or parking lot. The Contractor shall protect against the disruption of any of these items during the slurry sealing operation. Should a manhole, valve box, or survey marker be damaged, removed, or covered over, the Contractor shall repair, replace, or uncover that item in as good (or better) condition. All striping, ceramic markings, reflective markings, etc. shall be inventoried prior to slurry sealing and plotted on a site plan. All raised markings, ceramics, reflectors, and any other markers shall be removed and replaced identically and in the same location prior to construction unless specified or directed otherwise by the Owner, Agent, or Consultant. All striping and regulatory markings shall be installed to current local codes.

B-5 The existing bumper blocks shall be inventoried for existing breaks, cracks, etc. Any bumper blocks that are broken and/or have excessive cracking shall be removed and replaced with new blocks and pinned in place. All bumper blocks that are out of alignment with all other blocks or are not pinned shall be placed in alignment with the other blocks and pinned in place. All bumper blocks as well as any required new blocks shall be reinstalled in a straight configuration with each other and pinned to prevent movement. All broken bumpers shall be hauled off the site and disposed of.

B-6 All signs and poles shall be inventoried for condition and code requirements. All signs will be inventoried for damage, and any signs damaged afterwards shall be replaced at the Contractor's expense. All proper signage shall be installed according to current local codes and located at the direction of the Owner, Agent, or designated Consultant.

C. Application

C-1 Trimming around curbs, walls, walkways, sidewalks, and/or landscaped areas will be done by hand with necessary protection to prevent overapplication on curbs, sidewalks, walkways, etc. and slopping.

C-2 Application of the material shall be with slurry seal mixing equipment that has a continuous flow mixing unit, either an individual unit that returns to the stockpile for reloading or a continuous run unit that is resupplied on the job. All units must have suitable means of accurately metering each individual material being fed into the mixer. All feeding mechanisms must be continuous feed, and proportioning must remain constant at all times. The units shall be equipped with approved devices so the machine can be accurately calibrated and the quantities of material used during any one period can be estimated. In the event these metering devices stop working, the slurry unit will stop the application process until they are fixed. The mixer shall thoroughly blend all materials to form a homogeneous mixture before leaving the mixer.

C-3 The spreader box shall be equipped to prevent loss of slurry seal from all sides and with a flexible rear strike-off. It shall be capable of producing a uniform surface its full width. It shall have suitable means for side tracking to compensate for deviations in pavement geometry. Any type drag used shall be approved by the Owner, Representative, or Consultant and kept in complete flexible condition at all times. The box shall be kept clean, and buildup of asphalt and aggregate shall not be permitted.

C-4 The application rate of the slurry seal mixture shall be of proper consistency at all times in order to provide the amount of mixture required by the surface condition. The average application rate for type II slurry seal as measured by the Owner, Representative, or Consultant shall be approximately 18 pounds per square yard ±15 percent based on dry aggregate weight.

C-5 The existing surface should be pre-wet by fogging ahead of the slurry box when required by local conditions, usually when high air and pavement temperatures will cause the mix to break-set prematurely. Water used in pre-wetting the surface shall be applied such that the entire surface is damp with no apparent flowing water in front of the slurry box. The rate of application of the fog spray shall be adjusted during the day to suit temperatures, surface texture, humidity, and dryness of the pavement surface.

C-6   The slurry mixture shall be of the desired consistency upon leaving the mixer, and no additional materials shall be added. A sufficient amount of slurry shall be carried in all parts of the spreader at all times so that a complete coverage is obtained. Overloading of the spreader shall be avoided. No lumping, balling, or unmixed aggregates shall be permitted.

C-7   No streaks, such as those caused by oversized aggregate, will be left on the finished surface. If excess oversize develops, the job will be stopped until the Contractor proves to the Owner, Representative, or Consultant that the situation has been corrected.

C-8   No excessive buildup, uncovered areas, or unsightly appearance shall be permitted on longitudinal or transverse joints. An excessive overlap will not be permitted on longitudinal joints. The Contractor shall provide suitable width-spreading equipment to produce a minimum number of longitudinal joints throughout the project. When possible, longitudinal joints shall be placed on lane lines. Half passes and odd-width passes will be used only in minimum amounts. If half passes are used, they shall not be the last passes of any paved area.

C-9   The slurry mixture shall possess sufficient stability so that premature breaking of the slurry seal in the spreader box does not occur. The mixture shall be homogeneous during and following mixing and spreading; it shall be free of excess water or emulsion and free of segregation of the emulsion and aggregate.

C-10   Areas that cannot be reached with the slurry seal machine shall be surfaced using hand squeegees to provide complete and uniform slurry seal coverage. The area to be handworked shall be lightly dampened prior to mix placement and the slurry worked immediately. Care shall be required. Handwork shall be completed during the machine applying process.

C-11   Care shall be taken to ensure straight lines along curbs and shoulders. No runoff on these areas will be permitted. Lines at intersections will be kept straight to provide a good appearance.

C-12   All slurry sealed areas shall be rolled by a self-propelled 10-ton pneumatic roller with a tire pressure of 50 PSI (3.4 atm) and equipped with a water spray system or a 3- to 5-ton static steel wheel roller. The surfaced area shall be subjected to a minimum of five full passes by the roller. Rolling should not commence until the slurry seal has cured enough to prevent pickup on the tires of the roller.

C-13   The slurry seal shall not be applied if either the pavement or air temperature is below 55°F (15°C) and falling, but may be applied when both pavement and air temperature are above 45°F (7°C) and rising. No slurry seal shall be applied when there is danger that the finished product will freeze before twenty-four (24) hours. The mixture shall not be applied when weather conditions prolong opening to traffic beyond a reasonable time.

C-14   Allow recommended drying and curing time (at least twenty-four [24] hours depending upon local weather conditions, and air and pavement temperatures) before opening to local traffic.

C-15   When the surface temperature will remain above 100°F after the slurry seal is applied, a blotter material consisting of the slurry seal sand, without any emulsion or silica sand shall be applied to reduce power steering shear and pickup of slurry material by vehicles.

C-16   All areas, such as main ways, gutters, and intersections, shall have the slurry seal removed as specified by the Owner, Representative, or Consultant. The Contractor shall remove any and all debris associated with the performance of the work. This shall be done on a daily basis, and done fully at the completion of the project.

C-17 The Contractor shall clean all concrete structures (driveways, sidewalks, garage floors, curb cuts, valley gutters, spandrels or aprons, curbs and gutters, planters, etc.), by high-pressure steam/water or sandblasting.

C-18 Any material covering or spilled, placed, or running onto any of the above mentioned items will be cleaned by high-pressure hot water, or sandblasting, and/or methods acceptable to local agencies. All existing concrete spills, clay and soil deposits, and any other materials hardened to the pavement shall be removed prior to applying the slurry seal materials.

C-19 Rippling, also referred to as washboarding, shall be corrected by reapplication of the slurry material or by hot sand seal overlay over the entire area where the distress exists. The surface rippling condition shall be improved to provide an acceptable ride quality and no steering wheel vibration.

C-20 All speed bumps in a resurfaced sections that are covered with material and not visible shall be temporarily striped with white or yellow marking paint before any sealed sections are opened to traffic.

CHIP SEAL SPECIFICATIONS

A. Materials

A-1   CRS2 or CRS2P emulsified asphalt hot applied or specified (Medium Cure) MC70/250. The specifications for liquid grade and emulsified asphalts are listed in the "Uniform Standard Specifications for Public Works Construction."

A-2   The aggregate chip shall consist of an open graded aggregate with a maximum size grading of ½ inch, ⅜ inch, or ¼ inch, proportioned, mixed, and uniformly spread over a properly prepared surface as directed by the Owner, Representative, or Consultant. The completed chip seal shall leave a homogeneous mat and adhere firmly to the prepared surface. The specifications for aggregate chip materials are listed in the "Uniform Standard Specifications for Public Works Construction."

A-3   SS-1 or SS-1h emulsified asphalt and the specifications for liquid grade and emulsified asphalts are listed in the "Uniform Standard Specifications for Public Works Construction."

A-4   The Owner, Manager, or Consultant may take samples of materials to an independent laboratory for testing or require on-site inspection for application.

B. Preparation

B-1   Prior to application, ALL areas to be surfaced will be power swept with self-propelled pickup sweeper, power broom sweepers, or high-powered air brooms/blowers; vacuumed; or any combination of these methods until the pavement is completely clean of all loose debris and materials. All adjacent concrete structures (driveways, walks, gutters, garage floors, etc.) shall be cleaned of all dirt, dust, and debris prior to applying any materials to the designated sealing area. In urban areas, the surface shall be cleaned with a self-propelled pickup sweeper. In rural areas, power brooms may be used. When necessary, cleaning shall be supplemented by hand brooms. All existing concrete spills, clay and soil deposits, and any other materials hardened to the pavement shall be removed prior to applying the slurry seal materials. Oil spots will be manually scraped and cleaned with a mild detergent (if required) and a primer applied (Bond Seal or Petro Seal or approved equivalent) if required. All manholes, adjacent concrete structures (valley gutters, spandrels or aprons, walks, etc.), valve boxes, drop inlets, and other service entrances, as well as ALL survey markers, will be protected from the application of the product by a suitable method of covering and protection. Any material covering or spilled, placed, or running onto any of the above mentioned items will be cleaned by high-pressure steam and water, or sandblasting, and/or methods acceptable to local agencies.

B-2   Prior to chip sealing the existing pavement, the Contractor shall inventory all manholes, valve boxes, sewer clean-outs, survey markers, striping, and markers. The Contractor shall plot the locations of each of the manholes, valve boxes, sewer clean-outs, and survey markers on a site plan, leasing plan, or blueprints, as well as establish a marking on the curbing or edge of the street or parking lot. The Contractor shall protect against the disruption of any of these items during the chip sealing operation. Should a manhole, valve box, or survey marker be damaged, removed, or covered over, the Contractor shall repair, replace, or uncover that item in as good (or better) condition. All striping, ceramic markings, reflective markings, etc. shall be inventoried prior to chip sealing and plotted on a site plan. All raised markings, ceramics, reflectors, and any other markers shall be removed and replaced identically and in the same location prior to construction unless specified or directed otherwise by the Owner, Agent, or Consultant.

B-3 Prior to chip sealing, all cracks greater than ¼ inch shall be properly prepared before the slurry seal is applied by sealing with hot-poured elastomeric crack sealant.

B-4 Prior to chip sealing, any new skin patches and/or any new asphalt (R&R repairs) must be allowed a curing time of at least ten (10) days BEFORE the chip seal coating is applied. In desert climates, it is recommended to allow the full hot summer season for curing, especially when the patch material is placed in extremely hot weather conditions.

B-5 If required by the Owner, the existing bumper blocks shall be inventoried for existing breaks, cracks, etc. and removed and stacked off the construction site in such a manner as to not break or crack them. Any bumpers not marked during the inventory that are broken shall be replaced at the Contractor's expense.

B-6 All signs and poles shall be inventoried for condition and code requirements. All signs will be inventoried for damage, and any signs damaged afterwards shall be replaced at the Contractor's expense. All proper signage shall be installed according to current local codes and located at the direction of the Owner, Agent, or designated Consultant.

C. Application

C-1 The bituminous material shall be applied as soon as possible after preparation of surfaces. At the time of application, temperatures of the asphalt shall be within the ranges specified in the tables on pages 102 and 103 in Section 330 of the "Uniform Standard Specifications for Public Works Construction" MAG (Maricopa Association of Governments) specifications.

C-2 Apply the CRS2, CRS2P, or MC70/250 at a rate of 0.20 to 0.40 gallon per square yard spray applied. The emulsion shall be hot applied by means of a mechanical distributor capable of applying the desired amount of tack material. The entire area to be chip sealed shall be tack coated prior to installation of the aggregate chips. The distributor trucks shall be of the pressure type with insulated tanks. Gravity distributors will not be permitted. Spray bars and extensions shall be of the full circulating type. The spray bar shall be adjustable to permit varying height above the surface to be treated.

C-3 The nozzle spacings, center to center, shall not exceed 6 inches. The valves shall be operated so that one or all valves may be quickly opened or closed in one operation. The valves that control the flow from the nozzles shall be of a positive acting design so as to provide a uniform, unbroken spread of bituminous material on the surface.

C-4 The distributor shall be equipped with devices and charts to provide for accurate, rapid determination and control of the amount of bituminous material being applied. The distributor shall be equipped with a tachometer of the auxiliary wheel type registering speed in feet per minute. The distributor shall also be equipped with pressure gauges and an accurate thermometer for determination of the temperature of bituminous material. The spreading equipment shall be designed so that uniform application of a bituminous material can be applied in controlled amounts ranging from 0.05 to 2.00 gallons per square yard. Transverse variation rate shall not exceed 10 percent of the specified application rate. The distributor shall be equipped with a hose and nozzle attachment to be used for spotting skipped areas and areas inaccessible to the distributor. Distributor and booster tanks shall be maintained to prevent dripping of bituminous material from any part of the equipment.

C-5   Immediately after applying the bituminous material, the aggregate chip shall be applied with a self-propelled mechanical spreader. The chip spreading equipment shall be capable of applying a uniform application of chip cover material. The self-propelled equipment may be waived for projects less than 10,000 square yards. At the time of application, precoated aggregate shall be within the temperature range of 250°F and 350°F, measured at a point 6 to 12 inches below the top of the load; uncoated chips shall not contain moisture in excess of a saturated, surface dry condition when liquid or paving grade asphalt is used; and chips shall be surface wet but free from running water when emulsified asphalt is used as the seal coat binder. The precise rate for chip cover material shall be 30 to 35 pounds per square yard for ½-inch chips, 25 to 30 pounds per square yard for ⅜-inch chips, and 15 to 25 pounds per square yard for ¼-inch chips.

C-6   Trimming around curbs, walls, walkways, sidewalks, and/or landscaped areas will be done by hand with necessary protection to prevent overapplication on curbs, sidewalks, walkways, etc. and slopping.

C-7   Immediately following the application of the chip cover material, the surface shall be rolled with self-propelled pneumatic rubber-tired rollers. Three passes shall be made with a pneumatic roller. Each roller shall carry a minimum of 2,000 pounds on each wheel and minimum of 60 PSI in each tire. The roller shall not travel in excess of 12 MPH. A minimum of three rollers shall be required for projects over 10,000 square yards. On projects less than 10,000 square yards, one roller may be used, provided it performs the same amount of coverage.

C-8   No excessive buildup, uncovered areas, or unsightly appearance shall be permitted on longitudinal or transverse joints. An excessive overlap will not be permitted on longitudinal joints.

C-9   The chip seal shall be allowed to set for curing purposes for at least twenty-four (24) hours before sweeping up the unused portion of aggregates. Sweeping shall be accomplished with a broom sweeper designed for cleaning streets. The sweeper shall have rotating side brooms and a rotating rear broom.

C-10 After the project is complete, the Contractor shall apply a fog coat of SS-1h emulsion sealer to the chip seal surface at an application rate of 0.10 gallon per square yard.

C-11 The Contractor shall clean all concrete structures (driveways, sidewalks, garage floors, curb cuts, valley gutters, spandrels or aprons, curbs and gutters, planters, etc.) by high-pressure steam/water or sandblasting.

C-12 Any material covering or spilled, placed, or running onto any of the above mentioned items will be cleaned by high-pressure hot water, or sandblasting, and/or methods acceptable to local governments. All existing concrete spills, clay and soil deposits, and any other materials hardened to the pavement shall be removed prior to applying the chip seal materials.

C-13 All speed bumps in a resurfaced sections that are covered with material and not visible shall be temporarily striped with white or yellow marking paint before any sealed sections are opened to traffic.

ASPHALT PAVING OVERLAY SPECIFICATIONS

A. Materials

A-1   Asphalt pavement shall be ¾ inch maximum size aggregate hot-mixed material according to local government construction specifications (state highway, county, or municipality) for overlays 1½ inches thick or more. The asphalt pavement shall be ½ inch maximum size aggregate hot mixed material according to local government construction specifications for overlays 1½ inches thick or less. The gradation and asphalt contents are based on project mix designs as established by local agencies, project architects, or independent consultants.

A-2   The Owner, Manager, or Consultant may take samples of materials to an independent laboratory for testing of proper mix design or require on-site inspection for proper application of materials.

B. Preparation

B-1   The existing pavement shall be prepared before installing a new asphalt hot-mixed overlay by remove and replace patching of all areas that have structurally failed according to the attached Specification for R&R patching. Where the edges of the overlay area are confined by curbs, gutters, driveways, garages, sidewalks, building foundations, or any other permanent structure, the edges will be milled a minimum of 3 feet from the confining barrier prior to installing the overlay. Any materials covering or spilled, placed, or running onto any of the above mentioned items will be cleaned by high-pressure hot water, or sandblasting, and/or methods acceptable to local agencies. All existing concrete spills, clay and soil deposits, and any other materials hardened to the pavement shall be removed prior to applying the overlay materials.

B-2   All cracks ¼ inch and wider shall be filled with a hot elastomeric crack sealant according to the attached Specifications for crack sealing.

B-3   The existing surface shall be thoroughly cleaned of all loose debris and dirt particles by brooming, or use of high-pressure air brooms and blowers, or a combination of both until the existing pavement is completely clean.

B-4   All valve boxes, manhole covers and lids, and any other structures in the newly paved area shall be brought to the new-finished paved grade prior to installing the overlay.

B-5   Prior to paving the existing pavement, the Contractor shall inventory all manholes, valve boxes, survey markers, striping, and markers. The Contractor shall plot the locations of each of the manholes, valve boxes, and survey markers on a site plan, leasing plan, or blueprints, as well as establish a marking on the curbing or edge of the street or parking lot. The Contractor shall protect against the disruption of any of these items during the paving operation. Should a manhole, valve box, or survey marker be damaged, removed, or covered over, the Contractor shall replace or uncover that item in as good (or better) condition. All striping, ceramic markings, reflective markings, etc. shall be inventoried prior to paving and plotted on a site plan. All markings, ceramics, reflectors, and any other markings shall be removed and replaced identically and in the same location prior to construction unless specified or directed otherwise by the Owner, Agent, or Consultant. All striping and regulatory markings shall be installed to current local codes.

B-6   The existing bumper blocks shall be inventoried for existing breaks, cracks, etc. and removed and stacked off the construction site in such a manner as to not break or crack them. Any bumpers not marked during the inventory that are broken shall be replaced at the Contractor's expense. All bumper blocks as well as any required new blocks shall be reinstalled in a straight configuration with each other and pinned to prevent movement. All broken bumpers shall be hauled off the site and disposed of.

B-7   All signs and poles shall be inventoried for condition and code requirements. All signs will be inventoried for damage, and any signs damaged afterwards shall be replaced at the Contractor's expense. All proper signage shall be installed according to current local codes and located at the direction of the Owner, Agent, or designated Consultant.

B-8   MC70/250 or CRS2 emulsion tack coat shall be applied at 0.1 gallon per square yard before applying the new hot-mixed surfacing. This tack coat shall cover the entire overlay area to reduce slippage, movement, and separation of the new overlay from the existing pavement.

C. Application

C-1   The material shall be installed at the appropriate depth and compacted to yield a depth of 2 inches after compaction for ¾ inch maximum size aggregate mix or 1½ inches after compaction for ½ inch maximum size aggregate mix, or 1 inch after compaction for ⅜ inch maximum size aggregate mix.

C-2   The asphaltic concrete shall be compacted to 95 percent laboratory density (Marshall ASTM D-1559 or Hveem Mix Design ASTM D-1560) using a combination of steel wheel static roller and pneumatic rollers or vibratory rollers. Care shall be taken to not over-roll the new asphalt.

C-3   The initial rolling or "breakdown roll" will be accomplished by a steel wheel roller. This application will be performed as soon as the roller can move on the asphalt without picking up material, approximately 280°F but not less than 245°F. Any segregated areas or pockets shall have asphaltic material applied, raked to allow fines to fill surface voids, and made ready for further compaction passes. No segregated areas or pockets of the larger aggregates will be allowed in the finished surface.

C-4   A second rolling will be accomplished with a pneumatic rubber-tire roller. Any rough, overlapping, segregated joints shall be smoothed by raking and adding fine asphaltic materials to thoroughly blend the joint to the existing pavement and joint area. No visible or poor joint construction will be accepted.

C-5   A third or finish rolling will be accomplished with a steel wheel static roller.

C-6   A vibratory/static roller may be used in lieu of the above rolling procedure (C-3 through C-5) using the manufacturer's guidelines and proper roller patterns as established by local agency specifications to attain the required density. However, the roller should be in the static mode on the finished layer.

C-7   The roller pattern shall be established where the exposed or unconfined edge is left unrolled until the next or adjacent pass is installed, providing the adjacent pass will be laid within two (2) hours of the first pass. This roller pattern will create a better joint in the finished surface. Should the adjacent pass not be installed within two (2) hours or until the next workday, the edge shall be compacted during regular rolling.

C-8   Where there is a hot section placed next to a cold section, the breakdown rolling shall start from the hot section, with a 6-inch overlap of the roller wheel on the cold section.

C-9 When a transverse joint is created at the end of the workday, the proper method shall be applied to ensure a clean vertical edge at the beginning of the next workday. This can be accomplished by placing a wooden block or paper fold, or sanding so the cold asphalt can be easily removed and a flat, smooth, and vertical surface will be created, or the Contractor can saw cut the cold pavement to produce the straight, vertical edge.

C-10 Upon completion, the finished surface shall be smooth, dense, and of uniform texture and appearance. There shall be no large segregated areas or depressions holding water more than ¼ inch in depth at the deepest point. There will be no roller marks in the finished surface. All areas will drain and be free of standing water within four (4) hours of accumulation. The finished surface may not vary more than ¼ inch in a 12-foot distance as determined by a straightedge placed in any position on the finished surface, except across flow lines of drainways and inverted crowns. Paving joints shall be uniform and smooth at all locations and cold joints, and any areas where the pavement was raked to create a matching surface. Where there is segregation of larger aggregates on the surface (not from the use of mixes designed to yield a rocky surface, e.g., Superpave mix designs), poorly raked joints, or any rough surface, the location shall be skin patched using material with a maximum sized aggregate of ¼ inch or hot sand seal. Any puddle areas where the water is more than ¼ inch deep or lasts more than eight (8) hours before dissipation shall be patched level to eliminate the puddle in a manner to not relocate the puddle. Use of string line, leveling nails, or other grade markers will ensure proper level grade to eliminate water puddles. Should the collective or accumulative areas of corrective patches (used to remedy puddling, segregation, etc.) exceed 25 percent of the newly paved surface area, removal or resurfacing of the newly paved surface is required, utilizing proper leveling equipment, to ensure drainage integrity and reduce and/or eliminate surface discrepancies (puddling, segregated surface, etc.).

C-11 Asphalt concrete shall be placed only when the surface is dry and when the atmospheric temperature is 50°F (10°C) and rising. No asphalt concrete shall be placed when the weather is foggy or rainy, or when the base on which the asphalt is to be placed contains moisture in excess of the optimum.

C-12 The combination of particle sizes or sieve analysis of the asphalt material is according to local specifications and available materials.

C-13 All striping and signage shall be installed to the existing configuration prior to construction or as directed by the Owner, Agent, or Consultant. The striping shall be installed after the new paved surface has adequate curing time and and when there is no chance of color disfiguration.

C-14 All speed bumps in a resurfaced section that are constructed or overlaid and not visible shall be temporarily striped with white or yellow marking paint before any sealed sections are opened to traffic.

NEW PAVEMENT CONSTRUCTION SPECIFICATIONS

A. Materials

A-1 Aggregate base course as specified in the Architect's, Engineer's, or Consultant's design standards and the design Specifications.

A-2 Asphaltic concrete pavement as specified in the Architect's, Engineer's, or Consultant's design standards and the design Specifications.

A-3 The Owner, Manager, or Consultant may take samples of materials to an independent lab for testing of proper mix design or require on-site inspection for proper application of materials.

B. Preparation

B-1 The existing area shall be excavated to a depth of the planned new pavement section. The pavement section includes the design thickness of the pavement and aggregate base course. The area will have all vegetation, roots, and topsoil removed and discarded. All new utilities and trench excavation shall be accomplished prior to preparing the subgrade for final grading and compaction. All utility trenching shall be installed as specified in Trenching Specifications. The Contractor shall notify any local agency (e.g., Blue Stakes and Dig Alert) to have all utilities marked prior to excavating and grading. This may require more than one marking.

B-2 The subgrade shall be scarified, watered, and compacted to 95 percent Proctor (laboratory) density prior to installing the aggregate base course material. The requirements for the compacted density shall be 95 percent of standard Proctor (ASTM D-698) for clay soils or modified Proctor (ASTM D-1557) for aggregate bases, sandy soils, and sands.

B-3 All curbing, islands, light poles, dumpster pads, valley gutters, manholes, valve boxes, and all other structures shall be constructed prior to installing the aggregate base course. All disturbed subgrade next to the new structure shall be compacted to 95 percent Proctor density as specified above.

B-4 Aggregate base material, as specified, shall be installed to the designed depth and properly processed to reduce segregation by windrowing the base material prior to laying out and grading. The base material shall be graded over the area to the designed thickness.

B-5 The aggregate base shall be staked to proper finished and designed grade (blue top grading). The aggregate base shall be graded to the finished grade, watered, and compacted uniformly over the entire surface area to 95 percent modified Proctor density (ASTM D-1557).

B-6 The prepared base shall have AC-10 tack coat applied along curbings, planters, light pedestals, and any cold joint that will come in contact with new hot-mixed asphalt. Install 2 inches of hot asphaltic concrete (¾-inch maximum size aggregate), or 1½ inch of hot asphaltic concrete (½-inch maximum size aggregate) as designated in the design pavement section, and installed according to the Architect's, Engineer's, Consultant's, or local government construction specifications (state highway, county, or municipality).

C. Application

C-1 All topsoil and underlying soils shall be excavated to the depth as established by the design and new pavement section criteria. All excavated materials shall be hauled away and disposed of. Material that is deemed acceptable by the Engineer and/or the Consultant can be used as fill material in areas and locations where fill is required and specified. Topsoil material and material with vegetation, roots, or mulch shall not be used as fill material. This material shall be hauled away from the construction site to an acceptable dumpsite.

C-2   All acceptable fill materials shall be used from the project site or imported from an off-site source to areas designated and staked for fill. The material shall be deposited and installed in lifts of 6 inches, watered, and compacted to 95 percent Proctor density of standard Proctor (ASTM D-698) for clay soils or modified Proctor (ASTM D-1557) for aggregate bases, sandy soils, and sands.

C-3   All trenching shall be accomplished and utilities installed before the final preparation of the subgrade material and the final grading of the subgrade. The utilities shall be installed and the trench backfilled to the Trenching Specifications and tested for compliance. All vaults, manholes, valve boxes, drywells, and all other underground structures shall be installed and all structures backfilled according to the Backfill Specifications and tested for compliance.

C-4   The final lift of subgrade material shall be graded, watered, and uniformly compacted to 95 percent Proctor density as stated above.

C-5   All curbing, dumpster pads, island curbs, valley gutters, aprons, and all concrete flat work shall be installed according to the specifications established for installation of these structures.

C-6   Designated and specified aggregate base course shall be imported to project site. The base material shall be deposited by spreading the material in a manner to allow for proper processing by windrowing. The base material shall be windrowed a minimum of three times and tested for gradation specifications. Once the material is acceptable by gradation, the aggregate base material shall be graded and installed in lifts not to exceed 6 inches in depth, watered, and compacted to 95 percent modified Proctor density.

C-7   The hot asphalt material shall be installed at the appropriate depth as specified in the construction plans for design pavement section and compacted to yield a minimum depth of 2 inches for ¾-inch maximum size aggregate mix, and 1½ inches for ½-inch maximum size aggregate mix after compaction.

C-8   The asphaltic concrete shall be compacted uniformly over the entire surface area to 95 percent laboratory density (Marshall ASTM D-1559 or Hveem Mix Design ASTM D-1560) using a combination of steel wheel static rollers, pneumatic rollers, and/or vibratory rollers. Care should be taken to not over-roll the new asphalt surface, yet meet the required density.

C-9   The initial rolling or "breakdown roll" will be accomplished by a steel wheel roller. This application will be performed as soon as the roller can move on the asphalt without picking up material, approximately 280°F but not less than 245°F. Any segregated areas or pockets shall have asphaltic material applied, raked to allow fines to fill surface voids, and made ready for further compaction passes. No segregated areas or pockets of the larger aggregates will be allowed in the finished surface. Any rough, overlapping, segregated joints shall be smoothed by raking and adding fine asphaltic materials to thoroughly blend the joint to the existing pavement and joint area. No visible or poor joint construction will be accepted.

C-10 A second rolling will be accomplished with a pneumatic rubber-tire roller.

C-11 A third or finish rolling will be accomplished with a steel wheel static roller.

C-12 A vibratory/static roller may be used in lieu of the above rolling procedure using the manufacturer's guidelines and proper roller patterns to attain the required density. However, the roller should be in the static mode on the finished layer.

C-13 The roller pattern shall be established where the exposed or unconfined edge is left unrolled until the next or adjacent pass is installed, providing the adjacent pass will be laid within two (2) hours of the first pass. This roller pattern will create a better joint in the finished surface. Should the adjacent pass not be installed within two (2) hours or until the next workday, the edge shall be compacted during regular rolling.

C-14 Where there is a hot section placed next to a cold section, the breakdown rolling shall start from the hot section with a 6-inch overlap of the roller wheel on the cold section.

C-15 Upon completion, the finished surface shall be smooth, dense, and of uniform texture and appearance. There shall be no large segregated areas or depressions holding water more than ¼ inch in depth at the deepest point. There will be no roller marks in the finished surface. All areas will drain and be free of standing water within twelve (12) hours of accumulation. The finished surface may not vary more than ¼ inch in a 12-foot distance as determined by a straightedge placed in any position on the finished surface, except across flow lines of drainways and inverted crowns. Paving joints shall be uniform and smooth at all locations and cold joints, and any areas where the pavement was raked to create a matching surface. Where there are large deposits of segregation of larger aggregates on the surface (not from the use of mixes designed to yield a rocky surface, e.g., Superpave mix designs), poorly raked joints, or any rough surface, the location shall be skin patched using material with a maximum sized aggregate of ¼ inch or hot sand seal. Any puddle areas where the water is more than ¼ inch deep or lasts more than eight (8) hours before dissipation shall be patched level to eliminate the puddle in a manner to not relocate the puddle. Use of string line, leveling nails, or other grade markers will ensure proper level grade to eliminate water puddles. Should the collective or accumulative areas of corrective patches (used to remedy puddling, segregation, etc.) exceed 25 percent of the newly paved surface area, removal or resurfacing of the newly paved surface is required, utilizing proper leveling equipment to ensure drainage integrity and reduce and/or eliminate surface discrepancies (puddling, segregated surface, etc.).

C-16 When a transverse joint is created at the end of the workday, the proper method shall be applied to ensure a clean vertical edge at the beginning of the next workday. This can be accomplished by placing a wooden block or paper fold, or sanding so the cold asphalt can be easily removed and a flat, smooth, and vertical surface will be created, or the Contractor can saw cut the cold pavement to produce the straight, vertical edge.

C-17 Asphalt concrete shall be placed only when the surface is dry and when the atmospheric temperature is 50°F (10°C) and rising. No asphalt concrete shall be placed when the weather is foggy or rainy, or when the base on which the asphalt is to be placed contains moisture in excess of the optimum.

C-18 All valve boxes, manhole covers and lids, and any other structures in the newly paved area shall be brought to the new-finished paved grade.

C-19 The sieve or gradation of particles will be according to local available materials.

C-20 All striping and signage shall be installed to the existing configuration prior to construction or as directed by the Owner, Agent, or Consultant. The striping shall be installed after the new paved surface has adequate curing time, and the chance of any color disfiguration can occur.

C-21 All speed bumps in a resurfaced section that are constructed and not visible shall be temporarily striped with white or yellow marking paint before any sealed sections are opened to traffic.

RUBBERIZED ASPHALT PAVING OVERLAYS SPECIFICATIONS

A. Materials

A-1   Asphalt pavement shall be ¾-inch maximum size aggregate hot mixed material according to local government construction specifications (state highway, county, or municipality) for overlays 1½ inches thick or more. The asphalt pavement shall be ½ inch maximum size aggregate hot-mixed material according to local government construction specifications for overlays 1½ inches thick or less. The gradation and asphalt contents are based on project mix designs as established by local agencies, project architects, or independent consultants.

A-2   Asphalt rubber shall be according to local agency specifications.

A-3   The Owner, Manager, or Consultant may take samples of materials to an independent laboratory for testing of proper mix design or require on-site inspection for proper application of materials.

B. Preparation

B-1   The existing pavement shall be prepared before installing a new asphalt hot-mixed overlay by remove and replace patching of all areas that have structurally failed according to the attached Specification for R&R patching. Where the edges of the overlay area are confined by curbs, gutters, driveways, garages, sidewalks, building foundations, or any other permanent structure, the edges will be milled 6 feet from the confining barrier prior to installing the overlay. Any material covering or spilled, placed, or running onto any of the above mentioned items will be cleaned by high-pressure hot water, or sandblasting, and/or methods acceptable to local agencies. All existing concrete spills, clay and soil deposits, and any other materials hardened to the pavement shall be removed prior to applying the overlay materials.

B-2   All cracks ¼ inch and wider shall be filled with a hot elastomeric crack sealant according to the attached Specifications for crack sealing.

B-3   The existing surface shall be thoroughly cleaned of all loose debris and dirt particles by brooming, or use of high-pressure air brooms and blowers, or a combination of both until the existing pavement is completely clean.

B-4   All valve boxes, manhole covers and lids, and any other structures in the newly paved area shall be brought to the new-finished paved grade prior to installing the overlay.

B-5   Prior to paving the existing pavement, the Contractor shall inventory all manholes, valve boxes, survey markers, striping, and markers. The Contractor shall plot the locations of each of the manholes, valve boxes, and survey markers on a site plan, leasing plan, or blueprints, as well as establish a marking on the curbing or edge of the street or parking lot. The Contractor shall protect against the disruption of any of these items during the paving operation. Should a manhole, valve box, or survey marker be damaged, removed, or covered over, the Contractor shall replace or uncover that item in as good (or better) condition. All striping, ceramic markings, reflective markings, etc. shall be inventoried prior to paving and plotted on a site plan. All markings, ceramics, reflectors, and any other markings shall be removed and replaced identically and in the same location prior to construction unless specified or directed otherwise by the Owner, Agent, or Consultant. All striping and regulatory markings shall be installed to current local codes.

B-6   The existing bumper blocks shall be inventoried for existing breaks, cracks, etc. and removed and stacked off the construction site in such a manner as to not break or crack them. Any bumpers not marked during the inventory that are broken shall be replaced at the Contractor's expense. All bumper blocks as well as any required new blocks shall be reinstalled in a straight configuration with each other and pinned to prevent movement. All broken bumpers shall be hauled off the site and disposed of.

B-7   All signs and poles shall be inventoried for condition and code requirements. All signs will be inventoried for damage, and any signs damaged afterwards shall be replaced at the Contractor's expense. All proper signage shall be installed according to current local codes and located at the direction of the Owner, Agent, or designated Consultant.

B-8   MC70/250 or CRS2 emulsion tack coat shall be applied at 0.1 gallon per square yard before applying the new hot-mixed surfacing. This tack coat shall cover the entire overlay area to reduce slippage, movement, and separation of the new overlay from the existing pavement.

C. Application

C-1   The material shall be installed at the appropriate depth and compacted to yield a depth of 2 inches after compaction for ¾ inch maximum size aggregate mix or 1½ inches after compaction for ½ inch maximum size aggregate mix, or 1 inch after compaction for ⅜ inch maximum size aggregate mix.

C-2   The asphaltic concrete shall be compacted to 95 percent laboratory density (Marshall ASTM D-1559 or Hveem Mix Design ASTM D-1560) using a combination of steel wheel static rollers and pneumatic rollers or vibratory rollers. Care shall be taken to not over-roll the new asphalt.

C-3   The initial rolling or "breakdown roll" will be accomplished by a steel wheel roller. This application will be performed as soon as the roller can move on the asphalt without picking up material, approximately 280°F but not less than 245°F. Any segregated areas or pockets shall have asphaltic material applied, raked to allow fines to fill surface voids, and made ready for further compaction passes. No segregated areas or pockets of the larger aggregates will be allowed in the finished surface.

C-4   A second rolling will be accomplished with a pneumatic rubber-tire roller. Any rough, overlapping, segregated joints shall be smoothed by raking and adding fine asphaltic materials to thoroughly blend the joint to the existing pavement and joint area. No visible or poor joint construction will be accepted.

C-5   A third or finish rolling will be accomplished with a steel wheel static roller.

C-6   A vibratory/static roller may be used in lieu of the above rolling procedure using the manufacturer's guidelines and proper roller patterns as established by local agency specifications to attain the required density. However, the roller should be in the static mode on the finished layer.

C-7   The roller pattern shall be established where the exposed or unconfined edge is left unrolled until the next or adjacent pass is installed, providing the adjacent pass will be laid within two (2) hours of the first pass. This roller pattern will create a better joint in the finished surface. Should the adjacent pass not be installed within two (2) hours or until the next workday, the edge shall be compacted during regular rolling.

C-8   Where there is a hot section placed next to a cold section, the breakdown rolling shall start from the hot section with a 6-inch overlap of the roller wheel on the cold section.

C-9 When a transverse joint is created at the end of the workday, the proper method shall be applied to ensure a clean vertical edge at the beginning of the next workday. This can be accomplished by placing a wooden block or paper fold, or sanding so the cold asphalt can be easily removed and a flat, smooth, and vertical surface will be created, or the Contractor can saw cut the cold pavement to produce the straight, vertical edge.

C-10 Upon completion, the finished surface shall be smooth, dense, and of uniform texture and appearance. There shall be no large segregated areas or depressions holding water more than ¼ inch in depth at the deepest point. There will be no roller marks in the finished surface. All areas will drain and be free of standing water within four (4) hours of accumulation. The finished surface may not vary more than ¼ inch in a 12-foot distance as determined by a straightedge placed in any position on the finished surface, except across flow lines of drainways and inverted crowns. Paving joints shall be uniform and smooth at all locations and cold joints, and any areas where the pavement was raked to create a matching surface. Where there are large deposits of segregation of larger aggregates on the surface (not from the use of mixes designed to yield a rocky surface, e.g., Superpave mix designs), poorly raked joints, or any rough surface, the location shall be skin patched using material with a maximum sized aggregate of ¼ inch or hot sand seal. Any puddle areas where the water is more than ¼ inch deep or lasts more than eight (8) hours before dissipation shall be patched level to eliminate the puddle in a manner to not relocate the puddle. Use of string line, leveling nails, or other grade markers will ensure proper level grade to eliminate water puddles. Should the collective or accumulative areas of corrective patches (used to remedy puddling, segregation, etc.) exceed 25 percent of the newly paved surface area, removal or resurfacing of the newly paved surface is required, utilizing proper leveling equipment to ensure drainage integrity and reduce and/or eliminate surface discrepancies (puddling, segregated surface, etc.).

C-11 Asphalt concrete shall be placed only when the surface is dry and when the atmospheric temperature is 50°F (10°C) and rising. No asphalt concrete shall be placed when the weather is foggy or rainy, or when the base on which the asphalt is to be placed contains moisture in excess of the optimum.

C-12 The combination of particle sizes or sieve analysis of the asphalt material is according to local specifications and available materials.

C-13 All striping and signage shall be installed to the existing configuration prior to construction or as directed by the Owner, Agent, or Consultant. The striping shall be installed after the new paved surface has adequate curing time and the chance of any color disfiguration can occur.

C-14 All speed bumps in a resurfaced section that are constructed or overlaid and not visible shall be temporarily striped with white or yellow marking paint before any sealed sections are opened to traffic.

ASPHALT PAVING OVERLAYS WITH FABRIC INNER LAYER SPECIFICATIONS

A. Materials

A-1 Asphalt pavement shall be ¾ inch maximum size aggregate hot mixed material according to local government construction specifications (state highway, county, or municipality) for overlays 1½ inches thick or more. The asphalt pavement shall be ½-inch maximum size aggregate hot mixed material according to local government construction specifications for overlays 1½ inches thick or less. The gradation and asphalt contents are based on project mix designs as established by local agencies, project architects, or independent consultants.

A-2 The Owner, Manager, or Consultant may take samples of materials to an independent laboratory for testing of proper mix design or require on-site inspection for proper application of materials.

B. Preparation

B-1 The existing pavement shall be prepared before installing a new asphalt hot-mixed overlay by remove and replace patching of all areas that have structurally failed according to the attached Specification for R&R patching. Where the edges of the overlay area are confined by curbs, gutters, driveways, garages, sidewalks, building foundations, or any other permanent structure, the edges will be milled a minimum of 3 feet from the confining barrier prior to installing the fabric and the overlay.

B-2 All cracks ¼ inch and wider shall be filled with a hot elastomeric crack sealant according to the attached specifications for crack sealing.

B-3 The existing surface shall be thoroughly cleaned of all loose debris and dirt particles by brooming, or use of high-pressure air brooms and blowers, or a combination of both until the existing pavement is completely clean.

B-4 Prior to paving, all valve boxes, manholes, and utility boxes and vaults shall be marked with an ID locator (supplied by Seal Master). All valve boxes, manhole covers and lids, and any other structures in the newly paved area shall be brought to the newly finished paved grade prior to installing the overlay.

B-5 Prior to paving the existing pavement, the Contractor shall inventory all manholes, valve boxes, survey markers, striping, and markers. The Contractor shall plot the locations of each of the manholes, valve boxes, and survey markers on a site plan, leasing plan, or blueprints, as well as establish a marking on the curbing or edge of the street or parking lot. The Contractor shall protect against the disruption of any of these items during the paving operation. Should a manhole, valve box, or survey marker be damaged, removed, or covered over, the Contractor shall replace or uncover that item in as good (or better) condition. All striping, ceramic markings, reflective markings, etc., shall be inventoried prior to paving and plotted on a site plan. All markings, ceramics, reflectors, and any other markings shall be removed and replaced identically and in the same location prior to construction unless specified or directed otherwise by the Owner, Agent, or Consultant. All striping and regulatory markings shall be installed to current local codes.

B-6 The existing bumper blocks shall be inventoried for existing breaks, cracks, etc. and removed and stacked off the construction site in such a manner as to not break or crack them. Any bumpers not marked during the inventory that are broken shall be replaced at the Contractor's expense.

B-7   All signs and poles shall be inventoried for condition and code requirements. All signs will be inventoried for damage, and any signs damaged afterwards shall be replaced at the Contractor's expense. All proper signage shall be installed according to current local codes and located at the direction of the Owner, Agent, or designated Consultant.

B-8   MC70/250 or CRS2 emulsion tack coat shall be applied at 0.1 gallon per square yard before applying the geotectile fabric layer. This tack coat shall cover the entire overlay area to reduce slippage, movement, and separation of the new overlay from the existing pavement.

B-9   An approved fabric geotectile fabric shall be installed evenly and in such a manner to prevent any folding or spacing of the fabric and leave a finished smooth surface.

C. Application

C-1   The material shall be installed at the appropriate depth and compacted to yield a depth of 2 inches after compaction for ¾ inch maximum size aggregate mix or 1½ inch after compaction for ½-inch maximum size aggregate mix, or 1 inch after compaction for ⅜ inch maximum size aggregate mix.

C-2   The asphaltic concrete shall be compacted to 95 percent laboratory density (Marshall ASTM D-1559 or Hveem Mix Design ASTM D-1560) using a combination of steel wheel static rollers and pneumatic rollers or vibratory rollers. Care shall be taken to not over-roll the new asphalt.

C-3   The initial rolling or "breakdown roll" will be accomplished by a steel wheel roller. This application will be performed as soon as the roller can move on the asphalt without picking up material, approximately 280°F but not less than 245°F. Any segregated areas or pockets shall have asphaltic material applied, raked to allow fines to fill surface voids, and made ready for further compaction passes. No segregated areas or pockets of the larger aggregates will be allowed in the finished surface. Any rough, overlapping, segregated joints shall be smoothed by raking and adding fine asphaltic materials to thoroughly blend the joint to the existing pavement and joint area. No visible or poor joint construction will be accepted.

C-4   A second rolling will be accomplished with a pneumatic rubber-tire roller.

C-5   A third or finish rolling will be accomplished with a steel wheel static roller.

C-6   A vibratory/static roller may be used in lieu of the above rolling procedure using the manufacturer's guidelines and proper roller patterns as established by local agency specifications to attain the required density.

C-7   The roller pattern shall be established where the exposed or unconfined edge is left unrolled until the next or adjacent pass is installed, providing the adjacent pass will be laid within two (2) hours of the first pass. This roller pattern will create a better joint in the finished surface. Should the adjacent pass not be installed within two (2) hours or until the next workday, the edge shall be compacted during regular rolling.

C-8   Where there is a hot section placed next to a cold section, the breakdown rolling shall start from the hot section with a 6-inch overlap of the roller wheel on the cold section.

C-9   When a transverse joint is created at the end of the workday, the proper method shall be applied to ensure a clean vertical edge at the beginning of the next workday. This can be accomplished by placing a wooden block or paper fold, or sanding so the cold asphalt can be easily removed and a flat, smooth, and vertical surface will be created, or the Contractor can saw cut the cold pavement to produce the straight, vertical edge.

C-10 Upon completion, the finished surface shall be smooth, dense, and of uniform texture and appearance. There shall be no large segregated areas or depressions holding water more than ¼ inch in depth at the deepest point. There will be no roller marks in the finished surface. All areas will drain and be free of standing water within four (4) hours of accumulation. The finished surface may not vary more than ¼ inch in a 12-foot distance as determined by a straightedge placed in any position on the finished surface, except across flow lines of drainways and inverted crowns. Paving joints shall be uniform and smooth at all locations and cold joints, and any areas where the pavement was raked to create a matching surface. Where there are large deposits of segregation of larger aggregates on the surface (not from the use of mixes designed to yield a rocky surface, e.g., Superpave mix designs), poorly raked joints, or any rough surface, the location shall be skin patched using material with a maximum sized aggregate of ¼ inch or hot sand seal. Any puddle areas where the water is more than ¼ inch deep or lasts more than eight (8) hours before dissipation shall be patched level to eliminate the puddle in a manner to not relocate the puddle. Use of string line, leveling nails, or other grade markers will ensure proper level grade to eliminate water puddles. Should the collective or accumulative areas of corrective patches (used to remedy puddling, segregation, etc.) exceed 25 percent of the newly paved surface area, removal or resurfacing of the newly paved surface is required, utilizing proper leveling equipment to ensure drainage integrity and reduce and/or eliminate surface discrepancies (puddling, segregated surface, etc.).

C-11 Asphalt concrete shall be placed only when the surface is dry and when the atmospheric temperature is 50°F (10°C) and rising. No asphalt concrete shall be placed when the weather is foggy or rainy, or when the base on which the asphalt is to be placed contains moisture in excess of the optimum.

C-12 The combination of particle sizes or sieve analysis of the asphalt material is according to local specifications and available materials.

C-13 All striping and signage shall be installed to the existing configuration prior to construction or as directed by the Owner, Agent, or Consultant. The striping shall be installed after the new paved surface has adequate curing time, and the chance of any color disfiguration can occur.

C-14 All speed bumps in a resurfaced section that are constructed or overlaid and not visible shall be temporarily striped with white or yellow marking paint before any sealed sections are opened to traffic.

MILLING AND ASPHALT PAVING OVERLAYS SPECIFICATIONS

A. Materials

A-1 Asphalt pavement shall be ¾ inch maximum size aggregate hot mixed material according to local government construction specifications (state highway, county, or municipality) for overlays 1½ inches thick or more. The asphalt pavement shall be ½-inch maximum size aggregate hot-mixed material according to local government construction specifications for overlays 1½ inches thick or less. The gradation and asphalt contents are based on project mix designs as established by local agencies, project architects, or independent consultants.

A-2 The Owner, Manager, or Consultant may take samples of materials to an independent laboratory for testing of proper mix design or require on-site inspection for proper application of materials.

B. Preparation

B-1 The existing pavement shall be milled to the existing aggregate base. All the milled material shall be disposed of by hauling to a designated dumpsite for waste materials or to a stockpile area to be reused in recycling asphalt pavement.

B-2 The existing soft areas shall be stabilized before installing a new asphalt hot mixed overlay by removing and replacing all soft areas that have structurally failed.

B-3 Prior to paving, all valve boxes, manholes, and utility boxes and vaults shall be marked with an ID locator. All valve boxes, manhole covers and lids, and any other structures in the newly paved area shall be brought to the new-finished paved grade prior to installing the overlay.

B-4 The base material shall be graded to the existing drainage layout, and the base material compacted to 95 percent modified Proctor density.

B-5 Prior to paving, the Contractor shall inventory all manholes, valve boxes, survey markers, striping, and markers. The Contractor shall plot the locations of each of the manholes, valve boxes, and survey markers on a site plan, leasing plan, or blueprints, as well as establish a marking on the curbing or edge of the street or parking lot. The Contractor shall protect against the disruption of any of these items during the paving operation. Should a manhole, valve box, or survey marker be damaged, removed, or covered over, the Contractor shall replace or uncover that item in as good (or better) condition. All striping, ceramic markings, reflective markings, etc. shall be inventoried prior to paving and plotted on a site plan. All markings, ceramics, reflectors, and any other markings shall be removed and replaced identically and in the same location prior to construction unless specified or directed otherwise by the Owner, Agent, or Consultant. All striping and regulatory markings shall be installed to current local codes.

B-6 The existing bumper blocks shall be inventoried for existing breaks, cracks, etc. and removed and stacked off the construction site in such a manner as to not break or crack them. Any bumpers not marked during the inventory that are broken shall be replaced at the Contractor's expense. All bumper blocks as well as any required new blocks shall be reinstalled in a straight configuration with each other and pinned to prevent movement. All broken bumpers shall be hauled off the site and disposed of.

B-7 All signs and poles shall be inventoried for condition and code requirements. All signs will be inventoried for damage, and any signs damaged afterwards shall be replaced at the Contractor's expense. All proper signage shall be installed according to current local codes and located at the direction of the Owner, Agent, or designated Consultant.

B-8 MC70/250 or CRS2 emulsion tack coat shall be applied at 0.1 gallon per square yard before applying the geotectile fabric layer. This tack coat shall cover the entire overlay area to reduce slippage, movement, and separation of the new overlay from the existing pavement. The fabric layer shall be installed in such a manner to prevent any folding or spacing of the fabric and leave a smooth surface.

C. Application

C-1 The material shall be installed at the appropriate depth and compacted to yield a depth of 2 inches after compaction for ¾-inch maximum size aggregate mix or 1½ inches after compaction for ½-inch maximum size aggregate mix, or 1 inch after compaction for ⅜ inch maximum size aggregate mix.

C-2 The asphaltic concrete shall be compacted to 95 percent laboratory density (Marshall ASTM D-1559 or Hveem Mix Design ASTM D-1560) using a combination of steel wheel static rollers and pneumatic rollers or vibratory rollers. Care shall be taken to not over-roll the new asphalt.

C-3 The initial rolling or "breakdown roll" will be accomplished by a steel wheel roller. This application will be performed as soon as the roller can move on the asphalt without picking up material, approximately 280°F but not less than 245°F. Any segregated areas or pockets shall have asphaltic material applied, raked to allow fines to fill surface voids, and made ready for further compaction passes. No segregated areas or pockets of the larger aggregates will be allowed in the finished surface. Any rough, overlapping, segregated joints shall be smoothed by raking and adding fine asphaltic materials to thoroughly blend the joint to the existing pavement and joint area. No visible or poor joint construction will be accepted.

C-4 A second rolling will be accomplished with a pneumatic rubber-tire roller.

C-5 A third or finish rolling will be accomplished with a steel wheel static roller.

C-6 A vibratory/static roller may be used in lieu of the above rolling procedure using the manufacturer's guidelines and proper roller patterns as established by local agency specifications to attain the required density. However, the roller should be in the static mode on the finished layer.

C-7 The roller pattern shall be established where the exposed or unconfined edge is left unrolled until the next or adjacent pass is installed, providing the adjacent pass will be laid within two (2) hours of the first pass. This roller pattern will create a better joint in the finished surface. Should the adjacent pass not be installed within two (2) hours or until the next workday, the edge shall be compacted during regular rolling.

C-8 Where there is a hot section placed next to a cold section, the breakdown rolling shall start from the hot section with a 6-inch overlap of the roller wheel on the cold section.

C-9 When a transverse joint is created at the end of the workday, the proper method shall be applied to ensure a clean vertical edge at the beginning of the next workday. This can be accomplished by placing a wooden block or paper fold, or sanding so the cold asphalt can be easily removed and a flat, smooth, and vertical surface will be created, or the Contractor can saw cut the cold pavement to produce the straight, vertical edge.

C-10 Upon completion, the finished surface shall be smooth, dense, and of uniform texture and appearance. There shall be no large segregated areas or depressions holding water more than ¼ inch in depth at the deepest point. There will be no roller marks in the finished surface. All areas will drain and be free of standing water within four (4) hours of accumulation. The finished surface may not vary more than ¼ inch in a 12-foot distance as determined by a straightedge placed in any position on the finished surface, except across flow lines of drainways and inverted crowns. Paving joints shall be uniform and smooth at all locations and cold joints, and any areas where the pavement was raked to create a matching surface. Where there are large deposits of segregation of larger aggregates on the surface (not from the use of mixes designed to yield a rocky surface, e.g., Superpave mix designs), poorly raked joints, or any rough surface, the location shall be skin patched using material with a maximum sized aggregate of ¼ inch or hot sand seal. Any puddle areas where the water is more than ¼ inch deep or lasts more than eight (8) hours before dissipation shall be patched level to eliminate the puddle in a manner to not relocate the puddle. Use of string line, leveling nails, or other grade markers will ensure proper level grade to eliminate water puddles. Should the collective or accumulative areas of corrective patches (used to remedy puddling, segregation, etc.) exceed 25 percent of the newly paved surface area, removal or resurfacing of the newly paved surface is required, utilizing proper leveling equipment to ensure drainage integrity and reduce and/or eliminate surface discrepancies (puddling, segregated surface, etc.).

C-11 Asphalt concrete shall be placed only when the surface is dry and when the atmospheric temperature is 50°F (10°C) and rising. No asphalt concrete shall be placed when the weather is foggy or rainy, or when the base on which the asphalt is to be placed contains moisture in excess of the optimum.

C-12 The combination of particle sizes or sieve analysis of the asphalt material is according to local specifications and available materials.

C-13 All striping and signage shall be installed to the existing configuration prior to construction or as directed by the Owner, Agent, or Consultant. The striping shall be installed after the new paved surface has adequate curing time, and the chance of any color disfiguration can occur.

C-14 All speed bumps in a resurfaced section that are constructed or overlaid and not visible shall be temporarily striped with white or yellow marking paint before any sealed sections are opened to traffic.

PULVERIZING EXISTING BASE AND PAVEMENT OVERLAY SPECIFICATIONS

A. Materials

A-1   The existing asphalt pavement shall be pulverized to usable base with the maximum particle size of 2 inches. Any particle greater than 2 inches shall be removed and disposed of or broken down to meet size requirements.

A-2   The prepared base shall have AC-10 tack coat applied along curbings and any cold joint that exists where new hot mixed asphalt will come in contact, and 2 inches of hot asphaltic concrete (¾-inch maximum size aggregate), or 1½ inch of hot asphaltic concrete (½-inch maximum size aggregate) as designated in the Request for Bids, installed according to local government construction specifications (state highway, county, or municipality).

A-3   The Owner, Manager, or Consultant may take samples of materials to an independent lab for testing of proper mix design or require on-site inspection for proper application of materials.

B. Preparation

B-1   Prior to pulverizing the existing pavement, the Contractor shall inventory all manholes, valve boxes, survey markers, striping, and markers. The Contractor shall plot the locations of each of the manholes, valve boxes, and survey markers on a site plan, leasing plan, or blueprints, as well as establish a marking on the curbing or edge of the street or parking lot. The Contractor shall protect against the disruption of any of these items during the pulverizing or paving operation. Should a manhole, valve box, or survey marker be damaged or removed, the Contractor shall replace that item in as good (or better) condition. All striping, ceramic markings, reflective markings, etc. shall be inventoried prior to pulverizing and plotted on a site plan. All markings, ceramics, reflectors, and any other markings shall be replaced identically and in the same location prior to construction unless specified or directed otherwise by the Owner, Agent, or Consultant. All striping and regulatory markings shall be installed to current local codes. **The Contractor shall notify Blue Stakes and have all utilities marked prior to pulverizing, excavating, and grading.** This may require more than one marking by Blue Stakes.

B-2   The existing bumper blocks shall be inventoried for existing breaks, cracks, etc. and removed and stacked off the construction site in such a manner as to not break or crack them. Any bumpers not marked during the inventory that are broken shall be replaced at the Contractor's expense. All bumper blocks as well as any required new blocks shall be reinstalled in a straight configuration with each other and pinned to prevent movement. All broken bumpers shall be hauled off the site and disposed of.

B-3   All signs and poles shall be inventoried for damage and code requirements, and removed, as well as any concrete anchor ball. The signs, if in compliance with existing codes, will be removed and saved for reinstallation. All signs to be saved will be inventoried for damage, and any signs damaged afterwards shall be replaced at the Contractor's expense. All proper signage shall be installed according to current local codes and located at the direction of the Owner, Agent, or designated Consultant.

B-4   All in-place asphalt pavement not pulverized (e.g., close to or surrounding manholes, valve boxes, curbs, drives, or walks; the corners of intersecting streets; or pavement adjacent to or surrounding valley gutters where the pulverizing, milling, or crushing machinery cannot reach) shall be removed and hauled away or pulverized prior to preparing and grading base for paving.

B-5 The new pulverized base material shall be properly prepared by watering and compacting uniformly over the entire surface area to 95 percent of modified Proctor density ASTM D-1557 and acceptable for installation of hot-mixed asphalt pavement.

B-6 The perimeter edges of any curbing, dumpster pads, walks, etc. of the paving area shall have an MC tack coat applied 2 feet from the edge of any solid foundation or curbing as well as any adjacent edge of the curbing, dumpster pads, or walks that will come into contact with the new pavement surface.

C. Application

C-1 The hot asphalt material shall be installed at the appropriate depth and compacted to yield a depth of 2 inches for ¾-inch maximum size aggregate mix, and 1½ inch for ½-inch maximum size aggregate mix after compaction.

C-2 The asphaltic concrete shall be compacted uniformly over the entire surface area to 95 percent laboratory density (Marshall ASTM D-1559 or Hveem Mix Design ASTM D-1560) using a combination of steel wheel static rollers, pneumatic rollers, and/or vibratory rollers. Rolling shall be accomplished in such a manner as to not over-roll the new asphalt surface, yet meet the required density requirements.

C-3 The initial rolling or "breakdown roll" will be accomplished by a steel wheel roller. This application will be performed as soon as the roller can move on the asphalt without picking up material, approximately 280°F but not less than 245°F. Any segregated areas or pockets shall have asphaltic material applied, raked to allow fines to fill surface voids, and made ready for further compaction passes. No segregated areas or pockets of the larger aggregates will be allowed in the finished surface. Any rough, overlapping, segregated joints shall be smoothed by raking and adding fine asphaltic materials to thoroughly blend the joint to the existing pavement and joint area. No visible or poor joint construction will be accepted.

C-4 A second rolling will be accomplished with a pneumatic rubber-tire roller.

C-5 A third or finish rolling will be accomplished with a steel wheel static roller.

C-6 A vibratory/static roller may be used in lieu of the above rolling procedure using the manufacturer's guidelines and proper roller patterns as established by local agency specifications to attain the required density. However, the roller should be in the static mode on the finished layer.

C-7 The roller pattern shall be established where the exposed or unconfined edge is left unrolled until the next or adjacent pass is installed, providing the adjacent pass will be laid within two (2) hours of the first pass. This roller pattern will create a better joint in the finished surface. Should the adjacent pass not be installed within 2 hours or until the next workday, the edge shall be compacted during regular rolling.

C-8 Where there is a hot section placed next to a cold section, the breakdown rolling shall start from the hot section with a 6-inch overlap of the roller wheel on the cold section.

C-9 When a transverse joint is created at the end of the workday, the proper method shall be applied to ensure a clean vertical edge at the beginning of the next workday. This can be accomplished by placing a wooden block or paper fold, or sanding so the cold asphalt can be easily removed and a flat, smooth, and vertical surface will be created, or the Contractor can saw cut the cold pavement to produce the straight, vertical edge.

C-10 Upon completion, the finished surface shall be smooth, dense, and of uniform texture and appearance. There shall be no large segregated areas or depressions holding water more than ¼ inch in depth at the deepest point. There will be no roller marks in the finished surface. All areas will drain and be free of standing water within four (4) hours of accumulation. The finished surface may not vary more than ¼ inch in a 12-foot distance as determined by a straightedge placed in any position on the finished surface, except across flow lines of drainways and inverted crowns. Paving joints shall be uniform and smooth at all locations and cold joints, and any areas where the pavement was raked to create a matching surface. Where there are large deposits of segregation of larger aggregates on the surface (not from the use of mixes designed to yield a rocky surface, e.g., Superpave mix designs), poorly raked joints, or any rough surface, the location shall be skin patched using material with a maximum sized aggregate of ¼ inch or hot sand seal. Any puddle areas where the water is more than ¼ inch deep or lasts more than four (4) hours before dissipation shall be patched level to eliminate the puddle in a manner to not relocate the puddle. Use of string line, leveling nails, or other grade markers will ensure proper level grade to eliminate water puddles. Should the collective or accumulative areas of corrective patches (used to remedy puddling, segregation, etc.) exceed 25 percent of the newly paved surface area, removal or resurfacing of the newly paved surface is required, utilizing proper leveling equipment to ensure drainage integrity and reduce and/or eliminate surface discrepancies (puddling, segregated surface, etc.).

C-11 Asphalt concrete shall be placed only when the surface is dry and when the atmospheric temperature is 50°F (10°C) and rising. No asphalt concrete shall be placed when the weather is foggy or rainy, or when the base on which the asphalt is to be placed contains moisture in excess of the optimum.

C-12 All valve boxes, manhole covers and lids, and any other structures in the newly paved area shall be brought to the newly finished paved grade.

C-13 The sieve or gradation of particles will be according to local available materials.

C-14 All striping and signage shall be installed to the existing configuration prior to construction or as directed by the Owner, Agent, or Consultant. The striping shall be installed after the new paved surface has adequate curing time, and the chance of any color disfiguration can occur.

C-15 All speed bumps in a resurfaced section that are constructed or overlaid and not visible shall be temporarily striped with white or yellow marking paint before any sealed sections are opened to traffic.

PULVERIZING EXISTING PAVEMENT AND BASE
INSTALLING 3 PERCENT LIME STABILIZATION
AND PAVEMENT OVERLAY SPECIFICATIONS

A. Materials

A-1   The existing asphalt pavement shall be pulverized to usable base with the maximum particle size of 2 inches. Any particle greater than 2 inches shall be removed and disposed of or broken down to meet size requirement.

A-2   The prepared base shall have 3 percent (by volume) lime stabilization material added to the pulverized material.

A-3   The prepared base shall have AC-10 tack coat applied along curbings and any cold joint that exists where new hot mixed asphalt will come in contact, and 2 inches of hot asphaltic concrete (¾-inch maximum size aggregate), or 1½ inches of hot asphaltic concrete (½-inch maximum size aggregate) as designated in the Request for Bids, installed according to local government construction specifications (state highway, county, or municipality).

A-4   The Owner, Manager, or Consultant may take samples of materials to an independent lab for testing of proper mix design or require on-site inspection for proper application of materials.

B. Preparation

B-1   Prior to pulverizing the existing pavement, the Contractor shall inventory all manholes, valve boxes, survey markers, striping, and markers. The Contractor shall plot the locations of each of the manholes, valve boxes, and survey markers on a site plan, leasing plan, or blueprints, as well as establish a marking on the curbing or edge of the street or parking lot. The Contractor shall protect against the disruption of any of these items during the pulverizing or paving operation. Should a manhole, valve box, or survey marker be damaged or removed, the Contractor shall replace that item in as good (or better) condition. All striping, ceramic markings, reflective markings, etc. shall be inventoried prior to pulverizing and plotted on a site plan. All markings, ceramics, reflectors, and any other markings shall be replaced identically and in the same location prior to construction unless specified or directed otherwise by the Owner, Agent, or Consultant. All striping and regulatory markings shall be installed to current local codes. **The Contractor shall notify Blue Stakes and have all utilities marked prior to pulverizing, excavating, and grading.** This may require more than one marking by Blue Stakes.

B-2   The existing bumper blocks shall be inventoried for existing breaks, cracks, etc. and removed and stacked off the construction site in such a manner as to not break or crack them. Any bumpers not marked during the inventory that are broken shall be replaced at the Contractor's expense. All bumper blocks as well as any required new blocks shall be reinstalled in a straight configuration with each other and pinned to prevent movement. All broken bumpers shall be hauled off the site and disposed of.

B-3   All signs and poles shall be inventoried for damage and code requirements, and removed, as well as any concrete anchor ball. The signs, if in compliance with existing codes, will be removed and saved for reinstallation. All signs to be saved will be inventoried for damage, and any signs damaged afterwards shall be replaced at the Contractor's expense. All proper signage shall be installed according to current local codes and located at the direction of the Owner, Agent, or designated Consultant.

B-4 All in-place asphalt pavement not pulverized (e.g., close to or surrounding manholes, valve boxes, curbs, drives, or walks; the corners of intersecting streets; or pavement adjacent to or surrounding valley gutters where the pulverizing, milling, or crushing machinery cannot reach) shall be removed and hauled away or pulverized prior to preparing and grading base for paving.

B-5 The pulverized material shall have lime stabilization material applied at a rate of 3 percent per volume of pulverized material to the top 8 to 12 inches. The pulverized material and lime shall be blended or re-pulverized to provide a homogeneous mixture with the lime.

B-6 The new pulverized base material shall be properly prepared by watering and compacting uniformly over the entire surface area to 95 percent of modified Proctor density ASTM D-1557 and acceptable for installation of hot-mixed asphalt pavement.

B-7 The perimeter edges of any curbing, dumpster pads, walks, etc. of the paving area shall have an MC tack coat applied 2 feet from the edge of any solid foundation or curbing as well as any adjacent edge of the curbing, dumpster pads, or walks that will come into contact with the new pavement surface.

C. Application

C-1 The hot asphalt material shall be installed at the appropriate depth and compacted to yield a depth of 2 inches for ¾-inch maximum size aggregate mix, and 1½ inches for ½-inch maximum size aggregate mix after compaction.

C-2 The asphaltic concrete shall be compacted uniformly over the entire surface area to 95 percent laboratory density (Marshall ASTM D-1559 or Hveem Mix Design ASTM D-1560) using a combination of steel wheel static rollers, pneumatic rollers, and/or vibratory rollers. Rolling shall be accomplished in such a manner as to not over-roll the new asphalt surface, yet meet the required density requirements.

C-3 The initial rolling or "breakdown roll" will be accomplished by a steel wheel roller. This application will be performed as soon as the roller can move on the asphalt without picking up material, approximately 280°F but not less than 245°F. Any segregated areas or pockets shall have asphaltic material applied, raked to allow fines to fill surface voids, and made ready for further compaction passes. No segregated areas or pockets of the larger aggregates will be allowed in the finished surface. Any rough, overlapping, segregated joints shall be smoothed by raking and adding fine asphaltic materials to thoroughly blend the joint to the existing pavement and joint area. No visible or poor joint construction will be accepted.

C-4 A second rolling will be accomplished with a pneumatic rubber-tire roller.

C-5 A third or finish rolling will be accomplished with a steel wheel static roller.

C-6 A vibratory/static roller may be used in lieu of the above rolling procedure using the manufacturer's guidelines and proper roller patterns as established by local agency specifications to attain the required density. However, the roller should be in the static mode on the finished layer.

C-7 The roller pattern shall be established where the exposed or unconfined edge is left unrolled until the next or adjacent pass is installed, providing the adjacent pass will be laid within two (2) hours of the first pass. This roller pattern will create a better joint in the finished surface. Should the adjacent pass not be installed within two (2) hours or until the next workday, the edge shall be compacted during regular rolling.

C-8   Where there is a hot section placed next to a cold section, the breakdown rolling shall start from the hot section with a 6-inch overlap of the roller wheel on the cold section.

C-9   When a transverse joint is created at the end of the workday, the proper method shall be applied to ensure a clean vertical edge at the beginning of the next workday. This can be accomplished by placing a wooden block or paper fold, or sanding so the cold asphalt can be easily removed and a flat, smooth, and vertical surface will be created, or the Contractor can saw cut the cold pavement to produce the straight, vertical edge.

C-10   Upon completion, the finished surface shall be smooth, dense, and of uniform texture and appearance. There shall be no large segregated areas or depressions holding water more than ¼ inch in depth at the deepest point. There will be no roller marks in the finished surface. All areas will drain and be free of standing water within four (4) hours of accumulation. The finished surface may not vary more than ¼ inch in a 12-foot distance as determined by a straightedge placed in any position on the finished surface, except across flow lines of drainways and inverted crowns. Paving joints shall be uniform and smooth at all locations and cold joints, and any areas where the pavement was raked to create a matching surface. Where there are large deposits of segregation of larger aggregates on the surface (not from the use of mixes designed to yield a rocky surface, e.g., Superpave mix designs), poorly raked joints, or any rough surface, the location shall be skin patched using material with a maximum sized aggregate of ¼ inch or hot sand seal. Any puddle areas where the water is more than ¼ inch deep or lasts more than eight (8) hours before dissipation shall be patched level to eliminate the puddle in a manner to not relocate the puddle. Use of string line, leveling nails, or other grade markers will ensure proper level grade to eliminate water puddles. Should the collective or accumulative areas of corrective patches (used to remedy puddling, segregation, etc.) exceed 25 percent of the newly paved surface area, removal or resurfacing of the newly paved surface is required, utilizing proper leveling equipment to ensure drainage integrity and reduce and/or eliminate surface discrepancies (puddling, segregated surface, etc.).

C-11   Asphalt concrete shall be placed only when the surface is dry and when the atmospheric temperature is 50°F (10°C) and rising. No asphalt concrete shall be placed when the weather is foggy or rainy, or when the base on which the asphalt is to be placed contains moisture in excess of the optimum.

C-12   All valve boxes, manhole covers and lids, and any other structures in the newly paved area shall be brought to the new-finished paved grade.

C-13   The sieve or gradation of particles will be according to local available materials.

C-14   All striping and signage shall be installed to the existing configuration prior to construction or as directed by the Owner, Agent, or Consultant. The striping shall be installed after the new paved surface has adequate curing time, and the chance of any color disfiguration can occur.

C-15   All speed bumps in a resurfaced section that are constructed or overlaid and not visible shall be temporarily striped with white or yellow marking paint before any sealed sections are opened to traffic.

REMOVE EXISTING PAVEMENT
AND INSTALL NEW ASPHALT PAVEMENT SPECIFICATIONS

A. Materials

A-1   Asphalt pavement shall be ¾-inch maximum size aggregate hot mixed material according to local government construction specifications (state highway, county, or municipality) for overlays 1½ inches thick or more. The asphalt pavement shall be ½-inch maximum size aggregate hot-mixed material according to local government construction specifications for overlays 1½ inches thick or less. The gradation and asphalt contents are based on project mix designs as established by local agencies, project architects, or independent consultants.

A-2   The Owner, Manager, or Consultant may take samples of materials to an independent laboratory for testing of proper mix design or require on-site inspection for proper application of materials.

B. Preparation

B-1   The existing pavement shall be saw cut and removed to the existing base. All the removed material shall be disposed of by hauling to a designated dumpsite for waste materials or to a stockpile area to be reused in recycling asphalt pavement.

B-2   The existing aggregate base shall be prepared before installing a new asphalt hot mixed pavement by treating all soft areas that have structurally failed and grading to the original and existing drainage.

B-3   Prior to paving, the Contractor shall inventory all manholes, valve boxes, survey markers, striping, and markers. The Contractor shall plot the locations of each of the manholes, valve boxes, and survey markers on a site plan, leasing plan, or blueprints, as well as establish a marking on the curbing or edge of the street or parking lot. The Contractor shall protect against the disruption of any of these items during the paving operation. Should a manhole, valve box, or survey marker be damaged, removed, or covered over, the Contractor shall replace or uncover that item in as good (or better) condition. All striping, ceramic markings, reflective markings, etc. shall be inventoried prior to paving and plotted on a site plan. All markings, ceramics, reflectors, and any other markings shall be removed and replaced identically and in the same location prior to construction unless specified or directed otherwise by the Owner, Agent, or Consultant. All striping and regulatory markings shall be installed to current and local codes.

B-4   All valve boxes, manhole covers and lids, and any other structures in the newly paved area shall be brought to the new finished paved grade prior to installing the new pavement. Should the valve boxes, manholes, and any other structures be covered during the paving process, the structure shall be marked for adjustment after paving.

B-5   All signs and poles shall be inventoried for condition and code requirements. All signs will be inventoried for damage, and any signs damaged afterwards shall be replaced at the Contractor's expense. All proper signage shall be installed according to current local codes and located at the direction of the Owner, Agent, or designated Consultant.

B-6   MC-70/250, AC-20, or CRS emulsion tack coat shall be applied to the edges of all existing and adjacent paved and concrete structures.

C. Application

C-1   The material shall be installed at the appropriate depth and compacted to yield a depth of 2 inches after compaction.

C-2   The asphaltic concrete shall be compacted to 95 percent laboratory density (Marshall ASTM D-1559 or Hveem Mix Design ASTM D-1560) using a combination of steel wheel static rollers and pneumatic rollers or vibratory rollers. Care shall be taken to not over-roll the new asphalt.

C-3   The initial rolling or "breakdown roll" will be accomplished by a steel wheel roller. This application will be performed as soon as the roller can move on the asphalt without picking up material, approximately 280°F but not less than 260°F.

C-4   A second rolling will be accomplished with a pneumatic roller.

C-5   A third or finish rolling will be accomplished with a steel wheel static roller.

C-6   A vibratory/static roller may be used in lieu of the above rolling procedure using the manufacturer's guidelines and proper roller patterns as established by local agency specifications to attain the required density.

C-7   The roller pattern shall be established where the exposed or unconfined edge is left unrolled until the next or adjacent pass is installed, providing the adjacent pass will be laid within two (2) hours of the first pass. This roller pattern will create a better joint in the finished surface. Should the adjacent pass not be installed within two (2) hours or until the next workday, the edge shall be compacted during regular rolling.

C-8   Where there is a hot section placed next to a cold section, the breakdown rolling shall start from the hot section with a 6-inch overlap of the roller wheel on the cold section.

C-9   When a transverse joint is created at the end of the workday, the proper method shall be applied to ensure a clean vertical edge at the beginning of the next workday. This can be accomplished by placing a wooden block or paper fold, or sanding so the cold asphalt can be easily removed and a flat, smooth, and vertical surface will be created, or the Contractor can saw cut the cold pavement to produce the straight, vertical edge.

C-10 Upon completion, the finished surface shall be smooth, dense, and of uniform texture and appearance. There shall be no large segregated areas or depressions holding water more than ¼ inch in depth at the deepest point. There will be no roller marks in the finished surface. All areas will drain and be free of standing water within four (4) hours of accumulation. The finished surface may not vary more than ¼ inch in a 12-foot distance as determined by a straightedge placed in any position on the finished surface, except across flow lines of drainways and inverted crowns. Paving joints shall be uniform and smooth at all locations and cold joints, and any areas where the pavement was raked to create a matching surface. Where there are large deposits of segregation of larger aggregates on the surface (not from the use of mixes designed to yield a rocky surface, e.g., Superpave mix designs), poorly raked joints, or any rough surface, the location shall be skin patched using material with a maximum sized aggregate of ¼-inch or hot sand seal. Any puddle areas where the water is more than ¼-inch deep or lasts more than eight (8) hours before dissipation shall be patched level to eliminate the puddle in a manner to not relocate the puddle. Use of string line, leveling nails, or other grade markers will ensure proper level grade to eliminate water puddles. Should the collective or accumulative areas of corrective patches (used to remedy puddling, segregation, etc.) exceed

25 percent of the newly paved surface area, removal or resurfacing of the newly paved surface is required, utilizing proper leveling equipment to ensure drainage integrity and reduce and/or eliminate surface discrepancies (puddling, segregated surface, etc.).

C-11 Asphalt concrete shall be placed only when the surface is dry and when the atmospheric temperature is 50°F (10°C) and rising. No asphalt concrete shall be placed when the weather is foggy or rainy, or when the base on which the asphalt is to be placed contains moisture in excess of the optimum.

C-12 The combination of particle sizes or sieve analysis of the asphalt material is according to local specifications and available materials.

C-13 If the valve boxes, manholes, and other structures have been covered during paving, the asphalt pavement shall be saw cut and removed by 1 foot around valve boxes and 2 feet around manholes and drywells, exposing the top 6 inches of the structure. The area surrounding the valve boxes and manholes shall be concreted using 3,500 PSI concrete, smoothed and finished with a tool joint every 90° angle on center.

C-14 All striping and signage shall be installed to the existing configuration prior to construction or as directed by the Owner, Agent, or Consultant. The striping shall be installed after the new paved surface has adequate curing time, and the chance of any color disfiguration can occur.

INSTALL NEW ASPHALT PAVEMENT SECTION SPECIFICATIONS

A. Materials

A-1   Asphalt pavement shall be ¾-inch maximum size aggregate hot mixed material according to local government construction specifications (state highway, county, or municipality) for overlays 1½ inches thick or more. The asphalt pavement shall be ½-inch maximum size aggregate hot-mixed material according to local government construction specifications for overlays 1½ inches thick or less. The gradation and asphalt contents are based on project mix designs as established by local agencies, project architects, or independent consultants.

A-2   The Owner, Manager, or Consultant may take samples of materials to an independent laboratory for testing of proper mix design or require on-site inspection for proper application of materials.

B. Preparation

B-1   The existing subgrade shall be clear and grubbed of all vegetation, topsoil, and all vegetable root systems. The subgrade shall be scarified, watered, and compacted to 95% standard Proctor density for clayey soils and 95% modified Proctor density for granular materials.

B-2   Aggregate base course shall be prepared before installing the new asphalt hot mixed pavement by grading, watering, and compaction to 95 percent modified Proctor density. When the section is installed next to adjacent pavements, the new section must be graded to accommodate the existing drainage design.

B-3   Any and all valve boxes, manhole covers and lids, and any other structures in the newly paved area shall be brought to the new-finished paved grade prior to installing the new asphalt pavement. Should the valve boxes, manholes, and any other structures be covered during the paving process, the structure shall be marked for adjustment after paving.

B-4   Prior to paving, the Contractor shall inventory all manholes, valve boxes, survey markers, striping, and markers. The Contractor shall plot the locations of each of the manholes, valve boxes, and survey markers on a site plan, leasing plan, or blueprints, as well as establish a marking on the curbing or edge of the street or parking lot. The Contractor shall protect against the disruption of any of these items during the paving operation. Should a manhole, valve box, or survey marker be damaged, removed, or covered over, the Contractor shall replace or uncover that item in as good (or better) condition. All striping and regulatory markings shall be installed to current and local codes.

B-5   All signs and poles shall be inventoried for condition and code requirements. All signs will be inventoried for damage, and any signs damaged afterwards shall be replaced at the Contractor's expense. All proper signage shall be installed according to current local codes and located at the direction of the Owner, Agent, or designated Consultant.

B-6   MC-70/250, AC-20, or CRS emulsion tack coat shall be applied to the edges of all existing and adjacent paved and concrete structures.

C. Application

C-1   The material shall be installed at the appropriate depth and compacted to yield a depth of 2 inches after compaction.

C-2 The asphaltic concrete shall be compacted to 95 percent laboratory density (Marshall ASTM D-1559 or Hveem Mix Design ASTM D-1560) using a combination of steel wheel static rollers and pneumatic rollers or vibratory rollers. Care shall be taken to not over-roll the new asphalt.

C-3 The initial rolling or "breakdown roll" will be accomplished by a steel wheel roller. This application will be performed as soon as the roller can move on the asphalt without picking up material, approximately 250°F but not less than 225°F.

C-4 A second rolling will be accomplished with a pneumatic roller.

C-5 A third or finish rolling will be accomplished with a steel wheel static roller.

C-6 A vibratory/static roller may be used in lieu of the above rolling procedure using the manufacturer's guidelines and proper roller patterns as established by local agency to attain the required density.

C-7 The roller pattern shall be established where the exposed or unconfined edge is left unrolled until the next or adjacent pass is installed, providing the adjacent pass will be laid within two (2) hours of the first pass. This roller pattern will create a better joint in the finished surface. Should the adjacent pass not be installed within two (2) hours or until the next workday, the edge shall be compacted during regular rolling.

C-8 Where there is a hot section placed next to a cold section, the breakdown rolling shall start from the hot section with a 6-inch overlap of the roller wheel on the cold section.

C-9 When a transverse joint is created at the end of the workday, the proper method shall be applied to ensure a clean vertical edge at the beginning of the next workday. This can be accomplished by placing a wooden block or paper fold, or sanding so the cold asphalt can be easily removed and a flat, smooth, and vertical surface will be created, or the Contractor can saw cut the cold pavement to produce the straight, vertical edge.

C-10 Upon completion, the finished surface shall be smooth, dense, and of uniform texture and appearance. There shall be no large segregated areas or depressions holding water more than ¼ inch in depth at the deepest point. There will be no roller marks in the finished surface. All areas will drain and be free of standing water within four (4) hours of accumulation. The finished surface may not vary more than ¼ inch in a 12-foot distance as determined by a straightedge placed in any position on the finished surface, except across flow lines of drainways and inverted crowns. Paving joints shall be uniform and smooth at all locations and cold joints, and any areas where the pavement was raked to create a matching surface. Where there are large deposits of segregation of larger aggregates on the surface (not from the use of mixes designed to yield a rocky surface, e.g., Superpave mix designs), poorly raked joints, or any rough surface, the location shall be skin patched using material with a maximum sized aggregate of ¼ inch or hot sand seal. Any puddle areas where the water is more than ¼ inch deep or lasts more than eight (8) hours before dissipation shall be patched level to eliminate the puddle in a manner to not relocate the puddle. Use of string line, leveling nails, or other grade markers will ensure proper level grade to eliminate water puddles. Should the collective or accumulative areas of corrective patches (used to remedy puddling, segregation, etc.) exceed 25 percent of the newly paved surface area, removal or resurfacing of the newly paved surface is required, utilizing proper leveling equipment to ensure drainage integrity and reduce and/or eliminate surface discrepancies (puddling, segregated surface, etc.).

C-11 Asphalt concrete shall be placed only when the surface is dry and when the atmospheric temperature is 50°F (10°C) and rising. No asphalt concrete shall be placed when the weather is foggy or rainy, or when the base on which the asphalt is to be placed contains moisture in excess of the optimum.

C-12 If the valve boxes, manholes, and other structures have been covered during paving, the asphalt pavement shall be saw cut and removed by 1 foot around valve boxes and 2 feet around manholes and drywells, exposing the top 6 inches of the structure. The area surrounding the valve boxes and manholes shall be concreted using 3,500 PSI concrete, smoothed, and finished with a tool joint every 90° angle on center.

C-13 The combination of particle sizes or sieve analysis of the asphalt material is according to local specifications and available materials.

C-14 All striping and signage shall be installed to the existing configuration prior to construction or as directed by Owner, Agent, or Consultant. The striping shall be installed after the new paved surface has adequate curing time, and the chance of any color disfiguration can occur.

PAINTING AND STRIPING SPECIFICATIONS

A. Materials

A-1   DOT-approved latex white exterior paint shall be used for all parking stall lines. All lines shall be at least 4 inches wide.

A-2   DOT-approved latex yellow exterior paint shall be used for all speed bumps, four square courts, and basketball courts.

A-3   DOT-approved latex yellow or white exterior paint shall be used for all handicapped stalls, lines, and symbols.

A-4   DOT-approved acrylic RED for curbs in no-parking fire lanes.

B. Preparation

B-1   Prior to application, ALL areas will be power swept, vacuumed, and cleared of all loose material. Oil spots will be manually scraped and cleaned with mild detergent and a primer applied if required.

B-2   If the existing paint on the existing pavement prior to sealing could "bleed" through the new seal coating, the Contractor shall seal the existing paint by blacking out with flat black paint prior to sealing.

B-3   All existing painted curbs shall be protected and should not be affected by asphalt or painting work. If painted curbs are affected by work, the Contractor shall repaint to Property Manager's/Owner's satisfaction.

B-4   Where the red and yellow curbing is repainted, all loose, curled, flaking paint on the existing curb shall be removed by brushing with a stiff metal brush prior to applying any new paint. Any stenciling applied shall be centered and all letters shall be evenly spaced and level. On fire lane stencils, only white letters shall be allowed on red curbing.

C. Application

C-1   Restripe as existing, except as where noted on attached plot plan or changes necessary to accommodate the ADA handicapped parking stall codes as enforced by the local governing agency (at no time less than the federal standards).

C-2   Allow sealer and asphalt to cure for at least twelve (12) hours before painting, depending on local weather conditions. If painting is to be applied over an overlay, or other new asphalt, allow at least two (2) days for curing before painting.

C-3   The paint shall be applied according to the manufacturer's recommendations and shall not be applied below recommended temperatures or in such a way as to fade into the new asphaltic materials.

HANDICAPPED PARKING SPECIFICATIONS

Locations, Signage, and Codes

- The locations and quantity of handicapped stalls are shown on the attached drawing supplied to the Contractor by the Owner, Manager, or Consultant. The Contractor is responsible for reviewing these locations, sizes, and spacing, and installing all markings, signs, and any required bumper blocks as shown.

- The existing striping and paint for the handicapped parking stalls and symbols shall be blacked out with flat black-out latex paint before any asphalt work or repainting.

- The locations, quantity, and type of handicapped signs are available in detail through the local government agency and department where the property is located. If there is a new requirement, the Contractor shall conform to the most recent code and the federal ADA requirements. *

Handicap Stalls Required
ADA Standard

TOTAL PARKING SPACES IN LOT	REQUIRED MINIMUM NUMBER OF ACCESSIBLE SPACES
1 to 25	1
26 to 50	2
51 to 75	3
76 to 100	4
101 to 150	5
151 to 200	6
201 to 300	7
301 to 400	8
401 to 500	9
501 to 1000	2% of total
1,001 and over	20 plus 1 for each 100 over 1,000

*Page 35612, *Federal Register*, Vol. 56, No. 144, Friday, July 26, 1991. Rules and Regulations. Local governing agencies can have different requirements, and it is the responsibility of the Contractor to know all local codes.

Universal Parking Space

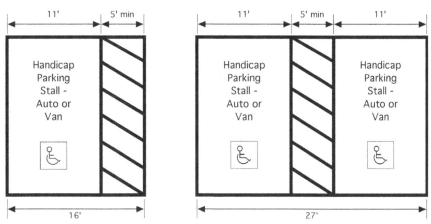

Figure 5.3: Universal Parking Stalls

The universal parking space design is an alternative to the provision of a percentage of spaces with a 96-inch side aisle, and the associated need to include additional signage. Under this design, *all* accessible spaces are 132 inches wide with a 60-inch access aisle. One advantage to this design is that no additional signage is needed because all spaces can accommodate a van with a side-mounted lift or ramp. Also, there is no competition between cars and vans for spaces, since all spaces can accommodate either. Furthermore, the wider space permits vehicles to park to one side or the other to allow persons to exit and enter the vehicle on either the driver or passenger side. This can eliminate some of the liability from having the driver back into the space. One space in eight is required to have a "Van Accessible" marking on the pole below the sign on the same pole.

Standard parking stall acceptable to ADA requirements

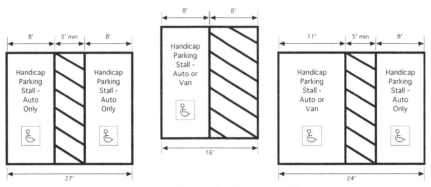

Figure 5.4: Standard ADA Stalls

Except as provided in the next paragraph, access aisles adjacent to accessible spaces shall be 60 inches wide minimum.

One in every eight accessible spaces, but not less than one, shall be served by an access aisle 96 inches wide minimum and shall be designated "Van Accessible" as required by the signage section of the ADA code and the design for van accessibility. The big disadvantage to this configuration is the liability because of tripping and falling over ramps, foliage, curbs, etc. that may be adjacent to the parking stall. When access aisles are shared with other handicap stalls, the vans may be required to back into the access space. This increases chances for accidents or injuries.

An essential consideration for any design is having the access aisle level with the parking space. Since a person with a disability, using a lift or ramp, must maneuver within the access aisle, the aisle cannot include a ramp or sloped area unless it complies with example C (ramp designs for existing parking). New construction must have an accessible ramp within the curb and walk, known as a "curb cut." Curb cut ramps can only be constructed within the design standards and must be accompanied with a minimum 5-feet by 5-feet landing area at the top. In the case of a crosswalk, the ramp cannot protrude into the traffic lane. The curb cut is the only option in this case. Also, the required dimensions of the access aisle cannot be restricted by planters, curbs, or wheel stops.

All curb ramps shall have a detectable warning complying with the ADA standard to warn of the existence of a ramp (the recommended standard is truncated domes). The visually impaired can detect the ramp with their cane, but the other disabled need to detect (visually and physically) a change in surface texture (an exposed aggregate or scored surface is an equivalent that may be safer than the truncated domes). As of this publication, equivalent specifications have yet to be defined that will accommodate all pedestrians.

Crosswalks shall be painted leading from the furthest most access aisle across any traffic lanes and into the curb cut or ramp of the sidewalk. All stalls on the opposite side of a drive lane from the ramp shall have a crosswalk leading to the ramp from the parking stalls.

All spaces, access aisles, and crosswalks shall be painted in a color to contrast with the existing striping (i.e., if the stripes are white, the contrast will be yellow, and vice versa). The universal symbol can have a background, but this is at the Owner's or Manager's discretion for aesthetic purposes. A background is not required by ADA.

Each parking space shall have the required and appropriate signage. All accessible parking spaces shall be identified by a sign on a stationary post or object. These signs shall not be obscured by a vehicle parked in the space. The bottom of the sign shall be located not less than 3 feet nor more than 6 feet above the grade and shall be visible directly in front of the parking space. If erected on a pole, the pole shall be installed in the ground in a 4-inch hole, 2 feet in the ground and permanently concreted. **Remember, before you drill or have the hole drilled, contact Blue Stakes for the location of underground utilities that may be in the direct location of the sign.** If any utility is damaged by the drilling, and Blue Stakes was not notified prior to drilling, the fines and repair costs are staggering. Accessible parking spaces shall be designated as reserved for the physically disabled by a sign showing the international wheelchair symbol and must meet local signage and enforcement requirements. The sign must have "Reserved Parking" and the international wheelchair symbol, "Van Parking," if appropriate, and the name of the city and city code number. Figure 5.5 illustrates that installation. Each parking stall shall have a corresponding sign. Each stall shall have a bumper block installed 3 feet from the sign pole to prevent damage from traffic. If required, each sign pole shall have a bumper block installed 3 feet from the pole on the opposite side for protection from exposure to traffic. An alternative is installing the sign on a structurally strong, steel-encased pole.

SIGN DESIGN FOR MARKING PARKING SPACE SPECIFICATIONS

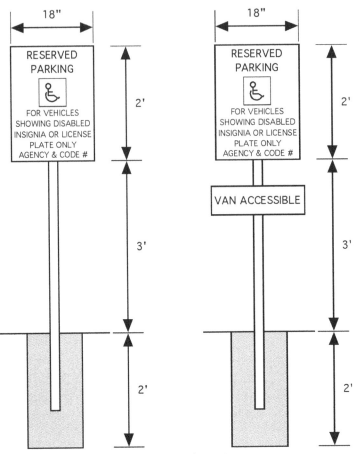

Figure 5.5: Handicap Signs

SIGN DESIGN FOR MARKING PARKING SPACE
WITH METAL SLEEVE AND CONCRETE FILLED SPECIFICATIONS

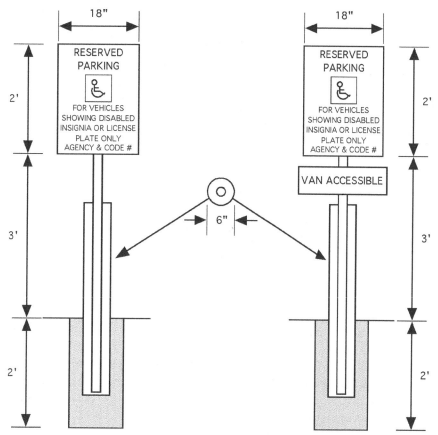

Figure 5.6: Handicap Signs with Metal Sleeves

All signs shall have a bumper block placed in front to protect the pole and sign from damage by parking vehicles. The only exception is when curb and gutter are substituted for the bumper block. Where an adjacent parking stall is located behind the sign and pole, two blocks will be placed at a distance of 42 inches.

Bumper Block Placement Configuration

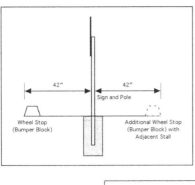
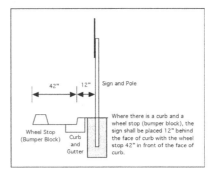
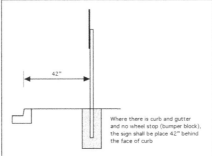

Figure 5.7 Handicap Sign Placement

Passenger loading zones shall provide an access aisle at least 60 inches wide and 20 feet (240 inches) long adjacent and parallel to the vehicle pull-up. If there are curbs between the access aisle and the vehicle pull-up space, then the appropriate curb ramp shall be provided. Vehicle standing spaces and access aisles shall be level with the surface slopes not exceeding 1:50 (2 percent in all directions.

RAMPS SPECIFICATIONS

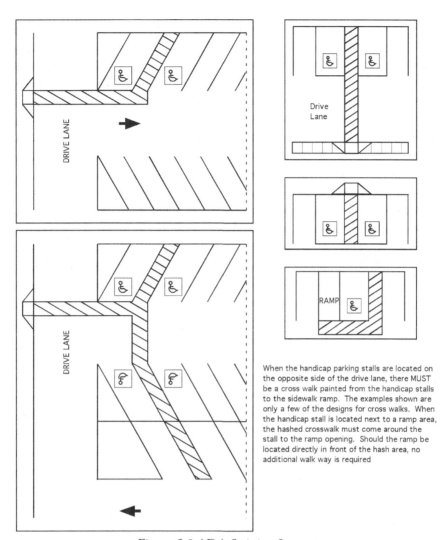

When the handicap parking stalls are located on the opposite side of the drive lane, there MUST be a cross walk painted from the handicap stalls to the sidewalk ramp. The examples shown are only a few of the designs for cross walks. When the handicap stall is located next to a ramp area, the hashed crosswalk must come around the stall to the ramp opening. Should the ramp be located directly in front of the hash area, no additional walk way is required

Figure 5.8 ADA Striping Layout

ADA Ramp Standards

When constructing a new ramp or replacing an existing ramp for handicap access, the following code requirements are to apply. The Americans with Disabilities Act (ADA) states that any accessible route with a slope greater than 1:20 shall be considered a ramp and shall comply with section 4.8 of "Ramps." The ratio 1:20 refers to 20 inches in depth or length for 1 inch in rise or height. When a new ramp is installed, whether in the sidewalk or inset or extruding from the curb, the code requires that the length or depth is related to the height of the curb. The ratio 1:12 means there shall be 12 inches of depth or length of an incline to 1 inch of height in the curb. If a curb is 4 inches in height, the depth or incline shall be 48 inches (4 × 12 = 48). If a curb is 5 inches in height, the depth or incline shall be 60 inches (5 × 12 = 60). If the ratio is 1:10, the depth or length shall be 10 inches for each 1 inch of curb height. When reading the code, keep this formula for depth or incline for ramps and flared sides in mind.

Refer to the latest ADA and ADAAG code for ramp details. However, the local government agencies (town, township, city, county, or state) can supersede the ADA code. Make sure you contact local agencies to find out what their requirements are. Usually, they will follow the guidelines established by ADA; however, they have the right to change ADA requirements as long as accessibility is attained. The following is the code for constructing ramps in parking lots for ADA compliance.

The slope and rise of the ramp shall be the least possible slope used for any ramp. The maximum slope of a ramp in new construction shall be 1:12. The maximum rise for any run shall be 30 inches. When constructing a ramp, the maximum rise for a 1:12 to 1:16 ramp shall be 30 inches and the maximum horizontal projection (length) shall be 30 feet. This means where the rise will exceed 30 inches or the ramp will exceed 30 feet in projection (length), a second ramp must be constructed, with a 5-foot × 5-foot landing between ramps. When constructing a ramp, the maximum rise for a 1:16 to 1:20 ramp shall be 30 inches and the maximum horizontal projection (length) shall be 40 feet. This means where the rise will exceed 30 inches or the ramp will exceed 40 feet in projection (length), a second ramp must be constructed with a 5-foot × 5-foot landing between ramps.

Where a ramp is constructed into an existing sidewalk and an obstruction (wall, garden, building, etc.) will not allow a landing at the top of the ramp that will fulfill the landing requirements, the ramp must have either a landing level with the parking lot (with the dimensions of 5 feet by 5 feet and the wings shall be 1:20), or a landing with the dimensions of 5 feet by 5 feet with sidewalk approach on each side with the slope of 1:12.

Where the ramp is bound by gardens, yard, or decorative areas on both sides of the ramp, the ramp shall be constructed from the paved area to the walkway on a slope of 1:12 as determined by the height of the curb. There shall be an opening of a minimum of 36 inches (3 feet) with return curbs on each side of the ramp.

Where the ramp is built from the curb into the parking lot, a ramp shall be constructed with the dimensions of a minimum of 36 inches wide on the ramp with a slope of 1:12. The wings shall be constructed with a slope of 1:10. The ramp shall *not* be constructed into a drive or drive lane with through traffic that can endanger the person using the ramp. The ramp shall be constructed where there is no exposure to passing vehicle traffic.

AGGREGATE BASE COURSE SPECIFICATIONS

A. Materials

A-1　Aggregate base course shall be ¾- to 1½-inch maximum size aggregate material according to local government construction specifications (state highway, county, or municipality) for the type and classification of the ABC used.

A-2　The Owner, Manager, or Consultant may take samples of materials to an independent laboratory for testing of proper gradation or design, or require on-site inspection for proper application and compaction of materials.

B. Preparation

B-1　The subgrade material shall be properly prepared with grading and compacting, and shall meet all compaction requirements for the type and classification of the existing soils prior to the base material being installed.

B-2　The base material shall be properly prepared, windrowed, and blended on-site before grading and compacting.

B-3　Upon completion of properly preparing the base material, it shall be processed and then spread by a motor blade or grader over the entire area to be covered to the specified design thickness before compaction. The base material shall be compacted with pneumatic, steel wheel static, or vibratory rollers, or any combination until the compacted material meets 95 percent compaction as determined by modified Proctor ASTM D-1557, and to the designed thickness for final grade and elevation. The maximum lift thickness for compaction shall be 6 inches at any one time. Where the designed finished thickness is greater than 6 inches, the material shall be installed in equal lifts and compacted as long as the lift being compacted does not exceed 6 inches. Each lift or layer shall be compacted to 95 percent of modified Proctor density as determined by ASTM D-1557 before another lift is placed.

C. Application

C-1　The material shall be installed at the appropriate depth and compacted to yield a designed depth (based on soil classification, Atterberg limits, and subgrade swell characteristic) in inches after compaction.

C-2　The base material shall be properly prepared by watering and compacting to 95 percent of modified Proctor density ASTM D-1557.

C-3　The material shall be rolled by pneumatic, static, or vibratory roller to accomplish required density, and water shall be added as needed to attain compaction and the required optimum moisture content within parameters established.

C-4　A second or finish rolling can be accomplished with a steel wheel static or vibratory roller.

C-5　The combination of particle sizes in the base material shall be according to local available materials as established by local governing agency specifications.

TRENCH AND SERVICE CUT EXCAVATION SPECIFICATIONS

A. Materials

A-1   Clean backfill material approved by the Owner, Manager, or Consultant to replace existing, native soils unsuitable for backfill material.

A-2   Asphaltic material for patching with ½-inch maximum size aggregate and a minimum of 5.5 percent asphalt by weight. Material shall meet all local government construction specifications for ½-inch hot-mixed asphalt mix.

A-3   The Owner, Manager, or Consultant may take samples of materials to an independent laboratory for testing of proper mix design or require on-site inspection for proper backfilling and compaction, subgrade preparation for patch installation, and application of patch materials. The gradation and asphalt contents of the patch material are based on project mix designs.

B. Preparation/APPLICATION

B-1   Saw cut or jackhammer the area to be excavated and remove the existing pavement. All materials removed shall be properly disposed of in an authorized disposal site. Be sure that all cuts are at right (90°) angles and as square as possible.

B-2   The subgrade and ABC shall be separately removed and stockpiled for use as backfill material. Should the material be saturated; contaminated; unsuitable from exposure to water, chemicals, or fuels; or unstable due to the soil composition (high swelling clays), the excavated material shall be removed, hauled away, and disposed of according to local regulations and laws. Where the trench excavation is over 3 feet in depth, shoring will be required before any workers can enter the trench.

B-3   The repairs to the utilities shall be completed, and the excavated areas shall be backfilled with the stockpiled or imported materials. The material shall be installed in lifts of not more than 8 to 12 inches and compacted to 95 percent Proctor density at optimum moisture content. This method shall hold for each lift until the excavated area is within 4 inches of the finished paved grade. The soils and imported materials shall be compacted to ASTM D-698 (standard Proctor) for clay soils and ASTM D-1557 (modified Proctor) for aggregate bases, sandy soils, and sands. The top surface of the backfilled material shall have a finished grade with consistent depth throughout the entire patched area.

B-4   The surface shall be replaced with a new patch consisting of hot mixed asphaltic concrete with a maximum size aggregate of ½ inch (MAG D 1/2). The patch material shall be installed ¼ inch above the existing pavement and compacted to 95 percent Marshall density. This method is required where the existing surface is asphalt or slurry seal. Should the existing surface be a chip seal surface, the patch shall be installed and compacted where the patch is slightly lower than the existing surface. The pavement shall then be covered with hot QS-1h asphalt emulsion and clean ⅜ inch stone chips installed over the entire surface, and rolled or compacted. The excess chips shall be swept away after the tack emulsion has cooled. The surface will then be sprayed with an SS-1h tack coat.

DUMPSTER PAD SPECIFICATIONS

A. Materials

A-1   5,000 pounds per square inch with added fibers when stated in the Scope of Work.

A-2   Reinforcing wire mesh (or type R rebar as required).

A-3   The Owner, Manager, or Consultant may take samples of materials to an independent laboratory for testing or require on-site inspection for application.

B. Preparation

B-1   Prior to installation, the areas designated for dumpster pad installation shall be saw cut and the asphalt removed approximately 2 feet beyond the edge of the designed dumpster pad on all asphalt-exposed sides. The existing base where the dumpster pad is to be installed shall be removed to a total depth of 8 inches, and all loose debris and materials shall be hauled off and disposed of.

B-2   The base shall be moistened and compacted before installation of forms.

B-3   The forms shall be installed to create straight edges and right angles.

B-4   Should the temperature be greater than 95°F, the base material shall be sprayed with water to eliminate rapid hydration.

C. Application

C-1   The existing pavement shall be saw cut and removed a minimum of 2 feet on either side of the dumpster pad. This removal shall be accomplished to the designed or recommended area to accommodate the existing pavement elevations.

C-2   The base material and subgrade material shall be removed to the width of the new dumpster pad and forms to the specified depth.

C-3   The subgrade material shall be compacted to 95 percent density as determined by the soil classification and type. Clay soils shall be installed by standard Proctor density ASTM D-698, and sandy or gravel soils shall be installed by modified Proctor density ASTM D-1557.

C-4   The concrete shall be poured in place, floated, and finished to accommodate proper and undisturbed drainage.

C-5   Tool joints shall be installed every 10 feet to reduce contraction cracking. Expansion joints shall be installed where the new concrete will be adjacent to any existing hard surfaces.

C-6   An approved curing compound shall be sprayed to manufacturer's recommendations over the entire surface area to retard the hydration process.

C-7   The base material surrounding the new valley gutter shall be moistened and compacted to 95 percent Proctor density ASTM D-698 or ASTM D-1557 prior to installing asphalt patch material.

C-8   The existing edges of the pavement and the edges of the new valley gutter shall be coated with tack coat prior to installing hot mixed asphalt.

C-9   Install hot mixed asphalt ½-inch to ¾-inch maximum size aggregate and compact to 95 percent Marshall stability ASTM D-1559 (refer to "Remove and Replace Patching Specifications").

VALLEY GUTTER REPAIRS SPECIFICATIONS

A. Materials

A-1   5,000 pounds per square inch with added fibers when stated in the Scope of Work.

A-2   The Owner, Manager, or Consultant may take samples of materials to an independent laboratory for testing or require on-site inspection for application.

B. Preparation

B-1   Prior to installation, the areas designated for valley gutter installation shall be saw cut and the asphalt removed approximately 5 feet beyond the edge of the designed valley gutter on all sides. The existing designated concrete valley gutter shall be removed and all debris hauled from the work site and disposed of in accordance to local government agency requirements.

B-2   The base shall be moistened and compacted before installation of forms.

B-3   The forms shall be installed to create straight edges and right angles to match the existing valley gutter on both ends of the new section.

B-4   Should the temperature be greater than 95°F, the base material shall be sprayed with water to eliminate rapid hydration.

C. Application

C-1   The existing pavement shall be saw cut and removed a minimum of 2 feet on either side of the designed valley gutter. This removal shall be accomplished to the designed or recommended area to accommodate the existing pavement elevations.

C-2   The existing damaged valley gutter section shall be removed.

C-3   The subgrade material shall be compacted to 95 percent density as determined by the soil classification and type. Clay soils shall be installed by standard Proctor density ASTM D-698, and sandy or gravel soils shall be installed by modified Proctor density ASTM D-1557.

C-4   The concrete shall be poured in place, floated, and finished to yield a clear flow line in the center (that matches the flow of any existing concrete valley gutter pieces on either or both ends of the new piece) to accommodate proper and undisturbed drainage.

C-5   Tool joints shall be installed every 12 feet to eliminate contraction cracking. Expansion joints shall be installed every 48 feet. Expansion joints shall be installed where the new concrete will be adjacent to any existing hard surfaces.

C-6   An approved curing compound shall be sprayed to manufacturer's recommendations over the entire surface area to retard the hydration process.

C-7   The base material surrounding the new valley gutter shall be moistened and compacted to 95 percent Proctor density ASTM D-698 or ASTM D-1557 prior to installing asphalt patch material.

C-8   The existing edges of the pavement and the edges of the new valley gutter shall be coated with tack coat prior to installing hot-mixed asphalt.

C-9   Install hot mixed asphalt ½-inch to ¾-inch maximum size aggregate and compact to 95 percent Marshall stability ASTM D-1559 (refer to "Remove and Replace Patching Specifications").

CONCRETE SIDEWALK, SLAB, AND DRIVEWAY APRON SPECIFICATIONS

A. Materials

A-1   2,500 pounds per square inch concrete shall be used in constructing the new concrete slab, sidewalk, and driveway.

A-2   Expansion joint materials.

A-3   The Owner, Manager, or Consultant may take samples of materials to an independent laboratory for testing or require on-site inspection for application.

B. Preparation

B-1   Prior to installation, the areas designated for removal shall be saw cut as required or broken evenly at the constructed tool joint and removed in such a manner as to not damage or break any adjacent concrete walks or slabs. The existing base where the walk or slab piece or pieces are to be installed shall be graded and all roots and debris removed. Any wet or soft spots shall be removed or treated to provide a solid base for the new concrete structure. All loose debris and materials shall be hauled off and disposed of.

B-2   The base shall be moistened and compacted before installation of forms.

B-3   The forms shall be installed to create straight edges and right angles and anchored or supported to prevent any bowing of the form from the weight of concrete mix.

B-4   Where the new concrete walk, slab, or driveway structure will come in contact with any poles, walls, other concrete slabs, or any other concrete structure, an acceptable expansion joint shall be installed using an acceptable expansion joint material.

B-5   Should the temperature be greater than 95°F, the base material shall be sprayed with water to eliminate rapid hydration and shrinkage of the wet concrete.

C. Application

C-1   The designated existing concrete sidewalk or slab shall be saw cut or broken at the adjacent tool joint and removed in such a manner as to not damage adjacent concrete walks or slab pieces. This removal shall be accomplished to the designed or recommended area.

C-2   The base material and subgrade material shall be graded and reshaped, and any wet or soft spots removed and filled with acceptable aggregate base material, and made ready for compaction. Where there is no adjacent concrete structure, a form will be required.

C-3   The subgrade material shall be compacted to 90 percent density as determined by the soil classification and type. Clay soils shall be installed by standard Proctor density ASTM D-698, and sandy or gravel soils shall be installed by modified Proctor density ASTM D-1557.

C-4   The concrete shall be poured in place, floated, and finished to accommodate proper and undisturbed drainage. The surface of the new concrete shall be treated, floated, or printed to match the existing concrete surfaces, or as outlined in the Scope of Work.

C-5 Tool joints shall be installed at every dimension to match the existing tool joints or as directed in the Scope of Work and plans to reduce contraction cracking. Expansion joints shall be installed where the new concrete will be adjacent to any existing hard surfaces.

C-6 An approved curing compound shall be sprayed to manufacturer's recommendations over the entire surface area to retard the hydration process.

C-7 The base material surrounding the new walk or slab shall be moistened and compacted to 95 percent Proctor density ASTM D-698 or ASTM D-1557 prior to installing any adjacent asphalt patch material.

C-8 The existing edges of the pavement and the edges of the new walk or slab shall be coated with tack coat prior to installing hot-mixed asphalt.

C-9 Install hot mixed asphalt ½-inch to ¾-inch maximum size aggregate and compact to 95 percent Marshall stability ASTM D-1559 (refer to "Remove and Replace Patching Specifications").

EXTRUDED CONCRETE CURB SPECIFICATIONS

A. Materials

A-1   3,500 pounds per square inch concrete or as stated in local government agency construction standards and specifications

A-2   Expansion joint materials.

A-3   The Owner, Manager, or Consultant may take samples of materials to an independent laboratory for testing or require on-site inspection for application.

B. Preparation

B-1   Prior to installation, the areas designated for removal shall be saw cut as required or broken evenly at the constructed tool joint and removed in such a manner as to not damage or break any adjacent concrete curb sections. The existing asphalt pavement base where the curb sections are to be installed shall be patched should any damage occur during curb removal. All loose debris and materials shall be hauled off and disposed of.

B-2   Forms shall be installed to create straight edges and right angles to match the existing curbing.

C. Application

C-1   The designated existing concrete curbing shall be saw cut or evenly broken at the adjacent tool joint and removed in such a manner to not damage adjacent curbing, concrete walks, or slab pieces.

C-2   The existing pavement where the extruded curb section shall be set shall be examined for damage from removal. Any damaged areas will be patched to smooth surface to match the elevation of the existing pavement. Should the asphalt pavement be too damaged to patch, the paved section shall be saw cut and removed and a new R&R patch installed to match the existing pavement.

C-3   The removed curb section shall be formed to match the existing concrete curbing; the new concrete shall be installed, formed, and finished to match the existing extruded curbing.

C-4   Tool joints shall be installed at every dimension to match the existing tool joints or as directed in the Scope of Work to reduce contraction cracking. Expansion joints shall be installed where the new concrete will be adjacent to any existing hard surfaces, as directed by the Owner or directed in the Scope of Work.

C-5   An approved curing compound shall be sprayed to manufacturer's recommendations over the entire surface area to retard the hydration process. Where the concrete is colored, the appropriate dye shall be used to match the existing color of adjacent curbing.

IN-PLACE CONCRETE CURB AND CURB AND GUTTER COMBINATION SPECIFICATIONS

A. Materials

A-1   3,500 pounds per square inch concrete or as stated in local government agency construction standards and specifications.

A-2   Expansion joint materials.

A-3   The Owner, Manager, or Consultant may take samples of materials to an independent laboratory for testing or require on-site inspection for application.

B. Preparation

B-1   Prior to installation, the areas designated for removal shall be saw cut as required or broken evenly at the constructed tool joint and removed in such a manner as to not damage or break any adjacent concrete curb sections. The adjacent asphalt pavement shall be saw cut and removed not less than 1 foot from the outer edge of the gutter. The existing base where the curb sections are to be installed shall be graded and all roots and debris removed. Any wet or soft spots shall be removed or treated to provide a solid base for the new concrete structure. All loose debris and materials shall be hauled off and disposed of.

B-2   The base shall be moistened and compacted before installation of forms.

B-3   The forms shall be installed to create straight edges and right angles.

B-4   Where the new concrete curb will come in contact with any poles, walls, other concrete slabs, or any other concrete structure, or where the curb section is more than 20 feet in length, an acceptable expansion joint shall be installed using an acceptable expansion joint material.

B-5   Should the temperature be greater than 95°F, the base material shall be sprayed with water to eliminate rapid hydration and shrinkage.

C. Application

C-1   The designated existing concrete curbing shall be saw cut or broken at the adjacent tool joint and removed in such a manner as to not damage adjacent curbing, concrete walks, or slab pieces. Also, the adjacent 1 foot of asphalt pavement shall be saw cut and removed.

C-2   The base material and subgrade material shall be graded and reshaped, and any wet or soft spots removed and filled with acceptable aggregate base material, and made ready for compaction. Where there is no adjacent concrete structure, a form will be required. A saw cut edge of existing asphalt pavement shall not be used as a form.

C-3   The subgrade material shall be compacted to 90 percent density as determined by the soil classification and type. Clay soils shall be installed by standard Proctor density ASTM D-698, and sandy or granular soils and the exposed aggregate base in the pavement shall be installed by modified Proctor density ASTM D-1557.

C-4   The concrete shall be poured in place, floated, and finished to accommodate proper and undisturbed drainage. The surface of the new concrete shall be treated, floated, washed, stained, or printed to match the existing concrete surfaces or as outlined in the Scope of Work.

C-5   Tool joints shall be installed at every dimension to match the existing tool joints or as directed in the Scope of Work to reduce contraction cracking. Expansion joints shall be installed where the new concrete will be adjacent to any existing hard surfaces, as directed by the Owner or directed in the Scope of Work.

C-6   An approved curing compound shall be sprayed to manufacturer's recommendations over the entire surface area to retard the hydration process. Where the concrete is colored, the appropriate dye shall be used to match the existing color of adjacent curbing.

C-7   The base material in the adjacent asphalt patch and next to the curbing shall be moistened and compacted to 95 percent Proctor density ASTM D-698 or ASTM D-1557 prior to installing any adjacent asphalt patch material. This will be accomplished after the form for the concrete structure has been removed.

C-8   The existing edges of the pavement and the edges of the curbing shall be coated with tack coat prior to installing hot-mixed asphalt.

C-9   Install hot mixed asphalt ½-inch to ¾-inch maximum size aggregate and compact to 95 percent Marshall stability ASTM D-1559 (refer to "Remove and Replace Patching Specifications").

REMOVAL OF EXISTING CONCRETE PAD AND REPLACE WITH ASPHALT PATCH
SPECIFICATIONS

A. Materials

A-1   Diamond blade saw

A-2   Jackhammer or power-driven hydraulic hammer

A-3   Loader and dump trucks

A-4   ½-inch or ¼-inch maximum size aggregate hot-mixed asphaltic

B. Preparation

B-1   Prior to removal of the areas designated, all adjacent curbing, sidewalks, structures, etc. shall be saw cut. The concrete shall be broken into pieces for safe moving with a loader and removed approximately to the existing subgrade or aggregate base course. The broken concrete shall be loaded on trucks, hauled off the project site, and disposed of according to local agency requirements.

B-2   The base shall be moistened and compacted to 95 percent standard Proctor density for clay materials, and 95 percent modified Proctor density for granular materials, before installation of aggregate base or full depth asphalt pavement.

B-3   Should aggregate base be installed, it shall be installed and compacted to 95 percent modified Proctor density and to a grade 2 inches below finished grade.

B-4   Asphaltic concrete shall be installed to an acceptable depth to yield a compacted depth of 2 inches according to the "Remove and Replace (R&R) Patching Specifications."

B-5   Should full-depth asphalt be used, it shall be installed in lifts of not more than 3 inches thick and according to the "Remove and Replace (R&R) Patching Specifications."

C. Application

C-1   The existing concrete shall be saw cut and removed. This removal shall be accomplished to the designed or recommended area to accommodate the existing pavement elevations.

C-2   The subgrade and/or aggregate base materials shall be compacted to 95 percent density as determined by the soil classification and type. Clay soils shall be installed by standard Proctor density ASTM D-698, and sandy or gravel soils shall be installed by modified Proctor density ASTM D-1557.

C-3   Install hot mixed asphalt ½-inch to ¾-inch maximum size aggregate (Superpave designation R19 or R12.5) and compact to 95 percent Marshall stability ASTM D-1559 (refer to "Remove and Replace Patching Specifications"). The pavement shall be installed and finished to accommodate the existing drainage and offer optimal drainage.

THREE-FOOT-WIDE SPEED BUMP SPECIFICATIONS

A. Materials

A-1   Asphaltic material shall be ½-inch maximum size aggregate and 5.5-6 percent asphalt by total weight. The material shall meet all local government construction specifications for ½-inch maximum size aggregate. The gradation and asphalt contents are based on project mix designs and local materials. When the speed bump is to be placed in an existing waterway (inverted street crown, asphalt valley gutter, etc.), an opening shall be constructed to allow for water flow at the low point of the inverted crown.

A-2   The Owner, Manager, or Consultant may take samples of materials to an independent laboratory for testing of proper mix design or require on-site inspection for proper application of materials.

B. Preparation

B-1   Prior to application, ALL areas will be power swept and cleared of all loose material. Oil spots will be manually scraped and cleaned with mild detergent if necessary to remove any petroleum residue.

B-2   A chalk line shall be strung on the existing pavement, outlining the outside edges of the speed bump at the correct width.

C. Application

C-1   Treat the entire area to be tack coated with AC-10 or an MC primer tack.

C-2   The edges of the speed bump shall be marked with a chalk line to ensure square even edges. The Contractor shall be responsible for installing all speed bumps with straight even edges. Any uneven, raveled, ragged edges will be corrected at the Contractor's expense.

C-3   Apply hot-mixed asphaltic material to a depth of not more than 2 inches on the center of the constructed speed bump. **All asphalt material shall have 1.5 percent Portland type II cement added to aid in strength and lessen tire deformation.** Make sure the edges are square and extend at least 1½ feet beyond the centerline of the finished speed bump and are as square or rectangular as possible. Care shall be exercised in raking the asphalt to provide a good-feathered edge and for blending of the new materials with the existing pavement and concrete curbs, openings, and walks. The pavement shall be installed in a manner to prevent segregation of the aggregate. The material shall be raked to feather the edges smoothly into the existing pavement to prevent an abrupt edge. This may require raking the edges to remove large stones from the mix approximately 6 to 8 inches from the edge and making an even, smooth transition from the speed bump to the existing pavement. The grade shall be ¼ inch for the first ½ foot, ¾ inch for the second ½ foot, and 1 inch for the remaining ½ foot (see Figure 5.9 for details). The ends of the speed bump shall be sloped on a gradual grade of 2:1.

C-4   Roll and compact with a 3- to 5-ton static roller, where the roller can move without interruption or sharp turning. Where the speed bump is in an inversion (especially in inverted crowns or asphalt valley gutters), or where the roller will require sharp turns, a vibratory plate will be used. In the case of an inverted crown or asphalt valley gutter, the speed bump shall be installed to fit the existing grade and not block the flow of water or cause damming of water behind it.

C-5   If a speed bump is installed when the ambient temperature is below 50°F and above 40°F for the daily high, the speed bump shall be coated with an asphalt-based emulsion to retard raveling.

Where the ambient and ground temperatures are below 40°F for the daily high, the speed bump shall not be installed.

C-6 The new speed bumps shall be allowed to thoroughly cure, based on local specifications, manufacturer's recommendations, and local weather conditions, before a seal coat is applied (approximately thirty [30] days). In desert climates, it is recommended to allow the full hot summer season for curing when the asphalt material is placed in extremely hot weather conditions.

C-7 The speed bump shall be painted a solid yellow or white, or yellow or white hash where specified. Also where specified, glass beads shall be installed over the fresh paint. **Box painting speed bumps will not be allowed. The crest of the speed bump must have some marking on it to be acceptable.**

C-8 The combination of particle sizes or sieve analysis of the asphalt material shall be according to local specifications and available materials.

THREE-FOOT SPEED BUMP

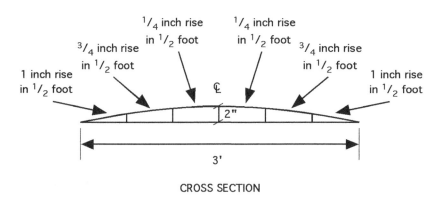

CROSS SECTION

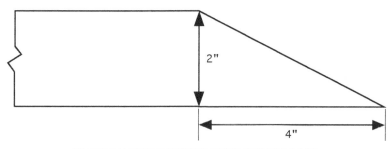

EDGE TRANSITION AT HIGH POINT (CENTER LINE)

Figure 5.9: Three-Foot-Wide Speed Bump

FOUR-FOOT-WIDE SPEED BUMP SPECIFICATIONS

A. Materials

A-1   Asphaltic material shall be ½-inch maximum size aggregate and 5.5-6 percent asphalt by total weight. The material shall meet all local government construction specifications for ½-inch maximum size aggregate. The gradation and asphalt contents are based on project mix designs and local materials. Where the speed bump is to be placed in an existing waterway (inverted street crown, asphalt valley gutter, etc.), an opening shall be constructed to allow for water flow at the low point of the inverted crown.

A-2   The Owner, Manager, or Consultant may take samples of materials to an independent laboratory for testing of proper mix design or require on-site inspection for proper application of materials.

B. Preparation

B-1   Prior to application, ALL areas will be power swept and cleared of all loose material. Oil spots will be manually scraped and cleaned with mild detergent if necessary to remove any petroleum residue.

B-2   A chalk line shall be strung on the existing pavement, outlining the outside edges of the speed bump at the correct width.

C. Application

C-1   Treat the entire area to be tack coated with AC-10 or an MC primer tack.

C-2   The edges of the speed bump shall be marked with a chalk line to ensure square even edges. The Contractor shall be responsible for installing all speed bumps with straight even edges. Any uneven, raveled, ragged edges will be corrected at the Contractor's expense.

C-3   Apply hot-mixed asphaltic material to a depth of not more than 2 inches on the center of the constructed speed bump. **All asphalt material shall have 1.5 percent Portland type II cement added to aid in strength and lessen tire deformation.** Make sure the edges are square and extend at least 2 feet beyond the centerline of the finished speed bump and are as square or rectangular as possible. Care shall be exercised in raking the asphalt to provide a good-feathered edge and for blending of the new materials with the existing pavement and concrete curbs, openings, and walks. The pavement shall be installed in a manner to prevent segregation of the aggregate. The material shall be raked to feather the edges smoothly into the existing pavement to prevent an abrupt edge. This may require raking the edges to remove large stones from the mix approximately 6 to 8 inches from the edge and making an even, smooth transition from the speed bump to the existing pavement. The grade shall be ¼ inch for the first ½ foot, ¾ inch for the second ½ foot, and 1 inch for the remaining ½ foot (see Figure 5.10 for details). The ends of the speed bump shall be sloped on a gradual grade of 2:1.

C-4   Roll and compact with a 3- to 5-ton static roller, where the roller can move without interruption or sharp turning. Where the speed bump is in an inversion (especially in inverted crowns or asphalt valley gutters), or where the roller will require sharp turns, a vibratory plate should be used. In the case of an inverted crown or asphalt valley gutter, the speed bump shall be installed to fit the existing grade and not block the flow of water or cause damming of water behind it.

C-5   If a speed bump is installed when the ambient temperature is below 50°F and above 40°F for the daily high, the speed bump shall be coated with asphalt-based emulsion to retard raveling. Where the ambient and ground temperatures are below 40°F for the daily high, the speed bump shall not be installed.

C-6   The new speed bumps shall be allowed to thoroughly cure, based on local specifications, manufacturer's recommendations, and local weather conditions, before a seal coat is applied (approximately thirty [30] days). In desert climates, it is recommended to allow the full hot summer season for curing when the asphalt material is placed in extremely hot weather conditions.

C-7   The speed bump shall be painted a solid yellow or white, or yellow or white hash where specified. Also where specified, glass beads shall be installed over the fresh paint. **Box painting speed bumps will not be allowed. The crest of the speed bump must have some marking on it to be acceptable.**

C-8   The combination of particle sizes or sieve analysis of the asphalt material shall be according to local specifications and available materials.

FOUR-FOOT SPEED BUMP

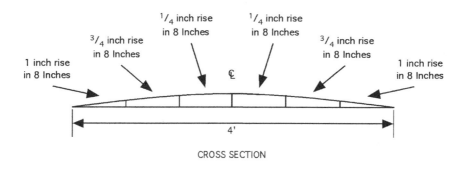

CROSS SECTION

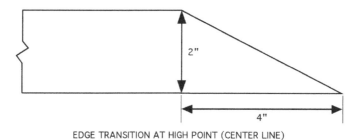

EDGE TRANSITION AT HIGH POINT (CENTER LINE)

Figure 5.10: Four-Foot-Wide Speed Bump

TWELVE-FOOT-WIDE SPEED HUMP SPECIFICATIONS

A. Materials

A-1 Asphaltic material shall be ½-inch maximum size aggregate and 5.5-6 percent asphalt by total weight. The material shall meet all local government construction specifications for ½-inch maximum size aggregate. The gradation and asphalt contents are based on project mix designs and local materials. Where the speed hump is to be placed in an existing waterway (inverted street crown, asphalt valley gutter, etc.), an opening shall be constructed to allow for water flow at the low point of the inverted crown.

A-2 The Owner, Manager, or Consultant may take samples of materials to an independent laboratory for testing of proper mix design or require on-site inspection for proper application of materials.

B. Preparation

B-1 Prior to application, ALL areas will be power swept and cleared of all loose material. Oil spots will be manually scraped and cleaned with mild detergent if necessary to remove any petroleum residue.

B-2 A chalk line shall be strung on the existing pavement, outlining the outside edges of the speed hump at the correct width.

C. Application

C-1 Treat the entire area to be tack coated with AC-10 or an MC primer tack.

C-2 The edges of the speed hump shall be marked with a chalk line to ensure square even edges. The Contractor shall be responsible for installing all speed humps with straight even edges. Any uneven, raveled, ragged edges will be corrected at the Contractor's expense.

C-3 Apply hot-mixed asphaltic material to a depth of not more than 4 inches on the center of the constructed speed hump. **All asphalt material shall have 1.5 percent Portland type II cement added to aid in strength and lessen tire deformation.** Make sure the edges are square and extend at least 6 feet beyond the centerline of the finished speed hump and are as square or rectangular as possible. Care shall be exercised in raking the asphalt to provide a good-feathered edge and for blending of the new materials with the existing pavement and concrete curbs, openings, and walks. The pavement shall be installed in a manner to prevent segregation of the aggregate. The material shall be raked to feather the edges smoothly into the existing pavement to prevent an abrupt edge. This may require raking the edges to remove large stones from the mix approximately 6 to 8 inches from the edge and making an even, smooth transition from the speed hump to the existing pavement. The grade shall be ½ inch per every ½ foot in width (see Figure 5.11 for details). The ends of the speed hump shall be sloped on a gradual grade of 2:1.

C-4 Roll and compact with a 3- to 5-ton static roller, where the roller can move without interruption or sharp turning. Where the speed hump is in an inversion (especially in inverted crowns or asphalt valley gutters), or where the roller will require sharp turns, a vibratory plate should be used. In the case of an inverted crown or asphalt valley gutter, the speed hump shall be installed to fit the existing grade and not block the flow of water or cause damming of water behind it.

C-5 If a speed hump is installed when the ambient temperature is below 50°F and above 40°F for the daily high, the speed hump shall be coated with an asphalt-based emulsion to retard raveling.

Where the ambient and ground temperatures are below 40°F for the daily high, the speed hump shall not be installed.

C-6 The new speed humps shall be allowed to thoroughly cure, based on local specifications, manufacturer's recommendations, and local weather conditions, before a seal coat is applied (approximately thirty [30] days). In desert climates, it is recommended to allow the full hot summer season for curing when the asphalt material is placed in extremely hot weather conditions.

C-7 The speed hump shall be painted a solid yellow or white, or yellow or white hash where specified, as shown in Figure 5.11. Also where specified, glass beads shall be installed over the fresh paint. **Box painting speed humps will not be allowed. The crest of the speed hump must have some marking on it to be acceptable.**

C-8 The combination of particle sizes or sieve analysis of the asphalt material shall be according to local specifications and available materials.

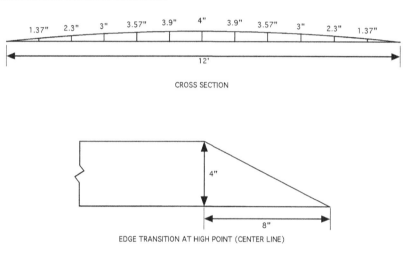

Figure 5.11 Twelve-Foot-Wide Speed Hump

SPEED TABLE SPECIFICATIONS

A. Materials

A-1. Asphaltic material shall be ½-inch maximum size aggregate and 5.5-6 percent asphalt by total weight. The material shall meet all local government construction specifications for ½-inch maximum size aggregate. The gradation and asphalt contents are based on project mix designs and local materials. Where the speed table is to be placed in an existing waterway (inverted street crown, asphalt valley gutter, etc.), an opening shall be constructed to allow for water flow at the low point of the inverted crown.

A-2. The Owner, Manager, or Consultant may take samples of materials to an independent laboratory for testing of proper mix design or require on-site inspection for proper application of materials.

B. Preparation

B-1. Prior to application, ALL areas will be power swept and cleared of all loose material. Oil spots will be manually scraped and cleaned with mild detergent if necessary to remove any petroleum residue.

B-2. A chalk line shall be strung on the existing pavement, outlining the outside edges of the speed table at the correct width.

C. Application

C-1. Treat the entire area to be tack coated with AC-10 or an MC primer tack.

C-2. The edges of the speed table shall be marked with a chalk line to ensure square even edges. The Contractor shall be responsible for installing all speed tables with straight even edges. Any uneven, raveled, ragged edges will be corrected at the Contractor's expense.

C-3. Apply hot-mixed asphaltic material to a depth of not more than 4 inches on the center of the constructed speed table . **All asphalt material shall have 1.5 percent Portland type II cement added to aid in strength and lessen tire deformation.** Make sure the edges are square and extend at least 6 feet beyond the centerline of the finished speed table and are as square or rectangular as possible. Care shall be exercised in raking the asphalt to provide a good-feathered edge and for blending of the new materials with the existing pavement and concrete curbs, openings, and walks. The pavement shall be installed in a manner to prevent segregation of the aggregate. The material shall be raked to feather the edges smoothly into the existing pavement to prevent an abrupt edge. This may require raking the edges to remove large stones from the mix approximately 6 to 8 inches from the edge and making an even, smooth transition from the speed table to the existing pavement (see Figure 5.12).

C-4. Roll and compact with a 3- to 5-ton static roller, where the roller can move without interruption or sharp turning. Where the speed table is in an inversion (especially in inverted crowns or asphalt valley gutters), or where the roller will require sharp turns, a vibratory plate should be used. In the case of an inverted crown or asphalt valley gutter, the speed table shall be installed to fit the existing grade and not block the flow of water or cause damming of water behind it.

C-5. If a speed table is installed when the ambient temperature is below 50°F and above 40°F for the daily high, the speed table shall be coated with an asphalt-based emulsion to retard raveling. Where the ambient and ground temperatures are below 40°F for the daily high, the speed table shall not be installed.

C-6. The new speed tables shall be allowed to thoroughly cure, based on local specifications, manufacturer's recommendations, and local weather conditions, before a seal coat is applied (approximately thirty [30] days). In desert climates, it is recommended to allow the full hot summer season for curing when the asphalt material is placed in extremely hot weather conditions.

C-7. The speed table shall be painted per Figure 5.12.

C-8. The combination of particle sizes or sieve analysis of the asphalt material shall be according to local specifications and available materials.

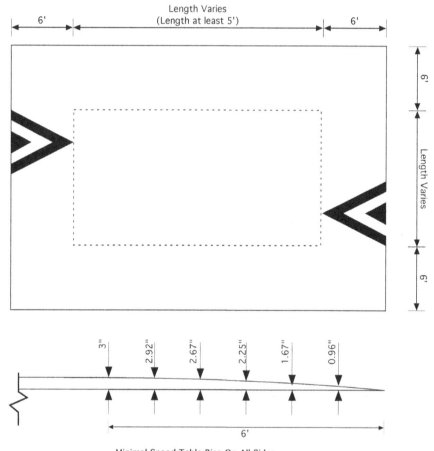

Minimal Speed Table Rise On All Sides

Figure 5.12 Speed Table

SWEEPING SPECIFICATIONS

A. Equipment Required

A-1   Mechanical broom (street) or rotary broom sweeper or vacuum mobile sweeper.

A-2   Air power packs, hand brooms, plastic disposal or garbage bags.

B. Preparation

B-1   Prior to sweeping the paved surfaces, all loose articles, cartons, glass, blown-in brush and large vegetation, and any other debris shall be picked up, placed in containers (plastic garbage bags, etc.), and disposed of properly. All debris such as cigarette butts, wrappers, cans, or any other small debris located in areas where a sweeper cannot reach shall be hand-broom swept or blown with power packs. The debris shall be collected, picked up, and placed in containers and disposed of properly.

B-2   Prior to sweeping the paved surface, behind and around parking bumper blocks, all corners, closed areas, behind and under containers and dumpsters, and any other surfaces not easily accessible for a sweeper (as described in A-1) shall be cleaned by use of power back packs or hand brooms, or a combination of both.

B-3   Any concrete walks, dumpster or storage pads, etc. requiring any steam cleaning shall be cleaned twenty-four (24) hours prior to sweeping to allow for the pavement and any dust or dirt debris to dry for removal.

B-4   Where the broom or street sweeper is used, all surfaces shall be clear of all standing water, and all muddy areas shall be dry. The Contractor shall spread any water puddles to aid in the drying process prior to sweeping.

B-5   The Contractor will accomplish the cleaning and sweeping of the paved surfaces during the off hours of company operations as directed by the property manager, facility manager, or maintenance engineer. The day and hours of cleaning will stay the same each week.

C. Application

C-1   All sidewalks, concrete loading areas, entryways to buildings, and any other surface inaccessible to a mechanical sweeper shall be blown clean with an air pack blower or swept with hand brooms to remove all loose debris, dirt, dust, cigarette butts, or any other trash, glass, or foreign objects. All debris will be swept to a location to be picked up by a mechanical sweeper, or packaged in acceptable containers and disposed of in proper disposal containers. Should these areas require steam or pressure washing, the Contractor shall allow twenty-four (24) hours before sweeping.

C-2   The paved parking area and/or streets will have all paper, cups, bottles, broken glass, brush, weeds, vegetation, and any other pieces of debris picked up and placed in acceptable containers and disposed of properly. All foliage, bushes, and flower gardens shall be inspected, and all disposable debris in, around, or under the plants shall be removed as previously stated.

C-3   All corners; areas behind, around, and under dumpsters and utility equipment; areas hard to get to with a mechanical device; and loading docks and delivery areas, where a rotary broom or air sweeper cannot access, shall be cleaned using hand brooms or air packs. The debris recovered shall be placed in the sweeper path for pickup, or will be picked up and placed in acceptable containers and disposed of properly.

C-4 Proper disposal of gathered material means hauling it to a designated disposal site off the property. The Contractor shall dispose of any and all material in a proper and *environmentally safe manner*. Swept debris shall not be disposed of in dumpsters or other trash receptacles on the property. The Contractor shall NOT dump or put any debris of any kind into dry wells or on landscaped areas at any time. The Contractor will not dispose of any debris in any dumpsters on any adjacent or neighboring properties, or on any other property adjacent to or neighboring to the work site, whether vacant or not. The Contractor shall clean up all areas where debris has been stored during the cleaning operation. The Contractor, under no circumstances, will leave any debris or piles of debris on the property.

C-5 Should the sweeping contractor notice a large deposit of sand and rock in the collected debris, or any potential or existing maintenance problems associated with the property, the sweeping contractor shall notify the Property Manager, Maintenance department, or Owner immediately.

C-6 The Contractor shall operate equipment in a safe manner to protect individuals, private property, and any special markings or barriers placed by the Owner. These barriers include ceramic pavement markers, ballards, delineators, signs, and poles. The Contractor shall notify the Owner, Property Manager, or Facility Manager of any and all damage to private and personal property or personal injury on the property immediately and in writing. The Contractor shall also, in writing, indicate the action taken, or to be taken, to correct any and all damage. This notification shall be within twelve (12) hours of the occurrence.

C-7 The Contractor shall perform the services at the agreed-to and designated hours and days of the week. Should weather prevent the normally scheduled sweeping, special effort will be taken to ensure that the sweeping is completed as soon as possible after the weather conditions change. The Contractor will be expected to pick up on the regular schedule following the substitution schedule.

STRUCTURE BACKFILL SPECIFICATIONS

A. Materials

A-1   Clean backfill material approved by the Owner, Manager, or Consultant to replace existing, native soils unsuitable for backfill material.

A-2   Asphaltic material for patching where required when a structure is built in an existing paved. The asphaltic material shall be a hot mixed asphalt material with ¾-inch maximum size aggregate and a minimum of 5 percent asphalt by weight or with ½-inch maximum size aggregate and a minimum of 5.5 percent asphalt by weight. Material shall meet all local construction specifications for ¾- and ½-inch hot-mixed asphalt mix.

A-3   The Owner, Manager, or Consultant may take samples of materials to an independent laboratory for testing of proper mix design or require on-site inspection for proper backfilling and compaction, subgrade preparation for patch installation, and application of patch materials. The gradation and asphalt contents of the patch material are based on project mix designs.

B. Preparation/Application

B-1   The structure shall be constructed and completed, and the excavated areas shall be backfilled with the stockpiled or imported materials. The material shall be installed in lifts of not more than 8 to 12 inches and compacted to 95 percent Proctor density at optimum moisture content. This method shall hold for each lift until the excavated area is within 6 inches of the finished paved grade. Should the structure be a wall or constructed in a non-asphalt surface, then the backfill will be filled to an elevation level with the surrounding soils. The soils and imported materials used as backfill shall be compacted to ASTM D-698 standards (standard Proctor) for clayey soils and ASTM D-1557 standards (modified Proctor) for aggregate bases, sandy soils, and sands. The top surface of the backfilled material shall have a finished grade with consistent depth throughout the entire area to be patched. Aggregate base material may be used to fill 4 inches of the remaining 6 inches and a 2-inch asphalt patch placed. Or, the entire depth of the surface (6 inches) can be patched back with full depth asphalt pavement placed in 2- to 3-inch-thick lifts and compacted.

B-2   Where the surface requires patching after the backfilling is completed, the surface shall be paved with a new patch consisting of hot mixed asphaltic concrete with a maximum size aggregate of ½ inch or ¾ inch. The patch material shall be installed ¼ inch above the existing pavement and compacted to 95 percent Marshall density. This method is required where the existing surface is asphalt or slurry seal. Should the existing surface be a chip seal surface, the patch shall be installed and compacted where the patch is slightly lower than the existing surface. The pavement shall then be covered with hot QS-h asphalt emulsion and clean ⅜-inch stone chips installed over the entire surface, and rolled or compacted. The excess chips shall be swept away after the tack emulsion has cooled. The chip seal surface will then be sprayed with an SS-1h tack coat.

ROOF DRAIN SPLASH PAD SPECIFICATIONS

A. Materials

A-1   2,500 pounds per square inch concrete as stated in local government specifications.

A-2   The Owner, Manager, or Consultant may take samples of materials to an independent laboratory for testing or require on-site inspection for application.

B. Preparation

B-1   Prior to installation, the areas at the outfall of the roof drain designated for splash pad installation shall be saw cut 3 feet wide and 4 feet in length. The asphalt and aggregate base shall be removed to a total depth of 4 inches and all loose debris and materials shall be hauled off and disposed of according to local government agency requirements.

B-2   The base shall be moistened and compacted before installation of concrete for the pad.

C. Application

C-1   The existing pavement shall be saw cut and removed 3 feet wide and 4 feet in length under the outfall (scupper) portion of roof drains.

C-2   The base material shall be removed to a total depth of 4 inches from the top of the pavement.

C-3   The subgrade material shall be compacted to 95 percent density as determined by the soil classification and type. Clay soils shall be installed by standard Proctor density ASTM D-698, and sandy or gravel soils shall be installed by modified Proctor density ASTM D-1557.

C-4   The concrete shall be poured in place, floated, and finished to accommodate proper and undisturbed drainage away from the building. If needed, the new splash pad will be finished with the rear of the pad (closest to the building) higher to allow for sloping away from the building. The pad shall be positioned to allow for roof drainage to strike the pad and flow away from the building.

C-5   An approved curing compound shall be sprayed to manufacturer's recommendations over the entire surface area to retard the hydration process.

Chapter 6

Developing Requests for Bids or Requests for Proposals

The previous chapters have taught you how to identify what distresses you have, how to measure your pavements and distresses, and how to repair those distresses or completely remove and repave your parking lot or street. In this chapter, we will discuss the request for bid (RFB). The Request for Bid (or request for proposal [RFP]), which includes the scope of work and specifications, is the most important communication you will have with the bidding contractors. This document contains the scope of work, the instructions to the bidders, the bidding documents listing the work tasks with associated quantities, and the specifications instructing the contractors how the tasks will be performed.

The RFB or RFP is divided into ten sections, as follows:

1. Scope of Work. The scope of work tells the bidding contractors what they are to do. It is a brief summary of each task. Scopes of work examples are included in chapter 5.
2. Specification Instructions. This section is brief and tells the contractors that the tasks listed in the scope of work are controlled by the attached specifications.
3. Instructions to Bidders. This section tells the bidding contractors what they need to perform the project. Also, in this section, the bidding contractors are required to list all subcontractors and material suppliers. This list is important because each subcontractor, sub-subcontractor, and material supplier has the right to lien your property for nonpayment by the contractor. Each listed subcontractor, sub-subcontractor, or material supplier will issue a notice to lien letter (a.k.a. 90-day letter). You will need this list so you can get a waiver of lien from the contractor you hired before releasing final payment.
4. Insurance Requirements. This section tells the bidding contractors what insurance is required (with minimum limits) to perform work on your property.
5. General Instructions. This section tells the contractors what submittals need to submitted and what bonding requirements (if applicable), work hours, scheduling requirements, additional bidding requirements, project requirements, work hours, and general rules and regulations the contractors must follow.
6. Submittal Requirements. This section tells the contractor the requirements for submitting the RFB or RFQ.
7. Bid by Task Submittal Sheet. This section or worksheet is set up for the contractors to insert their bids by task, sales tax (if applicable), and other requirements (license numbers, number of days to complete the project).
8. Patch Dimensions. This section is where the dimensions of the patches are listed. The measurements for skin and R&R patch areas from the inventory are listed with appropriate measurements.
9. Specifications. This section is where the specifications are attached. The specifications are as important as the scope of work in section 1. The specifications tell the contractors how to perform the work listed in the scope of work.
10. Attachments. In this section, all plans, drawings, maps or plats of the property, plotted patch locations, and so on are attached.

Regardless of the size or Scope of Work, this format should be followed to allow for proper bidding and proper responses. A short, vague RFB will only leave the owner or manager confused and unhappy with the final results.

Below is an example of an RFB. This RFB is an approved and acceptable document and can be used by filling in the blanks. Following this example is a filled-in RFB for your reference should you choose to use this blank RFB.

Another document required is the addendum to the RFB or RFP. Should any changes to the scope of work or specifications develop, the addendum is used to communicate that change to the contractors and formally change the RFB or RFP. This document will then become a part of the contract documents after all parties have agreed and signed them.

Contractors are requested to inspect the project and verify quantities of patching, crack sealing, project conditions, and so on. Should any one of the bidding contractors find a discrepancy or a problem that will cause additional costs to the contract, you can provide a change to the scope of work or quantities listed with the addendum.

Other documents included in the bidding package are the owner's terms and conditions, insurance requirements, and hold harmless. This document is also attached to the RFB or RFP. Some larger companies require their own terms and conditions to be included as an attachment to the RFB or RFP.

Figure 6.1: Blank Request For Bid Form

NAME OF OWNER OR MANAGEMENT CO OR OWNER
MANAGER ADDRESS

REQUEST FOR BID

FOR
NAME OF PROJECT
ADDRESS OF PROJECT

_____ requests bids for maintenance on the existing pavements and restriping at the property titled above. The project will be constructed as outlined in the Scope of Work and installed to the attached Specifications.

1. SCOPE OF WORK

The Contractor shall supply all required insurance certificates as shown with the appropriate named "additional insured(s)."

All work outlined within the Scope of Work shall be applied with the material, application rates, construction standards, and specifications as outlined in the Specifications section of this Request for Bids. All bidders will be responsible for reading and understanding each and every Specification for each maintenance application.

[INSERT SCOPE OF WORK BY TASK]

All materials, applications, quality of workmanship, designs, and schedules shall be in compliance with the attached Specifications. The Contractor shall abide by all the conditions of this Request for Bids, and provide all information requested. The quantities for bid are listed on the bid sheet, and it shall be the responsibility of the Contractor to verify all quantities, moves, and numbers of days to complete the project.

2. PROJECT SPECIFICATIONS

The Specifications for materials and applications are listed and attached with this Request for Bids. The bidder is required to read all specifications that apply to this project, and initial. The Specifications required for this project are highlighted with a red initial circle.

3. INSTRUCTIONS TO BIDDERS

In order to be considered, proposals must be made in accordance with these instructions to bidders.

The Property Manager and/or the Consultant will attempt to have all vehicles moved that are regarded as impeding work and being exposed to physical damage from blowing debris during construction. It is the responsibility of the Contractor to protect the well-being of this personal property. Therefore, any vehicles that are close to the cleaning and construction of the areas will be covered with a protective car cover supplied by the Contractor. This does not release the Contractor of responsibility for any damage caused by his/her vehicles or personnel to private property.

The Contractor shall keep all valve boxes, manholes, and survey markers clean of all maintenance materials. All fire hydrant "blue" markers shall be cleaned or replaced. Upon completion of the project, the Contractor shall clean all areas to the condition before construction, and all areas where materials were stored shall be restored to the existing condition before work began.

The Contractor will be responsible for all barricading and traffic control during the entire project. Care will be taken to direct all traffic in such a manner as to keep it moving safely and away from work areas. The Contractor shall be responsible for all cleanup including driveways, sidewalks, and any adjacent areas after cleaning cracks and all pavement surfaces. Where required, the Contractor shall notify all tenants with door hangers that detail the type of work to be performed and the date twenty-four (24) hours prior to doing any work. All tenants are to be kept completely aware of all activity at all times. **The Property Manager and/or the Consultant shall be notified at all times of all activity, changes in schedules, breakdowns, and/or any events that will slow down or stop the maintenance process. This is required in order to notify all the tenants and property manager in a timely manner.**

The Contractor shall advise the Property Manager or the Consultant of any and all damage to private and personal property or personal injury on the property to _____, his or her Consultant, or the Project Manager immediately and in writing. The Contractor shall also, in writing, indicate the action taken, or to be taken, to correct any and all damage. This notification shall be within twelve (12) hours of the occurrence.

The Contractor must list all the names of all qualified Subcontractors or Suppliers who will be employed for the various portions of the work indicated. Any changes to this list, either prior to or during construction, must be approved by _____ and/or their designated Consultant. Failure on the part of the Contractor to complete or properly complete this list will constitute sufficient grounds to reject this bid or cancel the Contract.

The Contractor may list himself/herself to perform one or more of the listed categories of work for which he/she has any requisite state licenses when required. In this case, all personnel performing such work at the site shall be carried on the payroll, except they may sublet those portions of the Work that were traditionally and commonly subletted by the representative Subcontractor in the community. If equipment is leased with operators, the operators need not be carried on the Contractor's payroll.

List only a single name for each listing. List names only for the bid as per your proposal.

_____   _____

_____   _____

_____   _____

_____   _____

Payment for all work will be paid net within fifteen (15) days from the completion of the project less fifteen percent (15%) for contingency reserves until final inspection has been completed and all punch list items have been corrected. The time period for completion of the punch list items shall be no more than sixty (60) days upon completion of the original project.

4. INSURANCE REQUIREMENTS

Entities

Owner: _____

Managing Agent: _____

Insurance

1. Certificates of Insurance must be submitted to _____ from successful bidder (including Subcontractors and Sub-subcontractors) before the time the Contract is signed.
2. Certificates of Insurance must be in _____ possession prior to the commencement of any work or service by the successful Bidder, Subcontractor, or Sub-subcontractor. Certificates must comply with the requirements outlined below. If not, certificates will be returned for correction. Work can begin only when all certificates have been reviewed by _____ and are in full compliance with these requirements.
    A. Comprehensive General Liability Insurance
    1. Including premises-operation, products/completed operations, personal injury blanket contractual liability, independent contractors, and broad form property damage.
    2. Include coverage pertaining to explosion, collapse, and underground hazards.
    3. Limits: Comprehensive general public liability and property damage insurance approved by Owner, naming _____ and _____ as additional insured(s) with minimum limits of not less than one million dollars ($1,000,000) for injury to one person, including death, in any one accident or occurrence, and also insurance in the sum of not less than one million dollars ($1,000,000) against claims for property damage. Such policy or policies of insurance shall insure against loss, injury, death, or damage to person or property of the public, _____ and _____ upon Notice to Proceed, and their respective board of directors, employees, residents, agents, patrons, consultants, or customers.
    B. Excess Liability Insurance
    1. Umbrella form.
    2. Required if primary comprehensive general liability limits are not adequate.
    C. Workers' Compensation Insurance
    1. According to statutory limits.
    2. Contractor shall also obtain and maintain workmen's compensation and employer's insurance liability covering Contractor and all personnel and agents directly or indirectly connected with the performance by Contractor of its services and activities.
    3. The Certificate of Insurance, regardless of completed Contract price, must include as additional insured(s):

    4. Certificates must provide thirty (30) days prior written notice of cancellation, material alteration, or non-renewal.

5. GENERAL INSTRUCTIONS

MSDS Certification

The Contractor shall supply with the Contract Documents all MSDS forms required for compliance to OSHA Safety and Health Standards as outlined in "State of _____ OSHA Safety and Health Standards for the Construction Industry" §1926.59. This includes any products containing asphaltines, coal tar pitch, and all paints used for striping and marking, whether oil or acrylic based. This information is supplied by the materials manufacturer and is available upon request.

Performance Bonds

NONE

Examination

Before submitting a proposal, Bidders shall carefully examine the proposal/bid package, visit the site of the Work, and fully inform themselves of all existing conditions and limitations. The proposal shall include a sum to cover the cost of all items included in the bid package. The Bidder, if awarded the Contract, shall not be allowed any extra compensation by reason of any matter or thing, which would have become known if the Bidder would have fully informed himself/herself prior to the bidding. Signing of this Contract shall indicate the Contractor has agreed to the existing specified Scope of Work and will complete the project as specified. The quantities listed in the bid documents are the measurements to be bid to, and all quantities are considered verified by the Bidders when submitting the signed proposal. When submitting the signed proposal, the Bidder agrees to complete the Work as stated in the Scope of Work.

Change Orders and Additions to Contract

Any additional Work or materials required above and beyond the Contract and Scope of Work shall be executed with an agreed change order between the Owner, Manager, Consultant, and Contractor. Any additional Work performed or materials used will not be compensated for if not approved prior to application and accompanied by an authorized change order. All labor and materials used for change orders must be verified, and any unauthorized labor and/or materials will not be compensated for.

Any tasks added to the Contract by change order shall be at the bid price and/or bid unit price as listed on the bid sheet in the Request for Bid. At no time shall the price for change order items be more than the listed price for similar tasks in the Request for Bid. Should the price for a change order task be in excess of the bid task and unit amount, the price will be adjusted accordingly.

Where a minimum amount is listed for a task and any additional Work is included which will keep the unit amount under the minimum charge, no additional compensation will be added.

Social Security Act

The Contractor agrees to comply with and to require all of his/her Subcontractors to comply with the provisions of the act of congress approved August 14, 1935, known and cited as the "Social Security Act," and also the provisions of the act of the state legislature approved, and known as the State Unemployment Compensation Law and all other laws and regulations pertaining to labor and workers and all amendments to such data, and the Contractor further agrees to indemnify and save harmless the _____ and

_____ of and from all claims and demands made against them by virtue of the failure of the Contractor or any Subcontractor to comply with the provisions of all said acts and amendments.

Sales and Use Tax

The Contractor agrees to comply with and to require all of his/her Subcontractors to comply with all the provisions of applicable state sales excise tax law and compensation use tax law and all amendments to same. The Contractor further agrees to indemnify and save harmless _____ and _____ from any and all claims and demands made against it by virtue of the failure of the Contractor or any Subcontractor to comply with the provisions of any or all said laws and amendments.

Qualifications of Contractors

Bidders to whom an award of a Contract is under consideration shall submit to the _____ _____ upon their request an executed Contractor's Qualification Statement listing license numbers, references, and any other information deemed necessary to determine the quality of the Contractor's workmanship.

Nondiscrimination

In connection with the performance of work under this Contract, the Contractor agrees (as prescribed in A.R.S. Section 23-373) not to discriminate against any employee or applicant for employment because of race, religion, color, or national origin. The aforesaid provision shall include, but not be limited to, the following: employment, upgrading, demotion, or transfer, recruitment or recruitment advertising, layoff or termination, rates of pay or other forms of compensation, and selection for training, including apprenticeship.

Quality Control and Quality Assurance

The _____ will inspect the materials being applied, the rates of application, and the method of application according to local specifications, manufacturer's recommendations, and the attached Specifications. The materials will be sampled and tested to ensure the mix designs are within the standards established at the acceptance of the bid documents and the Contract. Should the materials not be in the specifications parameters established by local agencies and manufacturers for quality or installation, the _____ reserves the right to adjust the material costs in the Contract price in direct proportion. The _____ shall have all materials tested by a licensed and practicing professional in the field of construction materials to ensure a third-party interpretation. **The attached Specifications are to be used and referred to during bidding and construction.**

Interpretations, Addenda

Should a Bidder find any ambiguity, inconsistency, or error in the proposal/bid package or should he/she be in doubt as to their meaning, he/she shall at once notify _____ in writing; a written addendum will be submitted to all Bidders. Neither _____ nor their designated consultant or agents will be responsible for oral instruction or information.

Where the Bidder is a corporation, the proposal must be signed by the legal names of the corporation followed by the name of state of incorporation and the legal signature of an officer authorized to bind the corporation to a Contract.

6. SUBMITTAL REQUIREMENTS

THE ENTIRE BID PACKAGE MUST BE RETURNED INTACT IN ORDER FOR THE BID TO BE CONSIDERED. The bid must be submitted in the original form plus *two copies.* The Bidder may attach his/her own proposal format, letterhead, specification fact sheets, etc. as an addendum to this proposal form. _____ reserves the right to reject any and all bids, and/or call for a re-bid. _____ **reserves the right to issue one complete Contract or several Contracts per each task.**

Project Work Hours

The Contractor will be allowed to work between the hours of 6:00 AM and 5:00 PM It is recommended that the Contractor bid according to his/her ability to complete the project in the designated number of days predicted by the Contractor in order to complete the project with diligence and prudence. **The Contractor shall not be allowed to work after 5:00 PM,** therefore, cleanup and shutdown time will need to be allowed in the workday.

Project Scheduling

The Contractor shall submit, with the proposal, the number of days it will take to complete this project in a reasonable and prudent manner, and maintain the quality of materials and workmanship specified. The Contractor will be required to submit a schedule of days it will take to complete the project in an acceptable manner and condition including cleanup. The _____ will assess a penalty of two-hundred fifty dollars ($250) per day the Contractor works beyond the submitted schedule of days. The exception to this rule is delays due to bad or inclement weather or other acts of God that are not the fault of the Contractor.

All proposals must be received (by hand delivery or mail) by 4:00 PM, _____ to _____. Proposals can also be mailed to _____. All proposals delivered by the mail must arrive before the deadline date and time. Judge your mailing accordingly in order to avoid having your bid disqualified for late delivery. You may fax a copy of the bid sheet to _____. However, the hard copy of the proposal must be received within five (5) days by mail or delivery. Any inquiries can be addressed by calling _____. Office hours are Monday-Friday 8:00 AM to 5:00 PM,

_____

* All paragraphs, sections, and specifications requiring initials of the Contractor must be signed for the bid to be accepted. Initialing of the paragraphs, sections, and specifications is required to indicate that the Contractor has read and understands each and its contents. These areas are marked with an INITIAL CIRCLE. There will be no exclusions or exceptions for not having read the required paragraphs, sections, and specifications.

_____

The Contractor is responsible for the amount of material needed to complete each task. The Contractor will have a current inventory of how much material was used and how much material will be required to complete the project at all times. Should the Contractor need additional materials, which will require an increase in the cost of a task and/or the Contract, this inventory will be required in order to request any change order or cost adjustments. There will be no change order for additional costs after the material is applied or installed and the task is completed. Should the Contractor need additional materials to complete the project, he/she will notify the Owner or Consultant prior to application and in a timely manner with substantial proof.

The Contractor will provide portable sanitary facilities (i.e., Portable Toilets) for the duration of the project. The Contractor will keep the facility clean and usable at all times.

Execution of this Request for Bid by the Contractor is a representation that the Contractor has visited the work site and has become completely familiar with all local conditions under which the Scope of Work is to be performed. The Contractor agrees to have examined all aspects of the work site, the Scope of Work, and the Bid Tasks as outlined in the Request for Bid for this project. Prior to submitting a completed proposal, the Contractor shall include in the proposal a sum to cover the cost of all items included in the Request for Bid and the bid package. Should the Contractor be awarded this Contract, any extra compensation shall not be allowed for any reason that would have become known if the Contractor would have fully informed himself/herself prior to submitting a completed bid. The quantities listed in this Request for Bid are measurements for bidding purposes only, and by signing this Request for Bid, the Contractor certifies all quantities have been verified and the project will be completed as stated in the Request for Bid and outlined in the Scope of Work and the attached Specifications.

This Request for Bid shall not be construed to create a contractual relationship of any kind between the Owner and the Contractor, the Consultant and the Contractor, the Owner and a Subcontractor or Sub-subcontractor, or between any persons or entities. This is only an offer by the Contractor to the Owner to perform the Work as stated in the Scope of Work, the Bid Tasks, and as specified and shall be only an offer from <u>the Contractor to the Owner.</u>

Should the Contractor be awarded a Contract based on this bid submittal, the Contractor shall supervise and direct the Work, using the Contractor's best skill and attention to ensure the project is being installed to the Scope of Work and Project Specifications. The Contractor shall be solely responsible for and have control over the construction means, methods, techniques, sequences and procedures, and coordinating all portions of the Work under the Scope of Work unless the Scope of Work gives other specific instructions concerning these matters. The Owner and/or the Owner's Agent or Consultant shall only advise and offer direction unless so specified in the Scope of Work.

As soon as the Work begins, the Owner, the Owner's Agent, or the Consultant will not have control over or charge of and will not be responsible for construction means, methods, techniques, sequences or procedures, or for safety precautions and programs in connection with the Work, since these are solely the Contractor's responsibility as provided and stated in the Scope of Work and the Specifications. The Owner, the Owner's Agent, and the Consultant will not be responsible for the Contractor's failure to carry out the Work in accordance with the Scope of Work and the Specifications.

When the Request for Bid and Scope of Work call for remove and replace (R&R) patching, milling or pulverizing the pavement, or any other Work function where the pavement is removed, the Contractor will notify the Consultant immediately should the removed pavement be thicker than bid upon or the estimated quantity of hot-mixed asphalt pavement is greater than bid upon. Failure to request an inspection of existing pavement or verification of material quantities and patch areas by the Contractor for extra thickness or tonnage prior to completing the Work shall render the change in price not acceptable, and will be denied for payment.

Contractor Company:_____**Date:**_____

BY:_____
      (Print Name and Title)

Signature:_____

7. BID BY TASK BREAKDOWN

UNIT PRICE

TASK I   _____   \$_____

TASK II   _____   \$_____

TASK III   _____   \$_____

TASK IV   _____   \$_____

TASK V   _____   \$_____

Sales tax if applicable \$_____

TOTAL BID \$_____

Contractor License No._____

Number of Days to Complete Project_____

Number of Moves to Complete Project_____

Contractor Name _____

Phone # _____ Fax # _____

Email Address _____

Contact Name _____

8. PATCHING DIMENSIONS

(for location, refer to Appendix A)

SKIN PATCHING   R&R (Remove and Replace)

9. ATTACHMENTS

REQUEST FOR BID FOR MY HOMEOWNERS ASSOCIATION 1234 W. BUCK ROAD ANY CITY, ANY STATE, ANY ZIP CODE

Normal Properties, Inc. requests proposals for maintenance on the existing pavements and restriping at the property titled above. The project will be constructed as outlined in the Scope of Work and installed to the attached Specifications.

 ## SCOPE OF WORK

The Contractor shall supply all required insurance certificates as shown with the appropriate named "additional insured(s)."

All Work outlined within the Scope of Work shall be applied with the material, application rates, construction standards, and specifications as outlined in the Specifications section of this Request for Bid. All Bidders will be responsible for reading and understanding each and every Specification for each maintenance application, verifying all quantities, and visiting the work site.

Crack Sealing
Fill all cracks with an opening of 3/16 inch or larger with hot pour elastomeric crack sealant. All cracks shall be thoroughly cleaned by blowing out all debris, dirt, dust, noncompressibles, and loose pavement with compressed air at a minimum of 200 PSI. Should any vegetation be present, that vegetation shall be removed, and the crack blown clean. All cracks shall be thoroughly dry of all moisture prior to applying sealant. Should the ambient and ground temperatures be excessive to cause the crack sealant to remain tacky and viscous, the Contractor shall apply a blotter material consisting of fine sand, silica sand, or other acceptable material which will prevent traffic from picking up the crack sealant and reduce tracking.

Skin Patching
Skin patch all areas marked on the pavement outlining the area to be patched and marked with the letter "S" and/or as plotted on the attached property plat, leasing plan, or map.

R&R Patching
Remove and replace (R&R) patch all areas marked on the pavement outlining the area to be patched and marked with the letter "R" and/or as plotted on the attached property plat, leasing plan, or map.

Seal Coat with Asphalt and Coal Tar Blend
2 Coats, Spray Applied
Seal coat all surface areas with asphalt emulsion and coal tar blend (80 percent x 20 percent (80 percent asphalt emulsion with 20 percent coal tar emulsion). First coat applied at an application rate of 0.14 gallon per square yard, and the second coat applied at a rate of 0.14 gallon per square yard, both coats spray applied, and diluted not more than 25 percent of concentrate or as specified by the manufacturer.

Restripe to the Existing Configuration
All materials, applications, workmanship, and schedule shall be as installed to the attached Specifications. The Contractor shall abide by all conditions of this Request for Bid and provide all information requested. The quantities for bid are listed on the bid sheet, and it shall be the responsibility of the Contractor to verify all quantities, moves, and number of days to complete the project.

PROJECT SPECIFICATIONS

The Specifications for materials and applications are listed and attached with this Request for Bid. The Bidder is required to read all Specifications that apply to this project, and initial. The Specifications required for this project are highlighted with a initial circle.

The Property Manager and/or the Consultant will attempt to have all vehicles moved that are regarded as impeding work and being exposed to physical damage from blowing debris during construction. However, it is the responsibility of the Contractor to protect the well-being of this personal property. Therefore, any vehicles that are close to the cleaning and construction of the areas will be covered with a protective car cover supplied by the Contractor. This does not release the Contractor of responsibility for any damage caused by his vehicles or personnel to private property.

The Contractor shall keep all valve boxes, manholes, and survey markers clean of all maintenance materials. All fire hydrant "blue" markers shall be cleaned or replaced. Upon completion of the project, the Contractor shall clean all areas to the condition before construction, and all areas where materials were stored shall be restored to the existing condition before work began.

 The Contractor will be responsible for all barricading and traffic control during the entire project. Care will be taken to direct all traffic in such a manner as to keep it moving safely and away from work areas. The Contractor shall be responsible for all cleanup including driveways, sidewalks, and any adjacent areas after cleaning cracks and all pavement surfaces. Where required, the Contractor shall notify all tenants with door hangers that detail the type of work to be performed and date twenty-four (24) hours prior to doing any work. All tenants are to be kept completely aware of all activity at all times. The Property Manager and/or his/her Consultant shall be notified at all times of all activity, changes in schedules, breakdowns, and/or any events that will slow down or stop the maintenance process. This is required in order to notify all the tenants and property manager in a timely manner.

The Contractor shall advise the Manager or the Consultant of any and all damage to private and personal property or personal injury on the property to Normal Properties, Inc., their Consultant, or the Project Manager immediately and in writing. The Contractor shall also, in writing, indicate the action taken, or to be taken, to correct any and all damage. This notification shall be within twelve (12) hours of the occurrence.

The Contractor must list all the names of the qualified Subcontractors or suppliers that will be employed for the various portions of the Work indicated. Any changes to this list, either prior to or during construction, must be approved by Normal Properties, Inc. and/or their designated Consultant. Failure on the part of the Contractor to complete or properly complete this list will constitute sufficient grounds to reject this bid or cancel the Contract.

The Contractor may list himself/herself to perform one or more of the listed categories of work for which he/she have any requisite state licenses when required. In this case, all personnel performing such work at the site shall be carried on the payroll, except that they may sublet those portions of the Work, which are traditionally and commonly sublet by the representative Subcontractor in the community. If equipment is leased with operators, the operators need not be carried on the Contractor's payroll.

List only a single name for each listing. List names only for the bid as per your proposal.

_____ _____

_____ _____

_____ _____

_____ _____

Payment for all work will be paid net within fifteen (15) days from the completion of the project less fifteen percent (15%) for contingency reserves until final inspection has been completed and all punch list items have been corrected. The time period for completion of the punch list items shall be no more than sixty (60_ days upon completion of the original project.

INSTRUCTIONS TO BIDDERS

In order to be considered, proposals must be made in accordance with these instructions to Bidders.

Entities
Owner: My Homeowners Association
Managing Agent: Normal Properties, Inc.

Insurance
1. Certificates of Insurance must be submitted to Normal Properties, Inc. from successful bidder (including Subcontractors and Sub-subcontractors) before the time the Contract is signed.
2. Certificates of Insurance must be in Normal Properties, Inc. possession prior to the commencement of Work or service by the successful Bidder, Subcontractor, or Sub-subcontractor and must comply with the requirements outlined below; if not, certificates will be returned for correction. Work can begin only when all certificates have been reviewed by Normal Properties, Inc. and are in full compliance with these requirements.
    A.   Comprehensive General Liability Insurance
       1.   Including premises-operation, products/completed operations, personal injury blanket contractual liability, independent contractors, and broad form property damage.
       2.   Include coverage pertaining to explosion, collapse, and underground hazards.
       3.   Limits: Comprehensive general public liability and property damage insurance approved by owner, naming Normal Properties, Inc. and My Homeowners Association as additional insured(s) with minimum limits of not less than one million dollars ($1,000,000) for injury to one person, including death, in any one accident or occurrence, and also insurance in the sum of not less than one million dollars ($1,000,000) against claims for property damage. Such policy or policies of insurance shall insure against loss, injury, death, or damage to person or property of the public, Normal Properties, Inc., and My Homeowners Association upon Notice to Proceed, and their respective boards of directors, employees, residents, agents, patrons, consultants, or customers.
    B.   Excess Liability Insurance
       1.   Umbrella form
       2.   Required if primary comprehensive general liability limits are not adequate.
    C.   Workers' Compensation Insurance
       1.   According to statutory limits

2. Contractor shall also obtain and maintain workers' compensation and employer's insurance liability covering Contractor and all personnel and agents directly or indirectly connected with the performance by Contractor of its services and activities.

3. The Certificate of Insurance, regardless of completed Contract price, must include as additional insured(s):
    A. The Owner upon Notice to Proceed and
    B. Normal Properties, Inc.

4. Certificates must provide thirty (30) days prior written notice of cancellation, material alteration, or non-renewal.

MSDS Certification

The Contractor shall supply with the Contract Documents all MSDS forms required for compliance to OSHA Safety and Health Standards as outlined in "local State OSHA Safety and Health Standards for the Construction Industry" §1926.59. This includes any products containing asphaltines, coal tar pitch, and all paints used for striping and marking whether oil or acrylic based. This information is supplied by the materials manufacturer and is available upon request.

Performance Bonds
NONE

Examination

Before submitting a proposal, Bidders shall carefully examine the proposal/bid package, visit the site of the Work, and fully inform themselves as to all existing conditions and limitations, and shall include in the proposal a sum to cover the cost of all items included in the bid package. The Bidder, if awarded the Contract, shall not be allowed any extra compensation by reason of any matter or thing, that would have become known if the bidder would have fully informed himself/herself prior to the bidding. Signing of this Contract shall indicate the Contractor has agreed to the existing specified Scope of Work and will complete the project as specified. The quantities listed in the bid documents are the measurements to be bid to, and all quantities are considered verified by the Bidders by submitting the signed proposal. When submitting the signed proposal, the bidder agrees to complete the Work as stated in the Scope of Work.

Change Orders and Additions to Contract

Any additional Work or materials required above and beyond the Contract and Scope of Work shall be executed with an agreed change order between the Owner, Manager, Consultant, and Contractor. Any additional Work performed or materials used will not be compensated for if not approved prior to application and accompanied by an authorized change order. All labor and materials used for change orders must be verified, and any unauthorized labor and/or materials will not be compensated for.

Any task(s) added to the Contract by change order shall be at the bid price and/or bid unit price as listed on the bid sheet in the Request for Bid. At no time shall the price for change order items be more than the listed price for similar tasks in the Request for Bid. Should the price for a change order task be in excess of the bid task and unit amount, the price will be adjusted accordingly.

Where a minimum amount is listed for a task and any additional Work is included which will keep the unit amount under the minimum charge, no additional compensation will be added.

Social Security Act

The Contractor agrees to comply with and to require all of his Subcontractors to comply with the provisions of the Act of Congress approved August 14, 1935, known and cited as the "Social Security Act," and also the provisions of the act of the state legislature approved, and known as the State Unemployment Compensation Law and all other laws and regulations pertaining to labor and workers and all amendments to such data, and the Contractor further agrees to indemnify and save harmless Normal Properties, Inc. and the Owner upon Notice to Proceed of and from all claims and demands made against them by virtue of the failure of the Contractor or any Subcontractor to comply with the provisions of all said acts and amendments.

Sales and Use Tax

The Contractor agrees to comply with and to require all of his/her Subcontractors to comply with all the provisions of applicable state sales excise tax law and compensation use tax law and all amendments to same. The Contractor further agrees to indemnify and save harmless Normal Properties, Inc. and the Owner upon Notice of Proceed from any and all claims and demands made against it by virtue of the failure of the Contractor or any subcontractor to comply with the provisions or any or all said laws and amendments.

Qualifications of Contractors

Bidders to whom an award of a Contract is under consideration shall submit to Normal Properties, Inc. upon their request an executed Contractor's Qualification Statement listing license numbers, references, and any other information deemed necessary to determine the quality of the Contractor's workmanship.

Nondiscrimination

In connection with the performance of work under this Contract, the Contractor agrees (as prescribed in A.R.S. Section 23-373) not to discriminate against any employee or applicant for employment because of race, religion, color, or national origin. The aforesaid provision shall include, but not be limited to, the following: employment, upgrading, demotion or transfer, recruitment or recruitment advertising, layoff or termination, rates of pay or other forms of compensation, and selection for training, including apprenticeship.

Quality Control and Quality Assurance

The Normal Properties, Inc. will inspect the materials being applied, the rates of application, and the method of application according to local specifications, manufacturer's recommendations, and the attached Specifications. The materials will be sampled and tested to ensure the mix designs are within the standards established at the acceptance of the bid documents and the Contract. Should the materials not be in the specification parameters established by local agencies and manufacturers for quality or installation, the Normal Properties, Inc. reserves the right to adjust the material costs in the Contract price in direct proportion. The Normal Properties, Inc. shall have all materials tested by a licensed and practicing professional in the field of construction materials to ensure a third-party interpretation. The attached Specifications are to be used and referred to during bidding and construction.

Interpretations, Addenda

Should a Bidder find any ambiguity, inconsistency, or error in the proposal/bid package or should he/she be in doubt as to their meaning, he/she shall at once notify Normal Properties, Inc. in writing, and a written addendum will be submitted to all Bidders. Neither Normal Properties, Inc. nor their designated Consultant or Agents will be responsible for oral instruction or information.

Where the Bidder is a corporation, proposal must be signed by the legal names of the corporation followed by the name of state of incorporation and the legal signature of an officer authorized to bind the corporation to a Contract.

THE ENTIRE BID PACKAGE MUST BE RETURNED INTACT IN ORDER FOR THE BID TO BE CONSIDERED. The bid must be submitted in the original form plus two copies. The Bidder may attach his/her own proposal format, letterhead, specification fact sheets, etc. as an addendum to this proposal form.

Normal Properties, Inc. reserves the right to reject any and all bids, and/or call for a re-bid. **Normal Properties, Inc. reserves the right to issue one complete Contract or several Contracts per each task.**

Project Work Hours

The Contractor will be allowed to work between the hours of 6:00 AM and 5:00 PM It is recommended that the Contractor bid according to his/her ability to complete the project in the designated number of days predicted by the Contractor in order to complete the project with diligence and prudence. The Contractor shall not be allowed to work after 5:00 PM; therefore, cleanup and shutdown time will need to be allowed in the workday.

Project Scheduling

The Contractor shall submit, with the proposal, the number of days it will take to complete this project in a reasonable and prudent manner, and maintain the quality of materials and workmanship specified. The Contractor will be required to submit a schedule of days it will take to complete the project in an acceptable manner and condition including cleanup. The Normal Properties, Inc. will assess a penalty of two-hundred fifty dollars ($250) per day the Contractor works beyond the submitted schedule of days. The exception to this rule is delays due to bad or inclement weather or other acts of God that are not the fault of the Contractor.

All proposals must be received (by hand delivery or mail) by 4:00 PM, May 27, 20__ to Normal Properties, Inc., 1234 E. Buck Road, Any City, Any State, Any Zip Code. Proposals can also be mailed to Normal Properties Inc. at P.O. Box 1234, Any City, Any State, Any Zip Code. All proposals delivered by the mail must arrive before the deadline date and time. Judge your mailing accordingly in order to avoid having your bid disqualified for late delivery. You may fax a copy of the bid sheet to 602-555-1234; however, the hard copy of the proposal must be received within five (5) days by mail or delivery. Any inquiries can be addressed by calling 602-555-1231. Office hours are Monday-Friday 8:00 AM to 5:00 PM.

 * All paragraphs, sections, and specifications requiring initials of the Contractor must be signed for the bid to be accepted. Initialing of the paragraphs, sections, and specifications is required to indicate that the Contractor has read and understands each and its contents. These areas are marked with an INITIAL CIRCLE. There will be no exclusions or exceptions for not having read the required paragraphs, sections, and specifications.

The Contractor is responsible for the amount of material needed to complete each task. The Contractor will have a current inventory of how much material was used and how much material will be required to complete the project at all times. Should the Contractor need additional materials, which will require an increase in the cost of a task and/or the Contract, this inventory will be required in order to request any change order or cost adjustments. There will be no change order for additional costs after the material is applied or installed and the task is completed. Should the Contractor need additional materials to complete the project, he/she will notify the Owner or Consultant prior to application and in a timely manner with substantial proof.

The Contractor will provide portable sanitary facilities (i.e., Porta Johns) for the duration of the project. The Contractor will keep the facility clean and usable at all times.

BID BY TASK BREAKDOWN

UNIT PRICE

TASK I   _____   Crack Seal all pavements, $_____
@ 33,333 LF

TASK II   _____   Skin Patch @ 4,887 Sq. Ft. $_____

TASK III   _____   R&R Patch @ 1,110 Sq. Ft. $_____

TASK IV   _____   Seal Coat @ 34,770 Sq. Yd. $_____

TASK V   _____   Restripe $_____

SALES TAX $_____

TOTAL BID $_____

Contractor License No._____

Number of Days to Complete Project_____

Number of Moves to Complete Project_____

Contractor Name _____

Phone # _____ Fax # _____

Email Address _____

Contact Name _____

Execution of this Request for Bid by the Contractor is a representation that the Contractor has visited the work site and has become completely familiar with all local conditions under which the Scope of Work is to be performed. The Contractor agrees to have examined all aspects of the work site, the Scope of Work, and the Bid Tasks as outlined in the Request for Bid for this project. Prior to submitting a completed proposal, the Contractor shall include in the proposal a sum to cover the cost of all items included in the Request for Bid and the bid package. Should the Contractor be awarded this Contract, any extra compensation shall not be allowed for any reason that would have become known if the Contractor would have fully informed himself/herself prior to submitting a completed bid. The quantities listed in this Request for Bid are measurements for bidding purposes only, and by signing this Request for Bid, the Contractor certifies all quantities have been verified and the project will be completed as stated in the Request for Bid and outlined in the Scope of Work and the attached Specifications.

This Request for Bid shall not be construed to create a contractual relationship of any kind between the Owner and the Contractor, the Consultant and the Contractor, the Owner and a Subcontractor or Sub-subcontractor, or between any persons or entities. This is only an offer by the Contractor to the Owner to perform the Work as stated in the Scope of Work, the Bid by Tasks, and as specified, and shall be only an offer from the Contractor to the Board of Directors of My Homeowners Association.

Should the Contractor be awarded a Contract based on this bid submittal, the Contractor shall supervise and direct the Work, using the Contractor's best skill and attention to ensure the project is being installed to the Scope of Work and Specifications. The Contractor shall be solely responsible for and have control over the construction means, methods, techniques, sequences, and procedures, and coordinating all portions of the Work under the Scope of Work unless the Scope of Work gives other specific instructions concerning these matters. The Owner and/or the Owner's Agent or Consultant shall only advise and offer direction unless so specified in the Scope of Work.

As soon as the Work begins, the Owner, the Owner's Agent, or the Consultant will not have control over or charge of and will not be responsible for construction means, methods, techniques, sequences, or procedures, or for safety precautions and programs in connection with the Work, since these are solely the Contractor's responsibility as provided and stated in the Scope of Work and the Specifications. The Owner, the Owner's Agent, and the Consultant will not be responsible for the Contractor's failure to carry out the Work in accordance with the Scope of Work and the Specifications.

When the Request for Bid and Scope of Work call for Remove and Replace (R&R) patching, milling or pulverizing the pavement, or any other work function where the pavement is removed, the Contractor will notify the Consultant immediately should the removed pavement be thicker than bid upon or the estimated quantity of hot-mixed asphalt pavement is greater than bid upon. Failure to request an inspection of existing pavement or verification of material quantities and patch areas by the Contractor for extra thickness or tonnage prior to completing the Work shall render the change in price not acceptable, and will be denied for payment.

Contractor Company:_____**Date:**_____

BY:_____
      (Print Name and Title)

Signature:_____

PATCHING DIMENSIONS
(for location, refer to Appendix A)

SKIN PATCHING	R&R (Remove and Replace)
1 - 5×6	5 - 10×22
2 - 10×12	6 - 28×25
3 - 9×9	
4 - 15×12	

ADDENDUM TO REQUEST FOR BID

_____   _____
(Name of Contractor)   (Date of Addendum)

Re: Addendum to Request for Bid to _____
                                     (Name of Project)

for _____
         (Describe Work)

located at the _____
              (Location of Work; parking lot, association streets, etc.)

and dated _____
          (Date of Request for Bid)

Addition or Change to Scope of Work and Bid by Task Breakdown to the Request for Bid dated _____.

Addendum Number 1   Date:_____ Pages:_____

Explanation for Addendum to Request for Bid:

Add or change Scope of Work by the following directions:

Change Bid by Task Breakdown on page _____of the Request for Bid to read: _____
_____

Description of Addition or Change

Task _____ $_____

Task _____ $_____

Task _____ $_____

Task _____ $_____

Task _____ $_____

Task _____ $_____

OWNER: CONTRACTOR:

_____   _____
(Signature)   (Signature)

_____   _____
(Printed Name and Title)   (Printed Name and Title)

Another helpful document is a pre-project checklist. This helps remind the Owner, Manager, board of directors, committee members, and the Consultant of the needed announcements and notifications.

RESIDENTIAL PRE-PROJECT CHECKLIST

1. **RESIDENT LETTERS AND NOTIFICATION OF PROJECT**

   Distributed and handed out to each resident   Y   N

2. **NOTIFY TRASH PICKUP**   Y   N

   AGENCY OR PRIVATE NAME_____

   PHONE #_____

   CONTACT NAME_____

3. **NOTIFY FIRE DEPARTMENT**   Y   N

   AGENCY NAME _____

   PHONE #_____

   CONTACT NAME_____

4. **NOTIFY POST OFFICE**   Y   N

   DISTRICT NAME_____

   PHONE #_____

   CONTACT NAME_____

5. **NOTIFY MEALS ON WHEELS FOR ELDERLY PROJECTS**   Y   N

   AGENCY NAME _____

   PHONE #_____

   CONTACT NAME_____

6. **NOTIFY LANDSCAPE COMPANY (WATERING CONTROL)**   Y   N

   COMPANY NAME_____

   PHONE # _____

   CONTACT NAME_____

7. **NOTIFY PAPER DELIVERY**   Y   N

8. **ALL SPRINKLERS TURNED OFF
    IN ALL AREAS AROUND PAVEMENT
    TO BE SERVICED.**   Y   N

9. **ALL REQUIRED PERMITS ISSUED
    (IF APPLICABLE) PARKING, STREET USE,
    MATERIALS, TIME LIMITS, ETC.**   Y   N

Any other notifications required or needed?

9. **NOTIFY**_____   Y   N

    NAME_____

    PHONE #_____

    CONTACT NAME_____

10. **NOTIFY** _____   Y   N

    NAME_____

    PHONE #_____

    CONTACT NAME_____

11. **NOTIFY** _____   Y   N

    NAME_____

    PHONE #_____

    CONTACT NAME_____

COMMERCIAL PRE-PROJECT CHECKLIST

1. **TENANT LETTERS AND NOTIFICATION**
   OF PROJECT   Y   N

2. **NOTIFY TRASH PICKUP**   Y   N

   AGENCY OR PRIVATE NAME _____

   PHONE #_____

   CONTACT NAME _____

3. **NOTIFY FIRE DEPARTMENT**   Y   N

   AGENCY NAME _____

   PHONE # _____

   CONTACT NAME _____

4. **NOTIFY POST OFFICE**   Y   N

   DISTRICT NAME _____

   PHONE # _____

   CONTACT NAME _____

5. **NOTIFY LANDSCAPE COMPANY**
   (WATERING CONTROL)   Y   N

   COMPANY NAME _____

   PHONE # _____

   CONTACT NAME _____

6. **ARE ALL SPRINKLERS TURNED**
   OFF IN ALL AREAS AROUND PAVEMENT
   TO BE SERVICED ?   Y   N

7. **NOTIFY ANY DELIVERY COMPANIES**   Y   N

   COMPANY NAME _____

   PHONE # _____

   CONTACT NAME _____

8. NOTIFY _____Y   N

    NAME _____

    PHONE # _____

    CONTACT NAME _____

9. NOTIFY _____Y   N

    NAME _____

    PHONE # _____

    CONTACT NAME _____

10. NOTIFY _____Y   N

    NAME _____

    PHONE # _____

    CONTACT NAME _____

When inviting contractors to bid on your project, there are ten items to make sure your bidding contractors measure up. Following is a list of ten tips for hiring a contractor. Use this list to help in your Contractor selection.

Figure 6.3: 10 TIPS
FOR MAKING SURE YOUR CONTRACTOR MEASURES UP

1. HIRE ONLY LICENSED, REPUTABLE CONTRACTORS.

2. CHECK CONTRACTOR'S LICENSE NUMBER WITH LOCAL REGISTRAR OF CONTRACTORS, Better Business Bureau, and if the contractor is a corporation or LLC, check the Corporation Commission and Secretary of State for filing status.

3. GET THREE REFERENCES with name of contact, phone number, and type of work completed. Make sure the work completed for the reference contact is the same scope of work you are requesting. Review past work by going to work site and looking at the completed project.

4. GET AT LEAST THREE BIDS. Make sure all bids have the same scope of work and specifications so you can compare the bids and pricing. It's not about comparing "apples to apples" but comparing the price quotes for a common Scope of Work and Specifications.

5. GET A WRITTEN CONTRACT AND DON'T SIGN ANYTHING UNTIL YOU COMPLETELY UNDERSTAND THE TERMS. You have the right to have the contract reviewed by an attorney or third-party consultant or both.

6. PAY NO MORE THAN 10 PERCENT DOWN, OR one thousand dollars ($1,000), WHICHEVER IS LESS. In some special cases only, the materials needed to complete a scope of work are more than the contractor can afford. Should a Contractor need more than one thousand dollars ($1,000) for materials, ask for a copy of his/her material costs from the material supplier and include the terms in the Contract. Also, make sure the contractor provides you with a waiver of lien from the Material Supplier to show the material supplier was paid.

7. DON'T LET PAYMENTS GET AHEAD OF WORK. KEEP RECORDS OF ALL PAYMENTS. In the case of homeowners associations, commercial projects, and industrial properties, a third-party consultant (who is very experienced in your scope of work) may be contracted to monitor your project and authorize proper payments. Include your payment schedule in the contract. It is imperative to receive a waiver of lien from the contractor for the amount being paid before releasing any payment.

8. DON'T MAKE FINAL PAYMENT UNTIL YOU'RE SATISFIED WITH THE JOB. In the contract documents, list a retention amount (usually 10 percent of the total price, which includes base bid and any change orders) to ensure final cleanup and any punch list items are completed. Once the contractor completes all cleanup and punch list items, you can release the retention amount. Make sure you get a written waiver from contractor for this payment.

9. DON'T PAY CASH. Always pay with check, money order, or credit card, where you receive a receipt. On the check, money order, or credit card receipt, make sure you list the contractor project number and scope of work. Do not pay the contractor until you receive a conditional waiver of lien for the amount paid (which will become an unconditional lien when your payment is posted in the contractor's account). Also make sure you have waivers of liens from the contractor's suppliers before releasing any payments.

10. KEEP A JOB FILE OF ALL PAPERS RELATING TO YOUR PROJECT. This includes proposal, contracts, any correspondence, change orders, payment receipts, waivers of liens, punch list items, and any project notes.

Courtesy of Contractors State License Board (CSLB) of California (with expanded comments by PMIS, Inc.).

Allow at least one to two weeks for the contractors to submit their proposals to you. Require the bidding contractors to visit the project site and become familiar with the project. The majority of property managers are faced with the task of reviewing the submitting quotes. A methodical way to review bids, other than the primary fact of cost, should be established. When reviewing submitted bids, a methodical way, other than cost or low bidder, should be established. Additional factors such as the following need to become part of the basis for comparison of all bids:

- Quality of materials the contractor is going to use
- Methods of application or installation
- Amount of work and quality of workmanship
- Reliability of the contractor

These factors can prevent the project from becoming one of marginal acceptability because you have many controls, and you will get what you paid for and what you requested to be completed. To avoid any pitfalls, the entire process of bidding should be structured as outlined.

Upon receipt of the submitted bids, it must be decided which contractor will do the work. This can be accomplished by the following steps:

1. Verify the contractor will meet all bid requirements including submittal of all subcontractors and material suppliers, insurance requirements, bonding (if required), and all items listed in the request for bid (RFB) and bidding terms and conditions.
2. Verify all quantities are compatible with your own estimates.
3. Use a worksheet to compare the factors you have outlined in your RFB or RFQ.
4. Make sure to get at least three references and contact them.
5. Look at a minimum of two of the winning bidders' jobs that match your scope of work.
6. Contact the local Registrar of Contractors and the Better Business Bureau and check if any complaints have been registered. If there are some complaints, check if they were cleared up satisfactorily.
7. Develop and incorporate a rating system for each of the tasks in your bid by tasks.

When controlling the bids by posting the tasks by types and quantities of work you desire, you can easily compare the dollar amount the contractors are submitting. The tasks and quantities are listed, and the contractors have filled in the amount they want to charge per task. A separate entry can be established for a unit cost for a specific task, or you can calculate the unit amount by dividing the quoted price by the measurement or quantity. It is easier if the distresses, the measurements, and a budget figure are already predetermined or developed. Should you ask for an open bid, make sure to compare the amount of each task requested (e.g., crack sealing, 88,888 linear feet). Should the bidding contractors be off in measurements by a substantial amount, there is something wrong. Same for patching, surface area, and even striping. By controlling the bids with stated measurements, you can compare pricing more easily. Some contractors will bid higher than others because of timing or their existing work schedule or maybe they do not want the project at this time. If the price is substantially lower than the other bidders, then there could be discounts from suppliers that the contractor is passing on, the contractor is bidding to get the project, there could be a mistake in calculations, or the contractor is providing an inferior material than specified. All these variables can be determined when comparing bids.

An example of a worksheet to compare bids is shown below:

	Contractor A	Contractor B	Contractor C
Crack Sealing			
Quantity LF	10,477	13,889	9,876
Price	$5,456.00	$6,876.00	$5,895.00
Unit Price	$0.52	$0.495	$0.597
Seal Coating 2 coats			
Quantity (Sq. Yd.)	30,878	29,456	31,668
Price	$26,246.00	$29,456.00	$23,766.00
	$0.85	$1.00	$.075

In our example, a substantial variation in quantities and prices can be seen. The unit prices for crack sealing show Contractor C is the most expensive, and there is quite a difference in the amount of crack sealing between Contractor A and B. Why do these differences exist?

• Are there going to be future change orders?
• Is there a bust in measurements?

The unit prices for seal coating show Contractor C is going to seal the pavement for less than the others, but his area is larger than A or B. Again, why do these differences exist?

• Are these contractors using the same material, and if so, are they diluting it the same?
• Are there going to be future change orders?
• Is there a bust in measurements?
• Is one contractor using an inferior material, overdiluting, or using a lesser application rate?

This table shows you a potential problem. Perhaps there is a good explanation, but you will need to check it out. With a premeasured quantity and budget, a more controlled bidding stage can be maintained. If in doubt, interview each contractor prior to awarding the contract. Above all, remember, "If it is too good to be true, then it must not be true."

Before anything else is done, you must make sure all of the bidding contractors are doing the same task. Look at quantities of each task first. This should have the least variance of difference between bidders (plus or minus 5 percent from your own takeoff is acceptable due to rounding off of quantities). If there is more than 5 percent variance, check all bids and determine where the average is. It is entirely possible the mistake may have been yours. Reevaluate the bids with a variance greater than 5 percent, and if they still do not meet the criteria, call the contractors and ask them to check their quantities for possible errors.

Should all bidders' quantities meet the acceptable parameters for bidding, then review the unit prices for each task to be completed. These prices may have a substantial difference due to different type and brands of materials that may be used. It is necessary to review the submittals to determine the manufacturer's recommendations on how to apply the material as well as the specifications (referred to as spec sheets). For example, seal coating materials are usually some form of an emulsion. There may well be different tolerances defining how much water may be added to their material to achieve the desired results. The bids will have to be grouped by the type of material specified and quoted to achieve any meaningful data. Once the bids are sorted, it can then be determined if any unit prices have any substantial differences. The higher unit prices may indicate a small contractor who has to pay more for material, a contractor with a large overhead, a contractor who charges a little more because they feel that they do better work, or someone who is deliberately bidding high because he/she doesn't want the project.

Conversely, low bids can often reflect contractors who are able to get discounts in purchasing material. Large contractors are able to disperse their overhead over a larger quantity of jobs, or a contractor can cut either their material or labor (quality of workmanship). The best way to determine this is to call the local distributor and ask what the cost of the material is. These types of determinations are much harder and more subjective with bids that only require a single task. Multitask bids will give you a better idea of contractors' trends in pricing. In turn, this will improve the quality of your analysis.

Using a concise and informative RFB or RFQ will aid in making sure the project is completed correctly and with the correct materials. Most bidding contractors will appreciate a bid document that explains what is desired to complete the project to make the bidding process equally competitive. Remember, it's not about comparing apples to apples, but more the exact scope of work, specifications, and proper and concise bid documents. The selected bidding contractor's bid document will become a major part of the contract; therefore, the more detailed and instructive the bid document, the better. Also, require the contractors to list their contractor's license numbers. Should this be a first-time contractor bidding with you or you have some doubt about a bidding contractor, contact the local state contractors' licensing agency (in most states, that is the Registrar of Contractors).

After you have selected a contractor to do your project, notify him/her in writing requesting his/her schedule to each task. This schedule will become a part of the contract and binds the contractor to start and completion dates (barring circumstances out of the contractor's control). In your notification, also list the correct names of the additional insureds to be named on all insurance documents. Take the time to also notify the other contractors, as they have spent time and resources to provide you with a bid.

Notify the chosen contractor with a formal notification of award of the project. Below is a suggested Notice of Award form to use. Have the contractor list his/her schedule to complete the project. This will give all parties a chance to review the schedule, and if there is any conflict of timing, a change can be made at that time and before a contract is issued.

Sample Notice of Award

Contractor Name _____

Contractor Address _____

_____

Dear _____ :

Your company has been awarded the project known as _____

_____

(NAME OF PROJECT).

This is a notice of award. A full Contract Document will be supplied as soon as you submit your dates and schedule for the project. You will be required to sign a full Contract prior to proceeding, which will be provided as soon as the information requested is submitted.

You are required to submit a certificate of your liability insurance and state compensation insurance as outlined in the Insurance section of the Request for Bid for the amounts stated. Also, all additional insured(s) as stated in the insurance section shall be shown and listed on the insurance certificates. MAKE THE ADDITIONAL INSURED(S) TO INCLUDE THE EXACT NAMES AS LISTED:

In order to expedite the Contract, you are required to schedule the dates to start and complete the project. These dates will be used to schedule all notifications and movement of residents or tenants, delivery vehicles, postal services, and all community service agencies. On the Request for Bid, the number of days were stated to complete the project. Please list the actual dates for the project below.

This form and all insurance certificates must be in receipt of _____ to expedite the Contract. The Contract will not be developed without this document and all required insurance documents. Please list your schedule to complete the project below by date and task (i.e., Task I crack seal 10/5 to 10/7, etc.), sign, and mail back to _____. Please refer to the Bid by Task form.

<u>Task #</u>   <u>Date</u>

_____   _____

_____   _____

_____   _____

_____   _____

Signature_____

Printed Name
_____Title_____

Company _____ Phone # _____

Fax # _____

Address _____

Sincerely
_____

Chapter 7

Developing the Proper Contract Documents

A Contract is an agreement between an owner, or board of directors of a homeowners association, and a contractor as outlined in the RFB or RFP, the scope of work, and the specifications for a specified amount. The Contract includes the contract document, the completely filled-out RFB signed by the contractor, all additional documents and certificates, any special instructions or changes, plans and details, specifications, and all permits (if required). The contract document contains all legal terms and conditions that execute the scope of work as outlined in the RFB or RFP. The following items, terms, and conditions are listed in the contract:

- Identification of all parties involved with the contract.
- Identification of the project location and type of work to be performed.
- The specific dates to begin and complete the contract and any penalties if applicable.
- The contract amount and any changes in the base contract price from any additions or deletions from the original scope of work and bid quantities.
- All terms and conditions of payment.
- All attachments to the contract, such as insurance riders, hold harmless agreements, the completed and signed RFB, specifications, and instructions to the bidder. All these items are brought together in the contract document to make the contract binding and enforceable.
- The general conditions of the contract that govern the entire process and project from start to finish.
- All drawings and details specific to the project.
- The signatures of all parties, which makes the contract official under the terms and conditions spelled out in the RFB or RFP, the general and special conditions stated in the contract documents, and all additional attachments, agreements, and changes affixed to the contract.

The contract should never be written between the contractor and the property manager or consultant. If the contract is written between the contractor and Consultant, then the Consultant is a broker and is considered a general contractor. The contract is a formal agreement that states the Contractor will perform the work as outlined in the scope of work, RFB or RFQ, and specifications. In return, the contract states the owner will pay the contractor for completing the work as outlined in the scope of work and specifications. The same holds for contracts as RFBs. Don't shorten the contractual agreement to accommodate any party or take a short way out. A brief contract can lead to trouble for both the contractor and the owner or manager.

The contract should be divided into nine main sections stating how the contractor is going to perform the work and all administrative tasks required. A contract is more than signing a proposal form and converting the proposal to a contract. A formal contract document protects the property owner and the contractor during the project, at the completion of the project, and closing out the project. An acceptable and correct contract should contain the following:

1. Introduction
2. Cover sheet
3. Contract scope of work and format: This section recaps the tasks in the Scope of Work and all rules and regulations for the completing the Contract.

4. Date of commencement and completion of the work.
5. The agreed contracted amount (base amount and any addendum prices and change order prices) and any agreed payment terms. Also listed is the schedule of payments and any retention that will be held for punch list work.
6. Final payment conditions to close out the contract.
7. Listing of any contract documents, attached drawings and plans, and addenda: This includes the RFB or RFQ, and any terms and conditions stipulated by the owner.
8. General conditions of the contract (which contains the insurance requirements required to work on the Owner's property) along with any supplementary conditions added by the owner.
9. Signature sheet, which contains any additional instructions

This format includes any formal contract (AIA, ASCE, or private contracts from the owner or a third party). If the AIA contract is used, there usually is a second attachment that is included for additional terms and conditions.

When completing a contract document, always make reference to and attach the proposal, RFB, or RFP, as this is the working part of the contract.

In review, there are three parts to a contract: the contract itself; any attachments (owner terms and conditions, blueprints, drawings and details, etc.); and the proposal, RFB, or RFQ, which contains the scope of work, job site instructions, and specifications. Figure 7.1 is an example of a contract with blank areas to be filled in, and Figure 7.2 is an example of a properly filled-in contract.

Should one of the bidding contractors bring up a point or issue pertaining to the bid documents or the project or supplies you with information you hadn't previously considered, this is the time and place to put it in writing and include it as part of the contract. Be very careful you do not change the amount of work to be done, alter the scope of work, or change the materials from any inquiry, since this would result in a change of conditions and would necessitate rebidding the project. If this does become the case, it is well advised to contact all bidding contractors and inform them of the change. Ask them if this will affect their price and whether or not they will want to reconsider and resubmit their bid.

The final area to discuss is the verification and notification processes. This is often said to be the area that will serve to keep everyone honest. In an open bidding situation, most contractors will include a statement in their bid eliminating the inclusion of any materials testing, drainage engineering, surveying and grading, and permitting. In many cases, they also will eliminate the inclusion of notification of residents or tenants, towing of vehicles, and any notifications to service entities. "Engineering" usually implies or refers to hiring a certified engineer to verify and certify materials being used, and establishing or certifying grades and compaction of the various material layers. "Testing" implies that an independent laboratory will be hired to sample and check materials and certify that the contractor is meeting the conditions and specifications of the contract. Notifications and towing are implied tasks that need to be taken care of by the owner, manager, or representative of the two. These are tasks of the project that will need to be taken care of by the owner or their designated representative (manager or consultant) and should not be a task of the contractor. Testing of materials can certainly be required; however, it is a conflict of interest for the contractor to be required to hire the laboratory. This is especially true if the project ends up in arbitration. The property owner should be the person who owns the test results. These tasks should be listed in the contract and shown as the responsibility of the owner, manager, (i.e., the owner or board of directors can or will be responsible for all testing and notifications). This will ensure these tasks are completed without any delays or conflicts.

Permitting should be a part of the contractor's responsibility and should be stated as such in the RFB or RFQ and the contract. The contractor needs to apply for and obtain all permits (right of way, parking, special work hours, dust, etc.) from all local agencies and should close out the permits once they have completed the project. All this is stated in the conditions of the contract. Many owners of commercial and industrial properties have a larger requirement for insurance coverage, which will be stated in a separate attachment known as terms and conditions or supplemental conditions of the contract. These supplemental conditions become the controlling terms of the contract and are attached directly to the contract documents and referred to in the body of the contract. These terms also override similar terms in the base contract.

Figure 7.1 Example of a Blank Contract

CONTRACTUAL AGREEMENT
FOR PAVEMENT AND CONCRETE REPAIR AND MAINTENANCE

BETWEEN
Owner/Homeowners Association
AND
THE CONTRACTOR

THIS DOCUMENT HAS IMPORTANT LEGAL CONSEQUENCES: CONSULTATION WITH AN ATTORNEY IS ENCOURAGED WITH RESPECT TO ITS COMPLETION OR MODIFICATION.

THIS DOCUMENT IS AN AGREEMENT BETWEEN THE HOMEOWNERS ASSOCIATION OR THE MANAGEMENT FIRM AS AGENT FOR THE HOMEOWNERS ASSOCIATION AND THE CONTRACTOR FOR PAVEMENT MAINTENANCE AND/OR CONCRETE MAINTENANCE PERFORMED AT THE DESCRIBED LOCATION AND PROJECT.

THIS AGREEMENT INCLUDES THE REQUEST FOR BID SIGNED AND AGREED TO BY THE CONTRACTOR AND THE HOMEOWNERS ASSOCIATION, OR THE AGENT FOR THE HOMEOWNERS ASSOCIATION. BY SIGNING THIS AGREEMENT AND CONTRACT, THE CONTRACTOR HAS VERIFIED ALL QUANTITIES LISTED IN THE REQUEST FOR BID, REVIEWED ALL THE ATTACHED SPECIFICATIONS, AND AGREED TO PERFORM THE WORK STATED AND IN THE TIME PERIOD AGREED TO.

THE CONTRACTOR, THE HOMEOWNERS ASSOCIATION'S AGENT, OR THE CONSULTANT SHALL NOT ALTER THIS CONTRACT OR ANY PART OF THIS CONTRACT AND/OR THE ATTACHED REQUEST FOR BID AND SPECIFICATIONS WITHOUT THE WRITTEN CONSENT OF ALL PARTIES AND THE HOMEOWNERS ASSOCIATION. AT NO TIME SHALL ANY PERSONS CROSS OUT, DELETE, OR WRITE IN ANY ITEMS OR PARAGRAPHS. THE CONTRACT AND AGREEMENT SHALL STAY INTACT AS IT IS WRITTEN. ANY CHANGES, CORRECTIONS, AND/OR DELETIONS SHALL BE SUBMITTED ON A SEPARATE LETTERHEAD AND ATTACHED TO THE CONTRACT PACKAGE.

IN ORDER FOR THE CONTRACTOR AND THE HOMEOWNERS ASSOCIATION TO ENTER INTO THIS CONTRACT, THE CONTRACTOR MAKES THE FOLLOWING REPRESENTATIONS:

1. THE CONTRACTOR HAS EXAMINED AND CAREFULLY STUDIED THE CONTRACT DOCUMENTS, THE SCOPE OF WORK, AND THE SPECIFICATIONS AND ANY OTHER RELATED DATA IDENTIFIED IN THE CONTRACT DOCUMENTS, SCOPE OF WORK, REQUEST FOR BID, AND THE SPECIFICATIONS.

2. THE CONTRACTOR HAS VISITED THE SITE AND BECOME FAMILIAR WITH AND IS SATISFIED WITH THE GENERAL, LOCAL, AND SITE CONDITIONS THAT MAY AFFECT THE COST, PROGRESS, AND PERFORMANCE OF COMPLETING THE WORK.

3. THE CONTRACTOR HAS VALIDATED THE QUANTITIES AND AMOUNTS OF WORK BY TASK TO BE PERFORMED AT THE PROJECT SITE.

4. THE CONTRACTOR IS FAMILIAR WITH AND IS SATISFIED WITH ALL FEDERAL, STATE, AND LOCAL LAWS AND REGULATIONS THAT MAY AFFECT THE COST, PROGRESS, AND PERFORMANCE OF THE WORK.

5. THE CONTRACTOR HAS CAREFULLY STUDIED ALL PROVIDED REPORTS OF EXPLORATIONS AND TESTS OF THE SUBGRADE CONDITION, AGGREGATE BASE THICKNESS/ CLASSIFICATION/CONDITION, THE TYPE AND THICKNESS AND CONDITION OF THE EXISTING ASPHALT PAVEMENT AND CONCRETE AT OR CONTIGUOUS TO THE PROJECT, AND ALL DRAWINGS OF PHYSICAL CONDITIONS IN OR RELATING TO THE EXISTING SURFACE OR SUBSURFACE STRUCTURES AT OR CONTIGUOUS TO THE PROJECT (EXCEPT FOR UNDERGROUND FACILITIES) WHICH HAVE BEEN IDENTIFIED IN ANY PLANS, DESIGNS, SCOPE OF WORK, REQUEST FOR BID, OR ANY SUPPLEMENTARY CONDITIONS AS PROVIDED IN THE GENERAL CONDITIONS. THE CONTRACTOR ACCEPTS THE DETERMINATIONS SET FORTH IN THE SUPPLEMENTARY CONDITIONS OF THE EXTENT OF THE TECHNICAL DATA CONTAINED IN SUCH REPORTS, MATERIAL TECHNICAL DATA SHEETS, AND DRAWINGS UPON WHICH THE CONTRACTOR IS ENTITLED TO RELY ON AS PROVIDED IN THE BID DOCUMENTS. THE CONTRACTOR ACKNOWLEDGES THAT SUCH REPORTS, MATERIAL TECHNICAL DATA SHEETS, AND DRAWINGS ARE NOT CONTRACT DOCUMENTS AND MAY NOT BE COMPLETE FOR THE CONTRACTOR'S PURPOSE. THE CONTRACTOR ACKNOWLEDGES THAT THE HOMEOWNERS ASSOCIATION AND THE CONSULTANT DO NOT ASSUME RESPONSIBILITY FOR THE ACCURACY OR COMPLETENESS OF INFORMATION AND DATA SHOWN OR INDICATED IN THE CONTRACT DOCUMENTS WITH RESPECT TO UNDERGROUND FACILITIES AT OR CONTIGUOUS TO THE PROJECT SITE. THE CONTRACTOR HAS OBTAINED AND CAREFULLY STUDIED (OR ASSUMES RESPONSIBILITY FOR HAVING DONE SO) ALL SUCH ADDITIONAL SUPPLEMENTARY EXAMINATIONS, INVESTIGATIONS, EXPLORATIONS, TESTS, STUDIES, AND DATA CONCERNING CONDITIONS (SURFACE, SUBSURFACE, AND UNDERGROUND FACILITIES) AT OR CONTIGUOUS TO THE SITE OR OTHERWISE WHICH MAY AFFECT THE COST, PROGRESS, PERFORMANCE, OR FURNISHINGS OF THE WORK OR WHICH RELATE TO ANY ASPECT OF THE MEANS, METHODS, TECHNIQUES, SEQUENCES, AND PROCEDURES OF CONSTRUCTION TO BE EMPLOYED BY THE CONTRACTOR AND ALL SAFETY PRECAUTIONS, EXPLORATIONS, TESTS, STUDIES, OR DATA NECESSARY FOR THE PERFORMANCE AND FURNISHING OF THE WORK AT THE CONTRACT PRICE, WITHIN THE CONTRACT TIMES, AND IN ACCORDANCE WITH THE OTHER TERMS AND CONDITIONS OF THE CONTRACT DOCUMENTS, SCOPE OF WORK, REQUEST FOR BID, OR THE SPECIFICATIONS.

6. THE CONTRACTOR IS AWARE OF THE GENERAL NATURE OF THE WORK TO BE PERFORMED BY THE HOMEOWNERS ASSOCIATION AND ALL OTHERS AT THE WORK SITE THAT RELATES TO THE WORK.

7. THE CONTRACTOR HAS CORRELATED ALL THE INFORMATION KNOWN TO THE CONTRACTOR, ALL INFORMATION AND OBSERVATIONS OBTAINED FROM VISITS TO THE SITE, ALL REPORTS AND DRAWINGS, AND ALL ADDITIONAL EXAMINATIONS, INVESTIGATIONS, EXPLORATIONS, TESTS, STUDIES, AND DATA PERTAINING TO THE WORK AND THE PROJECT SITE.

8. THE CONTRACTOR HAS GIVEN THE CONSULTANT WRITTEN NOTICE OF ALL CON-FLICTS, ERRORS, AMBIGUITIES, OR DISCREPANCIES THAT THE CONTRACTOR HAS DISCOVERED IN THE CONTRACT DOCUMENTS, THE SCOPE OF WORK, THE REQUEST FOR BID, AND THE SPECIFICATIONS, AND THE WRITTEN RESOLUTION THEREOF BY THE CONSULTANT IS ACCEPTABLE TO THE CONTRACTOR, AND THE CONTRACT DOCUMENTS, SCOPE OF WORK, THE REQUEST FOR BID, AND THE SPECIFICATIONS ARE GENERALLY SUFFICIENT TO INDICATE AND CONVEY UNDERSTANDING OF ALL TERMS AND CONDITIONS FOR PERFORMANCE AND FURNISHING OF THE WORK.

9. THE CONTRACTOR HAS SCHEDULED TO COMPLETE THE WORK IN A REASONABLE AMOUNT OF TIME WITHOUT CAUSING EXTREME HARDSHIP TO THE HOMEOWN-ERS ASSOCIATION, RESIDENTS AND VISITORS, OR VENDORS OF THE RESIDENTS. THE CONTRACTOR HAS REASONABLY SCHEDULED THE CORRECT AMOUNT OF DAYS AND MOVES TO COMPLETE THE WORK IN A SAFE AND PRUDENT MANNER.

10. THE CONTRACTOR WILL ONLY ATTEMPT OR COMPLETE ANY ADDITIONAL WORK OUTSIDE THE SCOPE OF WORK AND AGREED QUANTITIES AT THE WRITTEN ORDER OF THE HOMEOWNERS ASSOCIATION BY CHANGE ORDER. THE CONTRACTOR WILL ACCEPT A CHANGE ORDER FROM THE HOMEOWNERS ASSOCIATION'S AGENT OR CONSULTANT ONLY IF THE CHANGE ORDER IS AUTHORIZED AND SIGNED BY THE HOMEOWNERS ASSOCIATION. ONLY ADDITIONAL WORK AUTHORIZED IN WRITING BY THE HOMEOWNERS ASSOCIATION SHALL BE INVOICED AND PAID. THE ORIGI-NAL UNIT PRICE AS SHOWN IN THE REQUEST PER BID BY TASK SHALL BE THE UNIT PRICE FOR ALL CHANGE ORDER ITEMS OF LIKE KIND, UNLESS PREVIOUSLY AUTHO-RIZED IN WRITING BY THE HOMEOWNERS ASSOCIATION. THE CONTRACTOR WILL NEGOTIATE THE PRICE OF ANY ADDITIONAL WORK REQUESTED IN WRITING FROM THE HOMEOWNERS ASSOCIATION (THROUGH THE HOMEOWNERS ASSOCIATION'S AGENT OR CONSULTANT IF APPLICABLE) THAT IS OUTSIDE THE SCOPE OF WORK AND THE REQUEST FOR BID. THE CONTRACTOR WILL NOT ACCOMPLISH THE NEW WORK UNTIL A SIGNED CHANGE ORDER IS IN THE CONTRACTOR'S POSSESSION SIGNED BY THE HOMEOWNERS ASSOCIATION.

11. THE CONTRACTOR WILL KEEP THE WORK SITE, STAGING AREAS, AND ADJACENT STRUCTURES CLEAN OF ALL DEBRIS AND DIRT CAUSED FROM CONSTRUCTION, AND WILL REPAIR ANY DAMAGE TO AN EQUAL OR BETTER CONDITION AS IT WAS BEFORE INCIDENT OR DAMAGE.

THE FOLLOWING DEFINITIONS USED FOR CONTRACT PURPOSES, WHETHER USED IN THE CONTRACT DOCUMENTS, THE REQUEST FOR BID, THE SCOPE OF WORK, OR THE SPECIFICATIONS SHALL HAVE THE MEANINGS INDICATED WHICH ARE APPLICABLE TO BOTH THE SINGULAR AND PLURAL THEREOF.

ADDENDA/ADDENDUM: Written or Graphic additions or changes issued prior to the opening of the bids which will clarify, correct, or change the Bidding Requirements.

ADDITIONAL INSURED: Names of persons, entities, companies, homeowner/homeowners association, owner/homeowners associations, or board of directors who shall be listed as additional insureds on the contractor's insurance certificates for liability and employees' or workers' compensation insurance.

AGREEMENT: The written contract between the owner/homeowners association, owner/homeowners association's agent, and the contractor covering the work to be performed. Other contract documents attached to the original agreement shall be considered as part of the original agreement.

BOND, BID: A security instrument issued and delivered with the bid package to guarantee the Contractor's bid price.

BOND, PERFORMANCE: A security instrument issued to guarantee the performance of the contractor of the work. Should the contractor default, then the performance bond will be used to cover any expenses incurred in contracting the second bidder, including the bid price difference.

CHANGE ORDER: A document issued by the owner/homeowners association or the owner/homeowners association's agent with permission and authority of the owner/homeowners association by or through the consultant, which is signed by the contractor and the owner/homeowners association or the owner/homeowners association's agent and authorizes an addition, deletion, or revision in the work, the scope of work, the Specifications, or any adjustment in the contract price or the contract times. Change orders are issued on or after the effective date of the agreement.

CONDITIONAL WAIVER OF LIEN: An instrument issued by the contractor, subcontractor, sub-subcontractor, or materialman which will release the owner/homeowners association or homeowner/homeowners associations of an owner/homeowners association of all payment liabilities for the amount paid to the contractor. The term conditional shall mean the instrument does not release the owner/homeowners association or homeowner/homeowners associations of an owner/homeowners association from the contractors, subcontractor, sub-subcontractor, supplier, or materialman right to lien the property until all funds are transferred. This instrument shall become an unconditional waiver of lien upon transfer of funds to the Contractor, Subcontractor, Sub-subcontractor, Supplier, or Materialman for the agreed and paid amount.

CONSULTANT: A person, firm, or corporation having a contract with the owner/homeowners association or the owner/homeowners association's agent on behalf of the owner/homeowners association.

CONTRACT DOCUMENTS: The agreement, addenda (which pertain to the contract documents), the completed and signed original request for bid (RFB)by the contractor (including any documentation accompanying the request for bid (RFB) and any other bid documentation submitted prior to the award of the work) when attached as an exhibit to the agreement. Other documents considered part of the contract documents shall be the notice of award, any bonds if required, all general conditions, all specifications issued with the request for bid (RFB) or addenda or change orders, all drawings, all written amendments, all change orders, all work change directives, field orders, and all written interpretations and clarifications of and by the consultant or the consultant's engineer on or after the effective date of the agreement.

CONTRACT SUM: The monies payable by the owner/homeowners association to the contractor for completion of the work in accordance with the contract documents and the request for bid (RFB) and as stated in the agreement.

CONTRACTOR: The person, firm, or corporation with whom the owner/homeowners association or the owner/homeowners association's agent has entered into agreement to perform and complete the work as stated in the contract documents, the scope of work, and the specifications. Also considered the general contractor. The contractor is responsible for all subcontractors and sub-subcontractors.

DATE OF COMMENCEMENT OF WORK: The date agreed to by all parties of the agreement as the date the project is to begin.

DATE OF COMPLETION OF WORK: The date agreed to by all parties that the project shall be substantially completed as defined under substantial completion.

FINAL PAYMENT: The amount paid to the contractor by the owner/homeowners association, the homeowner/homeowners association's owner/homeowners association, or their agent on behalf of the owner/homeowners association for any outstanding sums, agreed retention, or additional work which pays the contractor in full for all work performed under the contract and for the exact contract price including any change orders or addenda.

GENERAL CONTRACTOR: The prime contractor with whom the owner/homeowners association is contracted to complete all tasks of the work listed in the contract documents, the scope of work, and the specifications. Also referred to as the contractor. The general contractor is responsible for all subcontractors and sub-subcontractors.

HOMEOWNERS ASSOCIATION: The public body or authority, corporation, association, firm, or person with whom the contractor has entered into the agreement and for whom the work is to be provided.

HOMEOWNERS ASSOCIATION'S AGENT OR REPRESENTATIVE: The person, firm, or corporation with whom the owner/homeowners association has contracted to represent them and their best interest.

HOMEOWNERS ASSOCIATION'S ASSOCIATION: Whether home, townhome, or condominium, this group shall represent a body of home or property owners who are collectively responsible for their community and hold residency in the same.

INSTRUCTIONS TO BIDDER: Those specific instructions given to the bidder or contractor outlining the requirements to work on the project site, the proper insurance and bond requirements, work hours and days, and all required forms and documents to be provided prior to bidding and starting the project.

LIABILITY INSURANCE: Insurance required by the owner/homeowners association to cover any losses caused by the contractor, the contractor's workers, and equipment, or from the negligence of the contractor, the contractor's workers, subcontractors, and sub-subcontractors.

LICENSED CONTRACTOR: Contractors who possess the proper category license required to perform the work as stated in the scope of work and the bid by task of the request for bid. A contractor who is current in fees with the licensing agency and whose license is current and active.

LIEN: Charges, security interests, or encumbrances upon real property or personal property.

MATERIALMAN: Manufacturer or supplier of materials to be used on the project.

NOTICE OF RIGHT TO LIEN (20-DAY LETTER): A document delivered to the owner/homeowners association to disclose the contractor's right to lien the property for nonpayment of the agreed contract amount within ninety (90) days of substantial completion. This letter is required by state lien laws to allow for a contractor's right to lien.

NOTICE TO PROCEED: A written notice given by the owner/homeowners association to the contractor fixing the date on which the contract times will commence to run and on which the contractor shall start to perform the contractor's obligations under the contract documents.

PENALTY CLAUSE: The clause that enables the owner/homeowners association to charge a contractor for failure to complete the work as set out and defined in the agreement, the contract documents, and the request for bid (RFB).

PROGRESS PAYMENT: A partial payment allowed to the contractor for work completed as agreed between the owner/homeowners association and the contractor.

PROJECT: The actual location and description of the work and referenced by physical name and/or number.

PUNCH LIST: The items identified as needing attention, repairs, or corrections upon completion of the work.

REGISTRAR OF CONTRACTORS: The government agency responsible for controlling and licensing contractors in their specific trades.

REQUEST FOR BID (RFB): Also refered to as request for quotes (RFQ). A document isuued to licensed, bonded and insured contractors containing the scope of work, Instructions to the bidders, insurance requirements and specifications.

RETENTION: The amount of monies agreed to between the owner/homeowners association and the contractor to be held from the full contract amount until the punch list items or damaged areas and property are corrected.

SCOPE OF WORK: The portion of the contract documents and the request for bid that explains to the contractor the specific type of work that is expected to occur on the project and the type of materials that is required to accomplish the work.

SPECIAL CONDITIONS: Any special instructions, materials, specifications, required documents, or additional work that is not listed or performed in the normal routine of pavement or concrete repair or maintenance. These conditions are unique to the project or work.

SPECIFICATIONS: The portion of the contract documents and the request for bid (RFB) which consists of written technical descriptions of materials, equipment, construction systems, standards, and workmanship as they apply to the project, work, and certain administrative details.

STATE COMPENSATORY INSURANCE: Insurance required by the owner/homeowners association to cover any work-related injuries to employees caused by the contractor, the contractor's workers, and equipment, or from the negligence of the contractor, the contractor's workers, subcontractors, and sub-subcontractors.

STOP WORK: A notice given to the contractor by the owner/homeowners association's agent or the consultant at the direction of the owner/homeowners association to stop the work in process due to material, workmanship, or equipment problems, or a violation of specifications.

SUBCONTRACTOR: The person, firm, or corporation with whom the contractor has entered into agreement to perform and complete a portion of the work as stated in the contract documents, the scope of work, and the specifications.

SUBSTANTIAL COMPLETION: The completion of all tasks and the scope of work as defined in Section D2-1 of this contract document.

SUB-SUBCONTRACTOR: The person, firm, or corporation with whom the subcontractor has entered into agreement to perform and complete a portion of the work which the subcontractor has contracted to perform as stated in the contract documents, the scope of work, and the specifications.

SUPPLIER: A manufacturer, fabricator, supplier, distributor, materialman, or vendor having a direct contract with the contractor, subcontractor, or sub-subcontractor to furnish materials or equipment to be incorporated in the work by the contractor, subcontractor, or sub-subcontractor.

TASK: A single item of the work as listed in the bid by task of the request for bid (RFB) and Section A1 of this contract.

TERMINATION OF CONTRACT: The contract shall be cancelled, stopped, discontinued, or ended by decision of the owner/homeowners association or the contractor for reasons of breach of contract or inability to complete the scope of work as stated in the request for bid (RFB), scope of work and the contract.

TERMS: The rules and regulations of performing the work. The agreed items as listed in the contract documents and the request for bid (RFB).

TIME SCHEDULE: The time line to complete the project from the start date to the date of substantial completion.

UNCONDITIONAL WAIVER: An instrument issued by the contractor, subcontractor, sub-subcontractor, supplier, or materialman which releases the owner/homeowners association or homeowner/homeowners associations of an owner/homeowners association of all payment liabilities for the amount paid to the contractor, subcontractor, sub-subcontractor, supplier, or materialman.

UNIT PRICE: Price per unit (sq ft, sq yd, linear ft, etc.) for a listed task.

WORK: The entire completed construction or the various separately identifiable parts thereof required to be furnished under the contract documents and the scope of work. Work includes and is the result of performing or furnishing labor and furnishing and incorporating materials and equipment into the construction and performing or furnishing services and furnishing documents as required by the contract documents and the scope of work.

THIS AGREEMENT IS MADE AS OF THE _____ OF _____ IN THE YEAR OF
_____

BETWEEN THE HOMEOWNERS ASSOCIATION:

AND THE CONTRACTOR:

THE PROJECT IS:

THE WORK IS:

THE CONSULTANT IS:

SECTION A
CONTRACT SCOPE OF WORK AND FORMAT

A1. The Contractor shall complete the project to the Scope of Work outlined in the attached Request for Bid for this project. The Contractor also agrees to install all structures according to any attached drawings without exception or changes unless authorized in writing by the Owner/Homeowners Association, the Owner/Homeowners Association's Agent, or the Consultant on behalf of the Owner/Homeowners Association. The Contractor agrees to complete the Scope of Work as listed in task form and to the quantities listed. By signing this Contract, the Contractor agrees with the Scope of Work and has verified all quantities as listed in the following task breakdown and the Request for Bid. The Contractor shall execute the entire Work described in the Contract Documents and the Request for Bid, except to the extent specifically indicated to be the responsibility of others. The Work is described as follows:

TASK I
TASK II
TASK III
TASK IV
TASK V
TASK VI
TASK VII
TASK VIII
TASK IX
TASK X
TASK XI
TASK XII

SECTION B
DATE OF COMMENCEMENT AND COMPLETION OF THE PROJECT

B1. The project shall commence on the date the Contractor has selected and is shown in the attached time schedule for this project as submitted by the Contractor. The start date shall be strictly adhered to once the Contractor has committed and scheduled the date. The Contractor agrees not to start the project earlier than the scheduled start date unless the Contractor has received written authorization from the Owner/Homeowners Association, the Owner/Homeowners Association's Agent, or the Consultant. The project shall last for the duration of the time established by the Contractor to complete all tasks of the project and the Scope of Work. The start date for the project shall be as agreed to by the Owner/Homeowners Association and the Contractor. Should the project be delayed by weather, equipment malfunction, labor problems of the Contractor, or any other reasons, the project shall be moved to the next reasonable and acceptable available date agreeable to the Owner/Homeowners Association and the Contractor. It shall be the responsibility of the Contractor to supply a new schedule in writing for approval by the Owner/Homeowners Association within twenty-four (24) hours of any cancellation of Work.

THE START DATE OF THIS CONTRACT SHALL BE: _____

The start date is submitted by the Contractor on the Notification of Award form attached and part of this Contract. This date shall become the contracted commencement date and shall be acceptable to the Owner/Homeowners Association and the Contractor UPON SIGNATURE OF THE CONTRACT BY BOTH PARTIES.

B2. The Contractor shall achieve Substantial Completion of the entire project not later than the date shown. Substantial Completion shall be defined as that portion of the Contract where all tasks listed in the Request for Bid (RFB) have been completed. This does not include any punch list items or future warranty work. Substantial Completion of the project shall be determined when all the Work outlined in the Scope of Work and each Task in the BID BY TASK BREAKDOWN in the attached Request for Bid (RFB) is completed and all change orders (if any) have been signed by the Contractor and received by the Owner/Homeowners Association.

**THE COMPLETION DATE OF THIS
CONTRACT SHALL BE**: _____

Subject to adjustments of this Contract Time as provided in the Contract Documents.

The Contractor and Contractor's surety, if any, shall be liable for and pay the Owner/Homeowners Association the sum hereinafter stipulated as liquidated damages for each calendar day until the Work is substantially completed. The penalty for late substantial completion shall be two hundred fifty dollars ($250) per day after the _____

One additional day for completion shall be added for each day of bad weather that prevents the Contractor from performing the Work covered by this agreement. Also, one day additional for completion shall be added for each day of any delay not deemed the Contractor's fault from performing the Work covered by this agreement as listed in Subsection B2. Moving off the project without just cause shall be considered delay as a result of the fault of the Contractor.

SECTION C
CONTRACTED AMOUNT TO PERFORM THE PROJECT

C1. THE HOMEOWNERS ASSOCIATION SHALL PAY THE CONTRACTOR IN CURRENT FUNDS FOR THE CONTRACTOR'S PERFORMANCE OF THE CONTRACT AND THE COMPLETION OF THE PROJECT AS OUTLINED IN THE SCOPE OF WORK AND THE BID TASKS IN THE REQUEST FOR BID AND THE TERMS SHOWN IN SUBSECTION B2. THE HOMEOWNERS ASSOCIATION SHALL PAY THE CONTRACTOR THE CONTRACT SUM OF

_____And _____/100-----------Dollars

($_____ TAX INCLUDED,
subject to all additions and deductions as provided in the Contract Documents and all change orders attached as part of this Contract.

C2. ANY ALTERNATIVES TO THE CONTRACT THAT WILL ALTER THE SCOPE OF WORK, THE QUANTITIES, OR THE PROJECT TASKS WILL HAVE TO BE LISTED BELOW. THIS WILL INCLUDE THE BASE BID FROM SUBSECTION C1 AND ANY AND ALL SUBSEQUENT AMOUNTS AS THEY WILL ALTER THE FINAL AMOUNT OF THE CONTRACT.

1- BASE BID $_____ TAX INCLUDED

C3. UNIT PRICES, IF ANY, ARE LISTED AS FOLLOWS SHOULD THE REQUEST FOR BID NOT BE ATTACHED. FOR UNIT PRICES ASSOCIATED WITH THIS BID, REFER TO THE ATTACHED REQUEST FOR BID AND THE TASK BREAKDOWN IN THE REQUEST FOR BID.

ITEM #   UNIT PRICE   TOTAL

SECTION D
PROGRESS PAYMENTS

D1. UNLESS SO AGREED, THE CONTRACTOR SHALL BE PAID UPON COMPLETION OF THE WORK AS STATED IN THE SCOPE OF WORK AND AS OUTLINED IN THE TASKS FOR BID OF THE ATTACHED REQUEST FOR BID. ANY AND ALL PROGRESS PAYMENTS AND TERMS WILL BE LISTED IN SECTION H.

D2. ALL PAYMENTS ON THE CONTRACT SHALL BE PAID TO THE FOLLOWING CONDITIONS AND SCHEDULE.

    a.   90 PERCENT PAYMENT WITHIN TWENTY (20) DAYS AFTER SUBSTANTIAL COMPLETION OF THE PROJECT. SUBSTANTIAL COMPLETION IS DETERMINED WHEN THE WORK OUTLINED IN THE SCOPE OF WORK IS COMPLETED, EACH TASK LISTED ON THE BID BY TASK BREAKDOWN OF THE REQUEST FOR BID IS COMPLETED, ALL CONDITIONAL WAIVERS ARE SUBMITTED, AND ALL SIGNED CHANGE ORDERS HAVE BEEN SUBMITTED BY THE CONTRACTOR.

    b.   10 PERCENT PAYMENT WITHIN TWENTY (20) DAYS UPON COMPLETION OF ALL PUNCH LIST ITEMS. COMPLETION OF THE PUNCH LIST ITEMS SHALL BE DETERMINED WHEN ALL ITEMS LISTED ON THE PUNCH LIST PROVIDED FROM A FINAL WALK-THROUGH ATTENDED BY THE CONSULTANT, HOMEOWNERS ASSOCIATION'S REPRESENTATIVE, AND THE CONTRACTOR ARE COMPLETED AND ACCEPTED BY THE HOMEOWNERS ASSOCIATION.

    c.   ALL SPECIAL CONDITIONS (IF ANY). THESE SPECIAL CONDITIONS WILL BE LISTED IN SECTION H OF THIS CONTRACT.

D3. THE 90-DAY PERIOD FOR THE TIME LIMIT FOR THE RIGHT TO LIEN SHALL BE SET FOR TWO PORTIONS OF THE CONTRACT IDENTIFIED AS THE SUBSTANTIAL COMPLETION PORTION (90 PERCENT) AND THE PUNCH LIST PORTION (10 PERCENT) OF THE CONTRACT. THE TIME LIMITS FOR ALL LIEN RIGHTS SHALL BE EXECUTED UPON ATTAINING THE SUBSTANTIAL COMPLETION OF THE CONTRACT FOR THE INITIAL CONTRACTED AMOUNT. THE REMAINDER PUNCH LIST ITEMS ARE CONSIDERED A SEPARATE CONTRACT, AND THE TIME LIMITS FOR ALL LIEN RIGHTS SHALL BE EXECUTED UPON COMPLETION OF ALL PUNCH LIST ITEMS, ONCE ALL PUNCH LIST ITEMS ARE COMPLETED AND SIGNED OFF AS COMPLETED BY THE HOMEOWNERS ASSOCIATION'S AGENT AND/OR THE CONSULTANT.

D4. INVOICES, DEMANDS, AND REQUESTS FOR PAYMENT WILL NOT BE SUBMITTED TO THE HOMEOWNERS ASSOCIATION OR MANAGER FOR PAYMENT UNTIL ALL WAIVERS ARE SUBMITTED FROM THE CONTRACTOR, SUBCONTRACTORS, SUB-SUBCONTRACTORS, AND MATERIALMEN. SUBMITTAL OF ALL WAIVERS (CONDITIONAL WAIVER

WHICH BECOMES UNCONDITIONAL UPON PAYMENT OF DEMAND BY THE CONTRAC-
TOR) IS REQUIRED AND IS CONSIDERED AN INTEGRAL PART OF THE CONTRACT AND
THE SCOPE OF WORK. FAILURE TO SUBMIT ALL DOCUMENTS BY THE CONTRACTOR
WILL DELAY THE TIMING FOR RIGHT TO LIEN SINCE THE SUBMITTAL OF ALL DOCU-
MENTS FOR PAYMENT IS CONSIDERED A PART OF THE REQUIREMENTS OF THE 90
PERCENT PAYMENT OF THE CONTRACT.

D5. PAYMENTS DUE AND UNPAID UNDER THE CONTRACT SHALL BEAR INTEREST FROM
THE DATE PAYMENT IS DUE AT THE RATE STATED BELOW. THIS RATE AND ALL TERMS
ARE TO BE AGREED TO BY THE HOMEOWNERS ASSOCIATION AND THE CONTRAC-
TOR PRIOR TO SIGNING OF THE CONTRACT.

(INSERT RATE OF INTEREST AGREED UPON, IF ANY.)

(Usury laws and requirement under the Federal Truth in Lending Act, similar state and local consumer
credit laws, and other regulations at the Owner/Homeowners Association's and Contractor's principal places
of business, the location of the project, and elsewhere may affect the validity of this provision. Legal advice
should be obtained with respect to deletions or modifications and also regarding requirements, such as written
disclosures or waivers.)

SECTION E
FINAL PAYMENT

E1. FINAL PAYMENT, CONSTITUTING THE ENTIRE UNPAID BALANCE OF THE CONTRACT
SUM, SHALL BE MADE BY THE HOMEOWNERS ASSOCIATION OR THE HOMEOWNERS
ASSOCIATION'S AGENT TO THE CONTRACTOR WHEN THE WORK HAS BEEN COM-
PLETED AND VERIFIED BY THE CONSULTANT, ALL PUNCH LIST ITEMS CORRECTED
AND VERIFIED BY THE CONSULTANT, THE CONTRACT FULLY PERFORMED AND VERI-
FIED BY THE CONSULTANT, AND ALL CHANGE ORDERS, ALL FINAL CERTIFICATES FOR
PAYMENT, AND WAIVERS OF LIENS BY THE CONTRACTOR, SUBCONTRACTOR, SUB-
SUBCONTRACTORS, AND MATERIALMEN HAVE BEEN ISSUED. NO FINAL PAYMENT
WILL BE SUBMITTED FOR PAYMENT UNTIL ALL THE ABOVE DOCUMENTS HAVE BEEN
SUBMITTED TO THE HOMEOWNERS ASSOCIATION, THE HOMEOWNERS ASSOCIA-
TION'S AGENT, OR THE CONSULTANT. FAILURE TO SUBMIT ALL DOCUMENTS BY THE
CONTRACTOR WILL DELAY THE TIMING FOR RIGHT TO LIEN SINCE THE SUBMITTAL
OF ALL DOCUMENTS FOR PAYMENT IS CONSIDERED A PART OF THE REQUIREMENTS
OF THE 10 PERCENT RETENTION CONTRACT.

SECTION F
LISTING OF CONTRACT DOCUMENTS

F1. ALL CONTRACT DOCUMENTS ARE LISTED IN ORDER IN SUBSECTION G1 AND
ATTACHED IN ORDER TO THIS CONTRACT DOCUMENT EXCEPT FOR MODIFICA-
TIONS, CHANGE ORDERS, AND ADDENDA ISSUED AFTER EXECUTION OF THIS AGREE-
MENT, AND ARE LISTED AS FOLLOWS.

F2. THIS CONTRACT DOCUMENT IS THE COVER DOCUMENT AND ATTACHED AS SECTION 1.

F3. ANY SUPPLEMENTARY AND/OR ANY OTHER CONDITIONS OF THIS CONTRACT ARE CONTAINED IN THE REQUEST FOR BID OR ANY APPLICABLE PROJECT MANUALS DATED _____

F4. ANY AND ALL SPECIFICATIONS LISTED IN ANY APPLICABLE PROJECT MANUALS DATED _____

F5. ANY AND ALL DRAWINGS, BLUEPRINTS, SKETCHES LISTED AND DATED _____

F6. THE ADDENDA AND CHANGE ORDERS TO THIS CONTRACT ARE LISTED AS FOLLOWS, WITH AN EXPLANATION OF THE CAUSE OF ANY ADDENDA AND CHANGE ORDERS.

ADDENDUM
CHANGE ORDER
NUMBER _____ **DATE** _____ **PAGES**

EXPLANATION:

ADDENDUM
CHANGE ORDER
NUMBER _____ **DATE** _____ **PAGES**

EXPLANATION:

ADDENDUM
CHANGE ORDER
NUMBER _____ **DATE** _____ **PAGES**

EXPLANATION:

ADDENDUM
CHANGE ORDER
NUMBER   **DATE**   **PAGES**

EXPLANATION:

F7. ALL OTHER DOCUMENTS, IF ANY, FORMING A PART OF THIS CONTRACT ARE LISTED IN ORDER AND AS FOLLOWS

    1.   REQUEST FOR BID DATED _____
    2.   INSTRUCTIONS TO BIDDER AS CONTAINED IN THE REQUEST FOR BID DATED _____
    3.   STANDARD SPECIFICATIONS
    4.   NOTICE OF SUCCESSFUL BID AND CONTRACTOR SCHEDULE

SECTION G
GENERAL CONDITIONS OF THIS CONTRACT

The Contractor is responsible for reading and understanding all the general conditions of this Contract. Signature of this Contract indicates the Contractor has read, completely understands, and acknowledges all the terms of Section G, which are the General Conditions of this Contract. There shall be no deviation or alteration of these terms as set forth in this Contract.

SECTION G1. Contract Documents

a. The Contract Documents consist of this Agreement with all the conditions of the Contract (General, Supplementary, and any other Conditions), Drawings, Blueprints, Sketches, Specifications, Addenda issued prior to the execution of this Agreement, Other Documents listed in this Agreement, and all Modifications issued after execution of this Agreement. The intent of the Contract Documents is to include all items necessary for the proper execution and completion of the Scope of Work by the Contractor. The Contract Documents are complementary, and what is required by one shall be as binding as if required by all parties. The performance of the Contractor shall be required only to the extent that is consistent with the Contract Documents, Scope of Work, Request for Bid, Specifications, and all other Contract Documents (Addenda, Change Orders, and Modifications) and reasonably inferable from them as being necessary to produce the intended results.

b. The Contract Documents shall not be construed to create a contractual relationship of any kind between the Consultant and the Contractor, the Owner/Homeowners Association and a Subcontractor or Sub-subcontractor, or between any persons or entities other than the Owner/Homeowners Association and the Contractor.

c. Execution of this Contract by the Contractor is a representation that the Contractor has visited the work site and has become completely familiar with the local conditions under which the Scope of Work is to be performed. The Contractor agrees to have examined all aspects of the work site, the Scope of Work, and the Bid Tasks as outlined and agreed to in the Request for Bid for this project. Prior to submitting a completed proposal, the Contractor certifies a complete and careful examination of the Proposal and Bid Documents, visited the work site, and fully informed themselves as to all existing conditions and limitations, and shall include in the Proposal a sum to cover the cost of all items included in the Request for Bid and the Bid package. The Contractor, when awarded this Contract, shall not be allowed any extra compensation by reason of any matter or thing, that would have become known if the Contractor would have fully informed himself/herself prior to submitting a completed bid. Signing this Contract shall indicate the Contractor has agreed to the existing specified Scope of Work and will complete the project as specified. The quantities listed in the Request for Bid documents are the measurements for bidding purposes only, and by signing this Contract, the Contractor has certified he/she has verified all quantities and will complete the project as outlined in the Scope of Work and as specified.

d. The term Work means the construction and services required by the Contract Documents, whether completed or partially completed, and includes all other labor, materials, equipment, and services provided or to be provided by the Contractor to fulfill the Contractor's obligations. The Work may constitute the whole or a part of the project.

SECTION G2. THE OWNER/HOMEOWNERS ASSOCIATION AND MANAGEMENT COMPANY

a.   The Owner/Homeowners Association may appoint an Agent (Manager, Management Company, or Asset Manager) to represent the Owner/Homeowners Association's best interest. Should the Owner/Homeowners Association appoint an Agent, the Agent shall be the contact person for the Owner/Homeowners Association during the entire project. All correspondence shall be directed to the Owner/Homeowners Association or the appointed Agent.

b.   The Owner/Homeowners Association or the Owner/Homeowners Association's Agent (Manager, Management Company, or Asset Manager) shall furnish and provide surveys, blueprints, and legal descriptions of the work site.

c.   Except for permits and fees, which are the responsibility of the Contractor under the Contract Documents, the Owner/Homeowners Association or the Owner/Homeowners Association's Agent shall secure and pay for necessary approvals, easements, assessments, and charges required for the construction, use, or occupancy of permanent structures or permanent changes in existing facilities.

d.   Should the Contractor fail to correct Work which is not in accordance with the requirements of the Contract Documents or persistently fail to carry out the Work in accordance with the Contract Documents, the Owner/Homeowners Association or the Owner/Homeowners Association's Agent, by a written order, may order the Contractor to stop the Work, or any portion thereof, until the cause for such order has been eliminated; however, the right of the Owner/Homeowners Association to stop the Work shall not give rise to a duty on the part of the Owner/Homeowners Association to exercise this right for the benefit of the Contractor or any other personal entity. The Owner/Homeowners Association or the Owner/Homeowners Association's Agent may hire or appoint a Consultant to represent their interests with the same authority as authorized by the Owner/Homeowners Association.

SECTION G3. THE CONTRACTOR

a.   The Contractor shall possess all required, proper licenses in the proper category for the type of work performed with the local licensing agency or the Registrar of Contractors. The Contractor shall also have the proper license bonding or reserve on record with the local licensing agency and/or the Registrar of Contractors.

b.   The Contractor shall have the capabilities of acquiring all required bonding or letters of credit necessary and as required by the Owner/Homeowners Association in the form of Bid Bonds and Performance Bonds.

c.   The Contractor shall supervise and direct the Work, using the Contractor's best skill and attention to ensure the project is being installed to the Scope of Work and Specifications. The Contractor shall be solely responsible for and have control over the construction means, methods, techniques, sequences, and procedures, and coordinating all portions of the Work under the Contract unless the Contract Documents or Scope of Work give other specific instructions concerning these matters. The Owner/Homeowners Association and/or the Owner/Homeowners Association's Agent or Consultant shall only advise and offer direction unless so specified in the Contract Documents.

d.   Unless provided in the Contract Documents, the Contractor shall provide and pay for all labor, material, equipment, tools, construction equipment and machinery, water, heat, utilities, transportation, and all other facilities and services necessary for the proper execution and completion of the Work, whether temporary or permanent and whether or not incorporated or to be incorporated into the Work.

e. The Contractor shall enforce strict discipline and good conduct among the Contractor's employees and other persons carrying out the Contract. There shall be no taunting or harassing of any persons employed or not employed by the Contractor that will make any person feel uncomfortable. The employees shall treat all persons in and around the project with complete respect. The Contractor's employees shall wear proper clothing and dress in a manner acceptable to the Owner/Homeowners Association. Any employee of the Contractor who is not dressed appropriately shall be asked to leave the project until he/she can return wearing proper clothing. The Contractor will ensure the employees are participating in a company drug awareness program under the Contractor's guidelines. At no time shall any employee be allowed on the work site who is intoxicated or under the influence of any drug, chemical, or alcohol. The Contractor shall have a competent and skilled leader and employees performing the tasks for this project. The leader and the majority of the skilled employees shall be on the project throughout the entire workday. All unskilled persons and trainees shall be used in a limited fashion and under the supervision of a skilled leader and other skilled employees.

f. The Contractor warrants to the Owner/Homeowners Association and Consultant that all materials and equipment furnished under the Contract will be of high quality and as specified in the Specification section of the Request for Bid unless otherwise required or permitted by the Contract Documents, that the Work will be free from defects not inherent in the quality required or permitted, and the Work will conform to the requirements of the Contract Documents, the Scope of Work, and the Specifications. Work not conforming to the requirements of the Request for Bid, Scope of Work, or the Specifications, including substitutions not properly approved and authorized, may be considered unacceptable. The Contractor's warranty excludes remedy for damage or defect caused by abuse, modifications not executed by the Contractor, improper or insufficient maintenance, improper operation, or normal wear and tear under normal usage. If required by the Owner/Homeowners Association or the Consultant, the Contractor shall furnish satisfactory evidence as to the kind and quality of materials and equipment.

g. Unless otherwise provided in the Contract Documents, the Scope of Work, or the Request for Bid, the Contractor shall pay sales, consumer, use, and other similar taxes which are legally enacted when bids are received or negotiations concluded, whether or not yet effective or merely scheduled to go into effect, and where applicable and so stated in the Request for Bid, shall secure and pay for all building, engineering, public works, and other permits and governmental fees, licenses, and inspections necessary for proper execution and completion of the Work.

h. The Contractor shall comply with and give notices required by laws, ordinances, rules, regulations, and lawful orders of public authorities bearing on performance of the Work. The Contractor shall promptly notify the Owner/Homeowners Association or Consultant if the Drawings and Specifications are observed by the Contractor to be at variance therewith.

i. The Contractor shall be responsible to the Owner/Homeowners Association for the acts and omissions of the Contractor's employees, Subcontractors, Sub-subcontractors, and their Agents and employees, and other persons performing portions of the Work under a Contract with the Contractor.

j. The Contractor shall review, approve, and submit to the Owner/Homeowners Association, Owner/Homeowners Association's Agent, or Consultant all Shop Drawings, Product Data, Samples, Materials Safety Data Sheets, and all similar submittals required by the Contract Documents, Scope of Work, or Request for Bid with reasonable promptness. The Work shall be in accordance with approved submittals. When professional certifications of performance criteria of materials, systems, or equipment is required by the Contract Documents or the Request for Bid, the Owner/Homeowners Association, the Owner/Homeowners Association's Agent, or Consultant shall be entitled to rely upon the accuracy and completeness of such certifications.

k. The Contractor shall keep the premises and surrounding area free from accumulation of waste materials or rubbish caused by operations under the Contract. The Contractor shall, before starting Work, remove or otherwise cover and/or protect all landscaping, curbing, equipment, accessories, signs, lighting fixtures, resident/tenant and customer personal property, and similar items or provide ample protection of such items. Upon completion of each work area, replace any of the above items if damaged during the maintenance or construction phase. Protect adjacent surfaces as required or directed. Any damage done shall be repaired by the Contractor at his/her expense. The Contractor shall observe all safety precautions as recommended by OSHA and the governing agency in which the project resides. All debris from the operation shall be removed daily from the job site and disposed of at an approved, suitable disposal site. The Contractor will not dispose of any debris or materials in the dumpsters. Should the Contractor dispose of any material in the dumpsters, the Contractor will pay all charges and fees to have the material hauled off and disposed of. The Contractor shall clean all sidewalks, driveways, building entrances, or any other surfaces adjacent to the work area of all dust, dirt, rocks, or any other debris by sweeping or air blowing immediately after the maintenance material has been applied. When the Contractor destroys, alters, defaces, or in any way damages any property adjacent to the project work area, the Contractor shall repair the damaged property to the condition equal to or better than it was before the Work began. This includes, but is not limited to, structures, walls, lawns, sprinkler systems, plumbing, irrigation, landscaping, all concrete structures, etc. The debris shall be cleaned away from all building entrances. The Contractor and his/her employees shall not track or walk back over the new Work, or track materials onto adjacent areas such as walks, patios, curbs, outdoor carpeting, etc. Final cleanup of the project shall consist of replacing any property disturbed by the Work; removing all debris, scraps, and containers from the work site; repairing, cleaning, painting, or replacing defaced or disfigured finishes and surfaces caused by Work performed, including curbs, sidewalks, walls, or any and all surfaces where construction materials have splashed or been oversprayed. The Contractor shall be responsible for cleaning all walks, floors, carpets, etc. should the material being used be tracked onto these surfaces. The Contractor shall be responsible for cleaning all property (autos, walls, windows, or any personal property or building fixtures) of material or residue as a result of the construction on all legitimate claims. These claims will be determined legitimate by the Owner/Homeowners Association, the Owner/Homeowners Association's Agent, or the Consultant.

The Contractor shall NOT dump or put construction materials, waste, debris, or anything into dry wells or dumpsters, or on landscaped areas at any time. The Contractor shall properly dispose of any and all material, either removed from the job site or leftover product, in a proper and environmentally safe manner.

l. The Contractor shall provide the Owner/Homeowners Association, the Owner/Homeowners Association's Agent, or the Consultant with access to the Work in preparation and progress wherever located.

m. The Contractor shall pay all royalties and license fees; shall defend suits or claims for infringement of patent rights; and shall hold the Owner/Homeowners Association, the Owner/Homeowners Association's Agent, or the Consultant harmless from loss on account thereof, but shall not be responsible for such defense or loss when a particular design, process, or product of a particular manufacturer or manufacturers is required by the Contract Documents unless the Contractor has reason to believe that there is an infringement of patent.

n. The Contractor agrees to indemnify and hold the Owner/Homeowners Association and its Agents, Consultants, employees, and residents harmless from claims, costs, suits, judgments, expenses, attorney's fees, and any and all professional fees on account of any damages which arise out of the respective acts or omissions of the Contractor, its Subcontractors, its Sub-subcontractors, its Consultants, its Agents, and its employees.

The Owner/Homeowners Association agrees to indemnify and hold the Contractor, its Subcontractors, its Sub-subcontractors, its Consultants, its Agents, and its employees harmless from claims, suits, judgments, expenses, attorney's fees, and any and all other professional fees on account of any damages which arise out of the respective acts or omissions of the Owner/Homeowners Association or its Consultants, Agents, and employees.

Any indemnification or hold harmless obligations of the Contractor shall extend only to the claims relating to bodily injury and property damage (and then only to that part or proportion) of any claims, damage, loss, or defect that results from the negligence or intentional act of the indemnifying party or someone for whom they are responsible. The Contractor shall not under any circumstance have a duty to defend any other persons or entities not connected to the Contractor by this Contract.

The Contractor shall not be required to indemnify any other parties from damages, attorney's fees, or personal or property damage for any amount exceeding the proportional amount of the Contractor's direct cause of such damages. The Contractor shall be liable for fines or assessments made against the Owner/Homeowners Association that are directly the fault or negligence of the Contractor. The Owner/Homeowners Association and the Contractor shall have the right of subrogation for claims to the proportion of their responsibility. The Contractor shall be responsible for all claims against their Subcontractors and Sub-subcontractors, their Agents, their Consultants, and their employees as it pertains to the Work. The Contractor shall be responsible for indemnifying the Owner/Homeowners Association of their actions and the actions of their Subcontractors, Sub-subcontractors, Agents, Consultants, and employees for negligence, personal property, and personal injury.

o. Claims against any persons or entities indemnified under this Paragraph by an employee of the Contractor, a Subcontractor, a Sub-subcontractor, or anyone directly or indirectly employed by them or anyone for whose acts they may be liable, the indemnification obligation under this Paragraph shall not be limited by a limitation on amount or type of damages, compensation, or benefits payable by or for the Contractor, a Subcontractor, and a Sub-subcontractor under workers' or workmen's compensation acts, disability benefit acts, or other employee benefit acts.

p. The obligations of the Contractor under this Paragraph shall not extend to the liability of the Consultant, the Consultant's Consultants, architects, Agents, and employees of any of them arising out of (1) the preparation or approval of maps, drawings, opinions, reports, surveys, Change Orders, Construction Change Directives, designs, or Specifications, or (2) the giving of or the failure to give directions or instructions by the Consultant, the Consultant's Consultants, architects, Agents, and employees of any of them provided such giving or failure to give is the primary cause of the injury or damage.

q. Should the completion of the project be delayed due to the fault of the Contractor by incomplete work, unacceptable work, poor or improper application of construction materials, or insufficient cleanup and corrective Work that will require additional services, testing, mediation, or arbitration of the Consultant, the Contractor, by the decision of the Owner/Homeowners Association, will be responsible for all additional fees charged by the Consultant for these additional services. These fees will be deducted from any retention left unpaid to the Contractor, and any balance left unpaid shall be invoiced by the Consultant to the Contractor.

SECTION G4. ADMINISTRATION OF THE CONTRACT

a. The Owner/Homeowners Association's Agent, and/or the Consultant will provide administration of the Contract and will be the Owner/Homeowners Association's Representative (1) during construction, (2) until final payment is due, and (3) with the Owner/Homeowners Association's concurrence, from time to time during the correction period described in Section G6.

b. The Owner/Homeowners Association's Agent and/or the Consultant will visit the site at intervals appropriate to the stage of construction to become generally familiar with the progress and quality of the completed Work and to determine in general if the Work is being performed according to the Scope of Work and the Specifications attached to the Request for Bid and attached with this Contract. The Owner/Homeowners Association's Agent and/or the Consultant will not be required to make exhaustive or continuous on-site inspections to check the quality or the quantity of the Work. On the basis of on-site observations, the Owner/Homeowners Association's Agent and/or the Consultant will keep the Owner/Homeowners Association informed of the progress of the Work and will endeavor to guard the Owner/Homeowners Association against defects and deficiencies in the Work.

c. The Owner/Homeowners Association's Agent and/or the Consultant will not have control over or charge of and will not be responsible for construction means, methods, techniques, sequences, or procedures, or for safety precautions and programs in connection with the Work, since these are solely the Contractor's responsibility as provided and stated in Sections G1a and G10a. The Owner/Homeowners Association's Agent and/or the Consultant will not be responsible for the Contractor's failure to carry out the Work in accordance with the Scope of Work, Specifications, Change Orders, or the Contract Documents.

d. Based on the Owner/Homeowners Association's Agent's and/or the Consultant's observations and evaluations of the Contractor's Applications for Payment, the Owner/Homeowners Association's Agent and/or the Consultant will review the amounts due the Contractor upon receipt of the Contractor's invoice, will compare the amounts invoiced to the amount of the Contract and any change orders, and after approval shall submit all invoices in the approved amounts, with appropriate waivers of liens from the Contractor, Subcontractor, Sub-subcontractors, and Materialmen, for payment from the Owner/Homeowners Association.

e. The Owner/Homeowners Association's Agent and/or the Consultant will interpret and will have input to the matters concerning performance under and requirements of the Scope of Work, Specifications, and Contract Documents on written request of either the Owner/Homeowners Association or Contractor. The Owner/Homeowners Association's Agent and/or the Consultant will make initial decisions on all claims, disputes, or other matters in question between the Owner/Homeowners Association and Contractor, but will not be liable for results of any interpretations or decisions rendered in good faith. The Owner/Homeowners Association shall have all final decisions in matters relating to the maintenance of his pavement or related concrete. The Owner/Homeowners Association's decisions in matters relating to any aesthetic effect will be final if consistent with the intent expressed in the Scope of Work, Specifications, or Contract Documents.

f. The Owner/Homeowners Association's Agent and/or the Consultant will have the authority to reject Work which does not conform to the Scope of Work, Specifications, or the Contract Documents, and will do so with the permission of the Owner/Homeowners Association or the Owner/Homeowners Association's Agent.

g. The Owner/Homeowners Association's Agent and/or the Consultant will review and approve, or take whatever appropriate action needed, upon the Contractor's submittals such as Shop Drawings, Product Data and Samples, Change Order requests, Material Data and Specification Sheets, and Laboratory Materials Test and Design Data Records, but only for the limited purpose of checking for conformance with information given and the design concept expressed in the Specifications and the Contract Documents.

h. <u>Arbitration of Disputes:</u> Any dispute or claim in law or equity arising out of this Contract, or relating to the Scope of Work described in this Contract that is not settled through mediation shall be determined by neutral, binding arbitration. The arbitration shall be conducted in accordance with Local Rules of Civil Procedure or the rules of the American Arbitration Association (AAA). The parties to the arbitration may agree in writing to use different rules and/or arbitrator(s). In all other respects, the arbitration shall be conducted in accordance with the State Rules of Civil Procedure. Judgment upon the award rendered by the arbitrator(s) may be entered in any court having jurisdiction thereof. Any action that is within the jurisdiction of a probate or small claims court is excluded from arbitration under this agreement. The filing of judicial action to enable the recording of a notice of pending action, for order of attachment, receivership, injunction, or other provisional remedy, shall not constitute a waiver of the right to arbitrate under this provision. Venue for any mediation, arbitration, or litigation relating to this Contract shall be in the local county.

i. <u>Attorney's Fees:</u> The prevailing party in any dispute (whether or not it is submitted to arbitration) shall be awarded its reasonable attorney's fees, any and all costs, and disbursements of counsel and expert witness fees, in addition to any other relief awarded as set forth and determined by said arbitrator.

SECTION G5. SUBCONTRACTS AND SUBCONTRACTORS

a. A Subcontractor is a person or entity who has a direct Contract with the Contractor to perform a portion of the Work at the work site. The Subcontractor must be a licensed contractor in the work category the Subcontractor performs. A Sub-subcontractor is a person or entity who has a direct Contract with the Contractor through the Subcontractor to perform a portion of the Work at the work site. The Sub-subcontractor must be a licensed contractor in the work category the Sub-subcontractor performs.

b. Unless otherwise stated in the Request for Bid, the Contract Documents, or the bidding requirements, the Contractor, as soon as practicable after award of the Contract, shall furnish in writing to the Owner/ Homeowners Association through the Consultant the names of all the Subcontractors, Sub-subcontractors, and Material Suppliers for each of the principal portions of the Work. The Contractor shall not contract with any Subcontractor or Sub-subcontractor about whom the Owner/Homeowners Association, Owner/Homeowners Association's Agent, or Consultant has made reasonable and timely objection. The Contractor shall not be required to contract with anyone about whom the Contractor has made reasonable objection. Contracts between the Contractor and Subcontractors, or Subcontractors and Sub-subcontractors, shall (1) require each Subcontractor and Sub-subcontractor, to the extent of the Work to be performed by the Subcontractor or Sub-subcontractor, to be bound to the Contractor by the terms of the Scope of Work, Request for Bid, Specifications, and the Contract Documents, and to assume toward the Contractor all the obligations and responsibilities which the Contractor, by the Scope of Work, Request for Bid, Specifications, and the Contract Documents, assumes toward the Owner/Homeowners Association and Consultant, and (2) allow the Subcontractor and Sub-subcontractor the benefit of all rights, remedies, and redress afforded to the Contractor by the Scope of Work, Request for Bid, Specifications, and the Contract Documents.

SECTION G6. CONSTRUCTION BY THE HOMEOWNERS ASSOCIATION AND ADDITIONAL CONTRACTORS

a. The Owner/Homeowners Association reserves the right to perform construction or operations related to the project with the Owner/Homeowners Association's own forces, and to award separate Contracts in connection with other portions of the project or other construction or operations on the site under conditions of the contract identical or substantially similar to these, including those portions related to insurance and waiver of subrogation. If the Contractor claims that delay or additional cost is involved because of such action by the Owner/Homeowners Association, the Contractor shall make such claim as provided elsewhere in the Contract Documents.

b. The Contractor shall afford the Owner/Homeowners Association and separate contractors reasonable opportunity for the introduction and storage of their materials and equipment and performance of their activities, and shall connect and coordinate the Contractor's construction and operations with theirs as required by the Contract Documents.

c. Costs caused by delays, improperly timed activities, or defective construction shall be the responsibility of the party responsible therefore.

SECTION G7. CHANGES IN THE SCOPE OF WORK AND CHANGE ORDERS TO WORK

a. The Owner/Homeowners Association, without invalidating the Contract, may order changes in the Work consisting of additions, deletions, or modifications, the Contract Sum and Contract Time being adjusted accordingly. Such changes in the Work shall be authorized by written Change Order signed by the Owner/Homeowners Association, Contractor, and Consultant, or by written Construction Change Directive signed by the Owner/Homeowners Association and Consultant.

b. The Contract Sum and Contract Time shall be changed only by a Change Order issued by the Owner/Homeowners Association signed by the Owner/Homeowners Association and the Contractor.

c. The cost or credit to the Owner/Homeowners Association from a change in the Work shall be determined by mutual agreement between the Owner/Homeowners Association and the Contractor.

d. Any additional Work or materials required above and beyond the Contract shall be executed with an agreed change order between the Owner/Homeowners Association, Manager, Consultant, and Contractor. Any additional Work performed or materials used will not be compensated for if not approved prior to application and accompanied by an authorized change order. All labor and materials used for change orders must be verified, and any unauthorized labor and/or materials will not be paid for.

e. Any tasks added to the Contract by change order shall be at the bid price and/or bid unit price as listed on the bid sheet in the Request for Bid. At no time shall the price for change order items be more than the listed price for similar tasks in the Request for Bid. Should the price for a change order task be in excess of the bid task amount and/or the unit amount, the price will be adjusted accordingly.

f. Where a minimum amount is listed for a task and any additional Work is included which will keep the unit amount under the minimum charge, no additional compensation will be added.

SECTION G8. CONTRACT AND WORK TIME, AND COMPLETION OF WORK TIME

a. Time limits stated in the Scope of Work and the Contract Documents are of the essence of the Contract. The Contractor is given the opportunity to set a reasonable schedule on the Bid by Tasks of the Request for Bid by declaring the number of days it will take to complete the project. Also, the Contractor has submitted a schedule in writing which is acceptable to the Owner/Homeowners Association and the Consultant This schedule has become part of the Contract Documents and will be referred to for any and all liquidated damages caused by over-running the scheduled time to complete the Work. By executing the Agreement, the Contractor confirms that the Contract Time is a reasonable period for performing the Work.

The Contractor will be allowed to work between the hours set out in the Request for Bid. The Contractor agrees to work only within the hours set out in the Request for Bid including cleanup time at the end of the workday. It is recommended that the Contractor bid according to his/her ability to complete the project in the designated number of days predicted by the Contractor in order to complete the project with diligence and prudence. The Contractor shall not be allowed to work after a set time as called out in the Request for Bid. Therefore cleanup and shutdown time will need to be allowed in the workday.

The Owner/Homeowners Association will assess an established penalty per day, as shown in the Request for Bid, if the Contractor works beyond the submitted schedule of days. The exception to this rule is delays due to bad or inclement weather, or other acts of God that are not the fault of the Contractor and further listed in item c below.

b. The date of Substantial Completion shall be designated when all tasks listed in the Bid by Tasks in the Request for Bid are completed, all change orders (if any) are signed by the Contractor and submitted to the Owner/Homeowners Association for approval, and all Conditional Waivers of Liens from the Contractor, Subcontractors, and Sub-subcontractors have been submitted with the Contractor's Invoice. Should the Work be completed and the Contractor has not submitted all signed change orders (if any) to the Contract to the Owner/Homeowners Association, the Owner/Homeowners Association's Agent, or the Consultant, then the date of substantial completion shall be the date the signed change orders are received from the Contractor.

c. If the Contractor is delayed at any time in progress of the Work by changes ordered in the Work, labor disputes, fire, unusual delay in deliveries, abnormal adverse weather conditions not reasonably anticipatable, unavoidable casualties, or any causes beyond the Contractor's control, or other causes which the Consultant determines may justify delay, then the Contract Time shall be extended by Change Order for such reasonable time as the Consultant may determine.

d. Should the Work be delayed for any reason, the Contractor shall continue Work on the project the first day following the delay.

SECTION G9. PAYMENTS ON CONTRACTS AND FINAL PAYMENT ON COMPLETION

a. Payments shall be made as provided in Sections D and E of this Contract.

b. Payments may be withheld on account of (1) defective Work not remedied, (2) claims filed by third parties, (3) failure of the Contractor to make payments properly to Subcontractors and Sub-subcontractors or for labor, materials, or equipment, (4) reasonable evidence that the Work cannot be completed for the unpaid balance of the Contract Sum, (5) damage to the Owner/Homeowners Association or another contractor, (6) reasonable evidence that the Work will not be completed within the Contract Time and that the unpaid balance would not be adequate to cover actual or liquidated damages for the anticipated delay, or (7) persistent failure to carry out the Work in accordance with the Scope of Work, Specifications, or the Contract Documents.

c. When the Consultant agrees that the Work is substantially complete and notifies the Owner/Homeowners Association, the Consultant shall provide the Owner/Homeowners Association with the invoices and Waivers of Liens from the Contractor for payment as outlined in Section D of this Contract.

d. Final payment shall not become due until the Contractor has delivered to the Owner/Homeowners Association all completed releases of all liens from the Contractor, Subcontractors, Sub-subcontractors, and Materialmen arising out of this Contract or receipts in full covering all labor, materials, and equipment for which a lien could be filed and a Preliminary Notice of Lien (20-day letter), or a bond satisfactory to the Owner/Homeowners Association to indemnify the Owner/Homeowners Association against any liens. If any liens remain unsatisfied after payments are made, the Contractor shall refund to the Owner/Homeowners Association all money that the Owner/Homeowners Association may be compelled to pay in discharging any liens, including all costs and reasonable attorney's fees.

e. The making of the final payment shall constitute a waiver of claims by the Owner/Homeowners Association except those arising from:

   1. Liens, claims, security interests, or encumbrances arising out of the Contract and unsettled;
   2. Failure of the Work to comply with the requirements of the Contract Specifications and Contract Documents; or
   3. Terms of special warranties required by the Contract Documents.

   Acceptance of final payment by the Contractor, a Subcontractor, a Sub-subcontractor, or Material Supplier shall constitute a waiver of claims by that payee except those previously made in writing and identified by that payee as unsettled at the time of final Application for Payment.

f. The Contractor shall provide, prior to submitting invoices for final payment, all waivers of liens from all Subcontractors, Sub-subcontractors, Material Suppliers, and any persons or entities that have submitted a Notice of Right to Lien (commonly referred to as a 20-day letter) to the Owner/Homeowners Association, the Owner/Homeowners Association's Agent, or the Consultant.

g. All invoices must be accompanied with all Conditional Waiver of Liens from the Contractor, Subcontractors, Sub-subcontractors, and Material Suppliers who have served notice with a Preliminary Notice of Lien (which will become an Unconditional Waiver of Lien upon receipt and endorsement of the demand deposit document) for the 90 percent payment portion.

   WITHOUT THE ACCOMPANYING CONDITIONAL WAIVERS, THE INVOICES WILL NOT BE SUBMITTED FOR PAYMENT. ONLY INVOICES WITH PROPER WAIVER DOCUMENTS WILL BE SUBMITTED FOR PAYMENT.

h. Upon completion of the Punch List Items, an invoice for final payment should be submitted and accompanied with a Conditional Waiver of Lien (which will become an Unconditional Waiver of Lien upon receipt and endorsement of the demand deposit document) for the 10 percent retention amount.

   WITHOUT THE ACCOMPANYING CONDITIONAL WAIVERS, THE INVOICES WILL NOT BE SUBMITTED FOR PAYMENT. ONLY INVOICES WITH PROPER WAIVER DOCUMENTS WILL BE SUBMITTED FOR PAYMENT.

SECTION G10. PROTECTION FROM BODILY INJURY AND DAMAGE

a. The Contractor shall be responsible for initiating, maintaining, and supervising all safety precautions and programs in connection with the performance of the Contract. The Contractor shall take reasonable precautions for safety of, and shall provide reasonable protection to prevent damage, injury, or loss to:

   1. The Contractors, Subcontractors, Sub-subcontractor's employees on the Work site, the Owner/ Homeowners Association and his/her Agent and Consultant, the Owner/Homeowners Association's employees, the Agent's employees, the Tenants, the Tenant's employees and customers, the Consultant's employees, and all other persons who may be affected thereby;
   2. The Work and materials and equipment to be incorporated therein; and
   3. All other property of the Owner/Homeowners Association, the Owner/Homeowners Association's Agent and Consultant, Tenants, and customers of the Owner/Homeowners Association and Tenants at the work site or adjacent thereto.

   The Contractor shall give notices and comply with applicable laws, ordinances, rules, regulations, and lawful orders of all public authorities bearing on safety of persons and property and their protection from damage, injury, or loss. The Contractor shall promptly remedy damage and loss to property at the site caused in whole or in part by the Contractor, a Subcontractor, a Sub-subcontractor, or anyone directly or indirectly employed by any of them, or by anyone for whose acts they may be liable and for which the Contractor is responsible except for damage or loss attributable to acts or omissions of the Owner/Homeowners Association, the Owner/Homeowners Association's Agent, or the Consultant or by anyone for whose acts either of them may be liable, and not attributable to the fault or negligence of the Contractor. The foregoing obligations of the Contractor are in addition to the Contractor's obligations under Subsection G3-n (The Contractor).

b. The Contractor shall advise the Owner/Homeowners Association, Owner/Homeowners Association's Agent, or the Consultant of any and all damage to private and personal property or personal injury on the project or the property immediately and in writing. The Contractor shall also, in writing, indicate the action taken, or to be taken, to correct any and all damage or damages and claims. This notification shall be within twelve (12) hours of the occurrence.

c. The Contractor shall not be required to perform without consent any Work relating to asbestos or polychlorinated biphenyl (PCB).

SECTION G11. INSURANCE, ADDITIONAL INSUREDS, INSURANCE RIDER

a. The Contractor shall purchase and maintain, from a company or companies lawfully authorized to do business in the jurisdiction in which the project is located, insurance for protection from claims under workers' or workmen's compensation acts and other employee benefit acts which are applicable, claims for damages because of bodily injury, including death, and claims for damages, other than to the Work itself, to property which may arise out of or result from the Contractor's operations under the Contract, whether such operations be by the Contractor or by a Subcontractor, Sub-subcontractor, or anyone directly or indirectly employed by any of them. This insurance shall be written for not less than limits of liability specified in the Insurance Rider listed in Subsection G11-g, the Request for Bid, and the Contract Documents or required by law, whichever coverage is greater, and shall include contractual liability insurance applicable to the Contractor's obligations. Certificates of such insurance shall be filed with the Owner/Homeowners Association prior to the commencement of the Work naming the Owner/Homeowners Association, the Owner/Homeowners Association's Agent, and the Consultant as additional insureds.

b. The Owner/Homeowners Association shall be responsible for purchasing and maintaining the Owner/Homeowners Association's usual liability insurance. Optionally, the Owner/Homeowners Association may purchase and maintain other insurance for self-protection against claims which may arise from operations under the Contract. The Contractor shall not be responsible for purchasing and maintaining this optional Owner/Homeowners Association's liability insurance unless specifically required and stipulated by the Contract Documents or in the Request for Bid.

c. Unless otherwise provided, the Owner/Homeowners Association shall purchase and maintain, from a company or companies lawfully authorized to do business in the jurisdiction in which the project is located, property insurance upon the entire Work at the site to the full insurable value thereof. This insurance shall be on an all risk policy form and shall include interests of the Owner/Homeowners Association, the Owner/Homeowners Association's Agent, the Consultant, the Contractor, Subcontractors, and Sub-subcontractors in the Work and shall insure against the perils of fire and extended coverage and physical loss or damage including, without duplication of coverage, theft, vandalism, and malicious mischief.

d. A loss insured under the Owner/Homeowners Association's property insurance shall be adjusted with the Owner/Homeowners Association and made payable to the Owner/Homeowners Association as fiduciary for the insureds, as their interests may appear, subject to the requirements of any applicable mortgagee clause.

e. The Owner/Homeowners Association shall file a copy of each policy with the Contractor before an exposure to loss may occur. Each policy shall contain a provision that the policy will not be cancelled or allowed to expire until at least thirty (30) days' prior written notice has been given to the Contractor.

f. The Owner/Homeowners Association and Contractor waive all rights against each other and Owner/Homeowners Association's Agent, the Consultant, Consultants to the Consultant as Associates, separate contractors described in Section H, if any, and any of their subcontractors, Sub-subcontractors, Agents, and employees, for damages caused by fire or other perils to the extent covered by property insurance obtained pursuant to Subsection G11 or any other property insurance applicable to the Work, except such rights as they may have to the proceeds of such insurance held by the Owner/Homeowners Association as fiduciary. The Contractor shall require similar waivers in favor of the Owner/Homeowners Association and the Contractor by Subcontractors and Sub-subcontractors. The Owner/Homeowners Association shall require similar waivers in favor of the Owner/Homeowners Association and Contractor by the Owner/Homeowners Association's Agent, Consultant, Consultants to the Consultant as Associates, separate contractors described in Section H, if any, and the Subcontractors, Sub-subcontractors, Agents, and employees of any of them.

g. Supplementing the terms of Section G11 of this Agreement, the Contractor shall obtain and keep in full force and effect the following insurance coverage (the "Insurance Coverage"):

A. Comprehensive General Liability Insurance in the form satisfactory to the industry and the insurance carrier as required by state law and with the following coverage: premises-operations, products/completed operations (the insured shall keep the insurance coverage in force until completion of the project and for ninety [90] days after), personal injury, blanket contractual liability, independent contractors, and broad form property damage (which shall include coverage for explosion and underground hazards). The limit of such coverage shall be in an amount not less than three million dollars ($3,000,000) combined single limit per occurrence for bodily injury and property damage.

B. Workers' Compensation Insurance in accordance with state and local statutory limits, in the form satisfactory to the industry and the insurance carrier as required by state law, including employer's liability in an amount not less than five hundred thousand dollars ($500,000) per accident.

C. Automobile Liability Insurance (owned, non-owned, and hired) in the form satisfactory to the industry and the insurance carrier as required by state law in the amount not less than one million dollars ($1,000,000) combined single limit per occurrence for bodily injury and property damage.

1. On or before the date of this agreement, the Contractor shall provide to the Owner/Homeowners Association Certificate(s) of Insurance in form and substance satisfactory to the Owner/Homeowners Association, evidencing the Insurance Coverage. In no event shall (a) the Contractor (or Subcontractors, or Sub-subcontractors) enter the property or (b) any Work or Service by the Contractor (or any Subcontractor, or Sub-subcontractor) commence prior to receipt of such certificates.

2. Certificates of Insurance and insurance policies set forth herein, regardless of completed Contract price, must include as additional insureds: the Owner/Homeowners Association, the Owner/Homeowners Association's Agent (if any), and the Owner/Homeowners Association's Consultant (if any).

3. All Subcontractors, Sub-subcontractors, Consultants, Architects, Materialmen, and Others involved in the project shall provide required Certificates of Insurance set forth herein, regardless of completed Contract price, and must include as additional insureds: the Owner/Homeowners Association, and the Owner/Homeowners Association's Agent (if any), or be listed as additional insureds of the Contractor.

4. All Certificates and Insurance Policies set forth herein must provide that the Owner/Homeowners Association be given thirty (30) days' notice in the event of cancellation and/or reduction of coverage limits, and ten (10) days' notice in the event of non-renewal of any of the insurance documents and coverage required and stipulated.

SECTION G12. CORRECTION OF WORK AND PUNCH LIST ITEMS

a. The Contractor shall promptly correct all Work rejected and all punch list items identified by the Owner/Homeowners Association, the Owner/Homeowners Association's Agent, and/or the Consultant or failing to conform to the requirements of the Scope of Work, the Specifications, and/or the Contract Documents, whether observed before or after Substantial Completion and whether or not fabricated, installed, or completed, and shall correct any Work found to be not in accordance with the requirements of the Scope of Work, Specifications, and/or the Contract Documents within a period of one year from the date of Substantial Completion of the Contract or by terms of an applicable special warranty required by the Request for Bid, and/or the Contract Documents. The provisions of this Section apply to Work done by Subcontractors, Sub-subcontractors, as well as to Work done by direct employees of the Contractor. THE WARRANTY PERIOD ON THIS PROJECT SHALL BE IN ACCORDANCE AND AS STATED IN THE REGULATIONS SET OUT BY THE STATE REGISTRAR OF CONTRACTORS. THE STATE LAW REQUIREMENT IS ONE (1) YEAR FROM SUBSTANTIAL COMPLETIONS AND/OR ACCEPTANCE OF THE PROJECT.

b. Nothing contained in this Section shall be construed to establish a period of limitation with respect to other obligations which the Contractor might have under the Contract Documents. Establishment of the time period of one year as described in the previous paragraph (a) relates only to the specific obligations of the Contractor to correct the Work and punch list items, and has no relationship to the time within which the obligation to comply with the Contract Documents may be sought to be enforced, nor to the time within which proceedings may be commenced to establish the Contractor's liability with respect to the Contractor's obligations other than specifically to correct the Work and all identified punch list items.

c. Correction of Work and punch list items are considered as a separate Work item, and any monies retained for punch list items is considered a separate portion of the Work, and substantial completion for correction to the Work and the punch list items will be the date the corrections to the Work and/or the punch list items are completed and accepted. This paragraph will refer to the 10 percent retention of the contracted payment for lien purposes. The right to lien on the retention or punch list items shall begin upon completion of the punch list items as determined by the Owner/Homeowners Association, the Owner/Homeowners Association's Agent, and/or the Consultant.

All punch list items will be completed within thirty (30) days of notification by the Owner/Homeowners Association, the Owner/Homeowners Association's Agent, and/or the Consultant. Should the Contractor not complete the punch list items within the thirty (30) days of notice, or decline to complete any of the punch list items, the Owner, the Owner's Agent, and/or the Consultant shall hire a contractor to complete the punch list items and the cost of completing the punch list items shall be subtracted from the 10 percent retention prior to release of the retention funds to the Contractor.

SECTION G13. OTHER DIRECTIONS AND OTHER ITEMS

a. The Contract shall be governed by the laws incorporated by the licensing state, the local licensing agency, and the local county and city where the project is located.

b. As between the Owner/Homeowners Association and the Contractor, any applicable statute of limitations shall commence to run and any alleged cause of action shall be deemed to have accrued:

   1. Not later than the date of Substantial Completion for acts or failures to act occurring prior to the relevant date of Substantial Completion;
   2. Not later than the date of issuance of the final Payment for acts or failures to act occurring subsequent to the relevant date of Substantial Completion and prior to issuance of the Payment; and
   3. Not later than the date of the relevant act or failure to act or acts by the Contractor for acts or failures to act occurring after the date of the final Payment.

SECTION G14. TERMINATION OF THE CONTRACT

a. If the Consultant fails to recommend payment for a period of thirty (30) days through no fault of the Contractor, or if the Owner/Homeowners Association fails to make payment thereon for a period of thirty (30) days, the Contractor may, upon seven additional days' written notice to the Owner/Homeowners Association and the Consultant, terminate the Contract and recover from the Owner/Homeowners Association payment for Work executed and for proven loss with respect to materials, equipment, tools, and construction equipment and machinery, including reasonable overhead, profit, and damages applicable to the project.

b. If the Contractor defaults or persistently fails or neglects to carry out the Work in accordance with the Request for Bid, the Specifications, and/or the Contract Documents or fails to perform a provision of the Contract, the Owner/Homeowners Association, after seven (7) days' written notice to the Contractor and without prejudice to any other remedy the Owner/Homeowners Association may have, may make good such deficiencies and may deduct the cost thereof, including compensation for the Consultant's services and expenses made necessary thereby, from the payment then or thereafter due the Contractor. Alternatively, at the Owner/Homeowners Association's option, and upon certification by the Consultant that sufficient cause exists to justify such action, the Owner/Homeowners Association may terminate the Contract and take possession of the site and of all materials, equipment, tools, and construction equipment and machinery thereon owned by the Contractor and may finish the Work by whatever method the Owner/Homeowners Association may deem expedient. If the unpaid balance of the Contract Sum exceeds costs of finishing the Work, including compensation for the Consultant's services and expenses made necessary thereby, such excess shall be paid to the Contractor, but if such costs exceed such unpaid balance, the Contractor shall pay the difference to the Owner/Homeowners Association.

SECTION H.
ALL ADDITIONAL ITEMS AND INSTRUCTIONS TO THE CONTRACT

The undersigned have read all portions of this Contract, and the attached Request for Bid and Specifications, and agree to all terms and conditions as set forward and outlined in the Contract Documents. By signature of this document, the Contractor agrees to complete the Work under the terms of the Contract, the Request for Bid, and the Specifications.

This Agreement and Contract is entered into as of the day and year written above and on page 1 of this Contract Document.

OWNER/HOMEOWNERS ASSOCIATION:

_____

(NAME OF HOMEOWNERS ASSOCIATION OR AGENT FOR THE HOMEOWNERS ASSOCIATION)

_____

(SIGNATURE)

_____

(PRINTED NAME AND TITLE)

_____

CONTRACTOR:

_____

(NAME OF CONTRACTOR)

_____

(SIGNATURE)

_____

(PRINTED NAME AND TITLE)

Figure 7.2 Example of a Completed Contract

**CONTRACTUAL AGREEMENT
FOR PAVEMENT AND CONCRETE REPAIR AND MAINTENANCE**

BETWEEN
ANY HOMEOWNERS ASSOCIATION
AND
ANY SEAL COATING AND PAVING COMPANY, INC.

THIS DOCUMENT HAS IMPORTANT LEGAL CONSEQUENCES: CONSULTATION WITH AN ATTORNEY IS ENCOURAGED WITH RESPECT TO ITS COMPLETION OR MODIFICATION.

THIS DOCUMENT IS AN AGREEMENT BETWEEN THE HOMEOWNERS ASSOCIATION OR THE MANAGEMENT FIRM AS AGENT FOR THE HOMEOWNERS ASSOCIATION AND THE CONTRACTOR FOR PAVEMENT MAINTENANCE AND/OR CONCRETE MAINTENANCE PERFORMED AT THE DESCRIBED LOCATION AND PROJECT.

THIS AGREEMENT INCLUDES THE REQUEST FOR BID SIGNED AND AGREED TO BY THE CONTRACTOR AND THE HOMEOWNERS ASSOCIATION, OR THE AGENT FOR THE HOMEOWNERS ASSOCIATION. BY SIGNING THIS AGREEMENT AND CONTRACT, THE CONTRACTOR HAS VERIFIED ALL QUANTITIES LISTED IN THE REQUEST FOR BID, HAS REVIEWED ALL THE ATTACHED SPECIFICATIONS, AND AGREED TO PERFORM THE WORK STATED AND IN THE TIME PERIOD AGREED TO.

THE CONTRACTOR, THE HOMEOWNERS ASSOCIATION'S AGENT, OR THE CONSULTANT SHALL NOT ALTER THIS CONTRACT OR ANY PART OF THIS CONTRACT AND/OR THE ATTACHED REQUEST FOR BID AND SPECIFICATIONS WITHOUT THE WRITTEN CONSENT OF ALL PARTIES AND THE HOMEOWNERS ASSOCIATION. AT NO TIME SHALL ANY PERSONS CROSS OUT, DELETE, OR WRITE IN ANY ITEMS OR PARAGRAPHS. THE CONTRACT AND AGREEMENT SHALL STAY INTACT AS IT IS WRITTEN. ANY CHANGES, CORRECTIONS, AND/OR DELETIONS SHALL BE SUBMITTED ON A SEPARATE LETTERHEAD AND ATTACHED TO THE CONTRACT PACKAGE.

IN ORDER FOR THE CONTRACTOR AND THE HOMEOWNERS ASSOCIATION TO ENTER INTO THIS CONTRACT, THE CONTRACTOR MAKES THE FOLLOWING REPRESENTATIONS:

1. THE CONTRACTOR HAS EXAMINED AND CAREFULLY STUDIED THE CONTRACT DOCUMENTS, THE SCOPE OF WORK, AND THE SPECIFICATIONS AND ANY OTHER RELATED DATA IDENTIFIED IN THE CONTRACT DOCUMENTS, SCOPE OF WORK, REQUEST FOR BID, AND THE SPECIFICATIONS.

2. THE CONTRACTOR HAS VISITED THE SITE AND BECOME FAMILIAR WITH AND IS SATISFIED WITH THE GENERAL, LOCAL, AND SITE CONDITIONS THAT MAY AFFECT THE COST, PROGRESS, AND PERFORMANCE OF COMPLETING THE WORK.

3. THE CONTRACTOR HAS VALIDATED THE QUANTITIES AND AMOUNTS OF WORK BY TASK TO BE PERFORMED AT THE PROJECT SITE.

4. THE CONTRACTOR IS FAMILIAR WITH AND IS SATISFIED WITH ALL FEDERAL, STATE, AND LOCAL LAWS AND REGULATIONS THAT MAY AFFECT THE COST, PROGRESS, AND PERFORMANCE OF THE WORK.

5. THE CONTRACTOR HAS CAREFULLY STUDIED ALL PROVIDED REPORTS OF EXPLORATIONS AND TESTS OF THE SUBGRADE CONDITION, AGGREGATE BASE THICKNESS/CLASSIFICATION/CONDITION, THE TYPE AND THICKNESS AND CONDITION OF THE EXISTING ASPHALT PAVEMENT AND CONCRETE AT OR CONTIGUOUS TO THE PROJECT, AND ALL DRAWINGS OF PHYSICAL CONDITIONS IN OR RELATING TO THE EXISTING SURFACE OR SUBSURFACE STRUCTURES AT OR CONTIGUOUS TO THE PROJECT (EXCEPT FOR UNDERGROUND FACILITIES) WHICH HAVE BEEN IDENTIFIED IN ANY PLANS, DESIGNS, SCOPE OF WORK, REQUEST FOR BID, OR ANY SUPPLEMENTARY CONDITIONS AS PROVIDED IN THE GENERAL CONDITIONS. THE CONTRACTOR ACCEPTS THE DETERMINATIONS SET FORTH IN THE SUPPLEMENTARY CONDITIONS OF THE EXTENT OF THE TECHNICAL DATA CONTAINED IN SUCH REPORTS, MATERIAL TECHNICAL DATA SHEETS, AND DRAWINGS UPON WHICH THE CONTRACTOR IS ENTITLED TO RELY ON AS PROVIDED IN THE BID DOCUMENTS. THE CONTRACTOR ACKNOWLEDGES THAT SUCH REPORTS, MATERIAL TECHNICAL DATA SHEETS, AND DRAWINGS ARE NOT CONTRACT DOCUMENTS AND MAY NOT BE COMPLETE FOR THE CONTRACTOR'S PURPOSE. THE CONTRACTOR ACKNOWLEDGES THAT THE HOMEOWNERS ASSOCIATION AND THE CONSULTANT DO NOT ASSUME RESPONSIBILITY FOR THE ACCURACY OR COMPLETENESS OF INFORMATION AND DATA SHOWN OR INDICATED IN THE CONTRACT DOCUMENTS WITH RESPECT TO UNDERGROUND FACILITIES AT OR CONTIGUOUS TO THE PROJECT SITE. THE CONTRACTOR HAS OBTAINED AND CAREFULLY STUDIED (OR ASSUMES RESPONSIBILITY FOR HAVING DONE SO) ALL SUCH ADDITIONAL SUPPLEMENTARY EXAMINATIONS, INVESTIGATIONS, EXPLORATIONS, TESTS, STUDIES, AND DATA CONCERNING CONDITIONS (SURFACE, SUBSURFACE, AND UNDERGROUND FACILITIES) AT OR CONTIGUOUS TO THE SITE OR OTHERWISE WHICH MAY AFFECT THE COST, PROGRESS, PERFORMANCE, OR FURNISHINGS OF THE WORK OR WHICH RELATE TO ANY ASPECT OF THE MEANS, METHODS, TECHNIQUES, SEQUENCES, AND PROCEDURES OF CONSTRUCTION TO BE EMPLOYED BY THE CONTRACTOR AND ALL SAFETY PRECAUTIONS, EXPLORATIONS, TESTS, STUDIES, OR DATA NECESSARY FOR THE PERFORMANCE AND FURNISHING OF THE WORK AT THE CONTRACT PRICE, WITHIN THE CONTRACT TIMES, AND IN ACCORDANCE WITH THE OTHER TERMS AND CONDITIONS OF THE CONTRACT DOCUMENTS, SCOPE OF WORK, REQUEST FOR BID, OR THE SPECIFICATIONS.

6. THE CONTRACTOR IS AWARE OF THE GENERAL NATURE OF THE WORK TO BE PERFORMED BY THE HOMEOWNERS ASSOCIATION AND ALL OTHERS AT THE WORK SITE THAT RELATES TO THE WORK.

7. THE CONTRACTOR HAS CORRELATED ALL THE INFORMATION KNOWN TO THE CONTRACTOR, ALL INFORMATION AND OBSERVATIONS OBTAINED FORM VISITS TO THE

SITE, ALL REPORTS AND DRAWINGS, AND ALL ADDITIONAL EXAMINATIONS, INVESTIGATIONS, EXPLORATIONS, TESTS, STUDIES, AND DATA PERTAINING TO THE WORK AND THE PROJECT SITE.

8. THE CONTRACTOR HAS GIVEN THE CONSULTANT WRITTEN NOTICE OF ALL CONFLICTS, ERRORS, AMBIGUITIES, OR DISCREPANCIES THAT THE CONTRACTOR HAS DISCOVERED IN THE CONTRACT DOCUMENTS, THE SCOPE OF WORK, THE REQUEST FOR BID, AND THE SPECIFICATIONS, AND THE WRITTEN RESOLUTION THEREOF BY THE CONSULTANT IS ACCEPTABLE TO THE CONTRACTOR, AND THE CONTRACT DOCUMENTS, SCOPE OF WORK, THE REQUEST FOR BID, AND THE SPECIFICATIONS ARE GENERALLY SUFFICIENT TO INDICATE AND CONVEY UNDERSTANDING OF ALL TERMS AND CONDITIONS FOR PERFORMANCE AND FURNISHING OF THE WORK.

9. THE CONTRACTOR HAS SCHEDULED TO COMPLETE THE WORK IN A REASONABLE AMOUNT OF TIME WITHOUT CAUSING EXTREME HARDSHIP TO THE HOMEOWNERS ASSOCIATION, RESIDENTS AND VISITORS, OR VENDORS OF THE RESIDENTS. THE CONTRACTOR HAS REASONABLY SCHEDULED THE CORRECT AMOUNT OF DAYS AND MOVES TO COMPLETE THE WORK IN A SAFE AND PRUDENT MANNER.

10. THE CONTRACTOR WILL ONLY ATTEMPT OR COMPLETE ANY ADDITIONAL WORK OUTSIDE THE SCOPE OF WORK AND AGREED QUANTITIES AT THE WRITTEN ORDER OF THE HOMEOWNERS ASSOCIATION BY CHANGE ORDER. THE CONTRACTOR WILL ACCEPT A CHANGE ORDER FROM THE HOMEOWNERS ASSOCIATION'S AGENT OR CONSULTANT ONLY IF THE CHANGE ORDER IS AUTHORIZED AND SIGNED BY THE HOMEOWNERS ASSOCIATION. ONLY ADDITIONAL WORK AUTHORIZED IN WRITING BY THE HOMEOWNERS ASSOCIATION SHALL BE INVOICED AND PAID. THE ORIGINAL UNIT PRICE AS SHOWN IN THE REQUEST PER BID BY TASK SHALL BE THE UNIT PRICE FOR ALL CHANGE ORDER ITEMS OF LIKE KIND, UNLESS PREVIOUSLY AUTHORIZED IN WRITING BY THE HOMEOWNERS ASSOCIATION. THE CONTRACTOR WILL NEGOTIATE THE PRICE OF ANY ADDITIONAL WORK REQUESTED IN WRITING FROM THE HOMEOWNERS ASSOCIATION (THROUGH THE HOMEOWNERS ASSOCIATION'S AGENT OR CONSULTANT IF APPLICABLE) THAT IS OUTSIDE THE SCOPE OF WORK AND THE REQUEST FOR BID. THE CONTRACTOR WILL NOT ACCOMPLISH THE NEW WORK UNTIL A SIGNED CHANGE ORDER IS IN THE CONTRACTOR'S POSSESSION SIGNED BY THE HOMEOWNERS ASSOCIATION.

11. THE CONTRACTOR WILL KEEP THE WORK SITE, STAGING AREAS, AND ADJACENT STRUCTURES CLEAN OF ALL DEBRIS AND DIRT CAUSED FROM CONSTRUCTION AND WILL REPAIR ANY DAMAGE TO AN EQUAL OR BETTER CONDITION AS IT WERE BEFORE INCIDENT OR DAMAGE.

DEFINITIONS USED FOR CONTRACT PURPOSES: WHEREVER USED IN THE CONTRACT DOCUMENTS, THE REQUEST FOR BID, THE SCOPE OF WORK, OR THE SPECIFICATIONS, THE FOLLOWING TERMS SHALL HAVE THE MEANINGS INDICATED WHICH ARE APPLICABLE TO BOTH THE SINGULAR AND PLURAL THEREOF.

ADDENDA/ADDENDUM: Written or graphic additions or changes issued prior to the opening of the bids which will clarify, correct, or change the bidding requirements.

ADDITIONAL INSURED: Names of persons, entities, companies, management company, owner/homeowners associations, or board of directors who shall be listed as additional insureds on the Contractor's insurance certificates for liability and employees' or workers' compensation insurance.

AGREEMENT: The written contract between the owner/homeowners association, owner/homeowners association's agent, and the contractor covering the work to be performed. Other contract documents attached to the original agreement shall be considered as part of the original agreement.

BOND, BID: A security instrument issued and delivered with the bid package to guarantee the contractor's bid price.

BOND, PERFORMANCE: A security instrument issued to guarantee the performance of the contractor of the work. Should the contractor default, then the performance bond will be used to cover any expenses incurred in contracting the second bidder, including the bid price difference.

CHANGE ORDER: A document issued by the owner/homeowners association or the owner/homeowners association's agent with permission and authority of the owner/homeowners association by or through the consultant, which is signed by the contractor and the owner/homeowners association or the owner/homeowners association's agent and authorizes an addition, deletion, or revision in the work, the scope of work, the specifications, or any adjustment in the contract price or the contract times. Change orders are issued on or after the effective date of the agreement.

CONDITIONAL WAIVER OF LIEN: An instrument issued by the contractor, subcontractor, sub-subcontractor, or materialman which will release the owner/homeowners association or homeowner/homeowners associations of an owner/homeowners association of all payment liabilities for the amount paid to the contractor. The term conditional shall mean the instrument does not release the owner/homeowners association or homeowner/homeowners associations of an owner/homeowners association from the contractors, subcontractor, sub-subcontractor, supplier, or materialman right to lien the property until all funds are transferred. This instrument shall become an unconditional waiver of lien upon transfer of funds to the contractor, subcontractor, sub-subcontractor, supplier, or materialman for the agreed and paid amount.

CONSULTANT: A person, firm, or corporation having a contract with the owner/homeowners association, or the owner/homeowners association's agent on behalf of the owner/homeowners association.

CONTRACTOR: The person, firm, or corporation with whom the owner/homeowners association or the owner/homeowners association's agent has entered into agreement to perform and complete the work as stated in the contract documents, the scope of work, and the specifications. Also considered the general contractor. The contractor is responsible for all subcontractors and sub-subcontractors.

CONTRACT DOCUMENTS: The agreement, addenda (which pertain to the contract documents), the completed and signed original request for bid (RFB)by the contractor (including any documentation accompanying the request for bid (RFB) and any other bid documentation submitted prior to the award of the work) when attached as an exhibit to the agreement. Other documents considered part of the contract documents shall be the notice of award, any bonds if required, all general conditions, all specifications issued with the request for bid or addenda or change orders, all drawings, all written amendments, all change orders, all work change directives, field orders, and all written interpretations and clarifications of and by the consultant or the consultant's engineer on or after the effective date of the agreement.

CONTRACT SUM: The monies payable by the owner/homeowners association to the contractor for completion of the work in accordance with the contract documents and the request for bid (RFB) and as stated in the agreement.

DATE OF COMMENCEMENT OF WORK: The date agreed to by all parties of the agreement as the date the project is to begin.

DATE OF COMPLETION OF WORK: The date agreed to by all parties that the project shall be substantially completed as defined under substantial completion.

FINAL PAYMENT: The amount paid to the Contractor by the owner/homeowners association, the homeowner/homeowners association's owner/homeowners association, or their agent on behalf of the owner/homeowners association for any outstanding sums, agreed retention, or additional work which pays the contractor in full for all work performed under the contract and for the exact contract price including any change orders or addenda.

GENERAL CONTRACTOR: The prime contractor with whom the owner/homeowners association is contracted to complete all tasks of the work listed in the contract documents, the scope of work, and the specifications. Also referred to as the contractor. The general contractor is responsible for all subcontractors and sub-subcontractors.

HOMEOWNERS ASSOCIATION: The public body or authority, corporation, association, firm, or person with whom the contractor has entered into the agreement and for whom the work is to be provided.

HOMEOWNERS ASSOCIATION'S AGENT OR REPRESENTATIVE: The person, firm, or corporation with whom the owner/homeowners association has contracted to represent them and their best interest.

HOMEOWNERS ASSOCIATION'S ASSOCIATION: Whether home, townhome, or condominium, this group shall represent a body of home or property owners who are collectively responsible for their community and hold residency in the same.

INSTRUCTIONS TO BIDDER: Those specific instructions given to the bidder or contractor outlining the requirements to work on the project site, the proper insurance and bond requirements, work hours and days, and all required forms and documents to be provided prior to bidding and starting the project.

LIABILITY INSURANCE: Insurance required by the owner/homeowners association to cover any losses caused by the contractor, the contractor's workers, and equipment, or from the negligence of the contractor, the contractor's workers, subcontractors, and sub-subcontractors.

LICENSED CONTRACTOR: Contractors who possess the proper category license required to perform the work as stated in the scope of work and the bid by task of the request for bid (RFB). A contractor who is current in fees with the licensing agency and whose license is current and active.

LIEN: Charges, security interests, or encumbrances upon real property or personal property.

MATERIALMAN: Manufacturer or supplier of materials to be used on the project.

NOTICE OF RIGHT TO LIEN (20-DAY LETTER): A document delivered to the owner/homeowners association to disclose the contractor's right to lien the property for nonpayment of the agreed contract amount within ninety (90) days of substantial completion. This letter is required by local state lien laws to allow for a contractor's right to lien.

NOTICE TO PROCEED: A written notice given by the owner/homeowners association to the contractor fixing the date on which the contract times will commence to run and on which the contractor shall start to perform the contractor's obligations under the contract documents.

PENALTY CLAUSE: The clause that enables the owner/homeowners association to charge a contractor for failure to complete the work as set out and defined in the agreement, the contract documents, and the request for bid (RFB).

PROGRESS PAYMENT: A partial payment allowed to the contractor for work completed as agreed between the owner/homeowners association and the contractor.

PROJECT: The actual location and description of the work and referenced by physical name and/or number.

PUNCH LIST: The items identified as needing attention, repairs, or corrections upon completion of the work.

REQUEST FOR BID (RFB): Also refered to as request for quotes (RFQ). A document isuued to licensed, bonded and insured contractors containing the scope of work, Instructions to the bidders, insurance requirements and specifications.

REGISTRAR OF CONTRACTORS: The government agency responsible for controlling and licensing contractors in their specific trades.

RETENTION: The amount of monies agreed to between the owner/homeowners association and the contractor to be held from the full contract amount until the punch list items or damaged areas and property are corrected.

SCOPE OF WORK: The portion of the contract documents and the request for bid (RFB) that explains to the contractor the specific type of work that is expected to occur on the project and the type of materials that is required to accomplish the work.

SPECIAL CONDITIONS: Any special instructions, materials, specifications, required documents, or additional work that is not listed or performed in the normal routine of pavement or concrete repair or maintenance. These conditions are unique to the project or work.

SPECIFICATIONS: The portion of the contract documents and the request for bid (RFB) which consists of written technical descriptions of materials, equipment, construction systems, standards, and workmanship as they apply to the project, work, and certain administrative details.

STATE COMPENSATORY INSURANCE: Insurance required by the owner/homeowners association to cover any work-related injuries to employees caused by the contractor, the contractor's workers, and equipment, or from the negligence of the contractor, the contractor's workers, subcontractors, and sub-subcontractors.

STOP WORK: A notice given to the contractor by the owner/homeowners association's agent or the consultant at the direction of the owner/homeowners association to stop the work in process due to material, workmanship, or equipment problems, or a violation of specifications.

SUBCONTRACTOR: The person, firm, or corporation with whom contractor has entered into agreement to perform and complete a portion of the Work as stated in the contract documents, the scope of work, and the specifications.

SUBSTANTIAL COMPLETION: The completion of all tasks and the scope of work as defined in Section D2-1 of this contract document.

SUB-SUBCONTRACTOR: The person, firm, or corporation with whom the subcontractor has entered into agreement to perform and complete a portion of the work which the subcontractor has contracted to perform as stated in the contract documents, the scope of work, and the specifications.

SUPPLIER: A manufacturer, fabricator, supplier, distributor, materialman, or vendor having a direct contract with the contractor, subcontractor, or sub-subcontractor to furnish materials or equipment to be incorporated in the work by the contractor, subcontractor, or sub-subcontractor.

TASK: A single item of the work as listed in the bid by task of the request for bid (RFB) and Section A1 of this contract.

TERMINATION OF CONTRACT: The contract shall be cancelled, stopped, discontinued, or ended by decision of the owner/homeowners association or the contractor for reasons of breach of contract or inability to complete the contract.

TERMS: The rules and regulations of performing the work. The agreed items as listed in the contract documents and the request for bid (RFB).

TIME SCHEDULE: The time line to complete the project from the start date to the date of substantial completion.

UNCONDITIONAL WAIVER: An instrument issued by the contractor, subcontractor, sub-subcontractor, supplier, or materialman which releases the owner/homeowners association or homeowner/homeowners associations of an owner/homeowners association of all payment liabilities for the amount paid to the contractor, subcontractor, sub-subcontractor, supplier, or materialman.

UNIT PRICE: Price per unit (sq ft, sq yd, linear ft, etc.) for a listed task.

WORK: The entire completed construction or the various separately identifiable parts thereof required to be furnished under the contract documents and the scope of work. Work includes and is the result of performing or furnishing labor and furnishing and incorporating materials and equipment into the construction and performing or furnishing services and furnishing documents as required by the contract documents and the scope of work.

THIS AGREEMENT IS MADE AS OF THE <u>17TH DAY</u> OF <u>APRIL</u> IN THE YEAR TWO THOUSAND NINE

BETWEEN THE HOMEOWNERS ASSOCIATION:

> MY HOMEOWNERS ASSOCIATION
> C/O THEIR MANAGEMENT COMPANY, INC.
> 123 AJAX ST.
> PHOENIX, AZ

AND THE CONTRACTOR:
> ANY SEAL COATING & PAVING, INC.
> 567 MOON AVE.
> PHOENIX, AZ

THE PROJECT IS:
> MY HOMEOWNERS ASSOCIATION
> 890 JONES WAY
> SCOTTSDALE, AZ

THE WORK IS:

CRACK SEALING, PATCHING (SKIN/SURFACE AND R&R), SEAL COATING 2 COATS, RESTRIPE TO EXISTING CONFIGURATION, REMOVE & REPLACE CONCRETE CURB.

THE CONSULTANT IS:
> PMIS, INC.
> DBA PAVEMENT MAINTENANCE INFORMATION SOURCE
> 18521 E. QUEEN CREEK RD.
> SUITE #105-435
> QUEEN CREEK, AZ 85242
> 602-854-7070

SECTION A
CONTRACT SCOPE OF WORK AND FORMAT

A1. The Contractor shall complete the project to the Scope of Work outlined in the attached Request for Bid for this project. The Contractor also agrees to install all structures according to any attached drawings without exception or changes unless authorized in writing by the Owner/Homeowners Association, the Owner/Homeowners Association's Agent, or the Consultant on behalf of the Owner/Homeowners Association. The Contractor agrees to complete the Scope of Work as listed in task form and to the quantities listed. By signing this Contract, the Contractor agrees with the Scope of Work and has verified all quantities as listed in the following task breakdown and the Request for Bid. The Contractor shall execute the entire Work described in the Contract Documents and the Request for Bid, except to the extent specifically indicated to be the responsibility of others. The Work is described as follows:

TASK I	CRACK SEAL ALL CRACKS WITH AN OPENING OF ¼ IN. OR MORE @ 24,658 LF
TASK II	SKIN/SURFACE PATCH @ 2,456 SQ. FT. (15 LOCATIONS)
TASK III	R&R PATCHING @ 35,456 SQ. FT. (23 LOCATIONS)
TASK IV	SEAL COAT, 2 COATS @ 30,560 SQ. YD.
TASK V	RESTRIPE TO EXISTING CONFIGURATION @ 3,256 LF STRIPE, 6 STOP BARS, 8 SPEED BUMPS (SOLID YELLOW WITH GLASS BEADS)
TASK VI	REMOVE AND REPLACE 6 CONCRETE CURB SECTIONS @ 72 LF

SECTION B
DATE OF COMMENCEMENT AND COMPLETION OF THE PROJECT

B1. The project shall commence on the date the Contractor has selected and is shown in the attached time schedule for this project as submitted by the Contractor. The start date shall be strictly adhered to once the Contractor has committed and scheduled the date. The Contractor agrees not to start the project earlier than the scheduled start date unless the Contractor has received written authorization from the Owner/ Homeowners Association, the Owner/Homeowners Association's Agent, or the Consultant. The project shall last for the duration of the time established by the Contractor to complete all tasks of the project and the Scope of Work. The start date for the project shall be as agreed to by the Owner/Homeowners Association and the Contractor. Should the project be delayed by weather, equipment malfunction, labor problems of the Contractor, or any other reasons, the project shall be moved to the next reasonable and acceptable available date agreeable to the Owner/Homeowners Association and the Contractor. It shall be the responsibility of the Contractor to supply a new schedule in writing for approval by the Owner/ Homeowners Association within twenty-four (24) hours of any cancellation of work.

THE START DATE OF THIS CONTRACT SHALL BE MAY 12, 2009.

The start date is submitted by the Contractor on the Notification of Award form attached and part of this Contract. This date shall become the contracted commencement date and shall be acceptable to the Owner/Homeowners Association and the Contractor UPON SIGNATURE OF THE CONTRACT BY BOTH PARTIES.

B2. The Contractor shall achieve Substantial Completion of the entire project not later than the date shown. Substantial Completion shall be defined as that portion of the Contract where all tasks listed in the Request for Bid (RFB) have been completed. This does not include any punch list items or future warranty work. Substantial Completion of the project shall be determined when all the Work outlined in the Scope of Work and each Task in the Bid by Task Breakdown in the attached Request for Bid is completed and all change orders (if any) have been signed by the Contractor and received by the Owner/Homeowners Association.

THE COMPLETION DATE OF THIS CONTRACT SHALL BE MAY 22, 2009

subject to adjustments of this Contract Time as provided in the Contract Documents.

The Contractor and Contractor's surety, if any, shall be liable for and pay the Owner/Homeowners Association the sum hereinafter stipulated as liquidated damages for each calendar day until the Work is substantially completed. The penalty for late substantial completion shall be two hundred fifty dollars ($250) per day after MAY 23, 2009.

One additional day for completion shall be added for each day of bad weather that prevents the Contractor from performing the Work covered by this agreement. Also, one day additional for completion shall be added for each day of any delay not deemed the Contractor's fault from performing the Work covered by this agreement as listed in Section B2. Moving off the project without just cause shall be considered delay as a result of the fault of the Contractor.

SECTION C
CONTRACTED AMOUNT TO PERFORM THE PROJECT

C1. THE HOMEOWNERS ASSOCIATION SHALL PAY THE CONTRACTOR IN CURRENT FUNDS FOR THE CONTRACTOR'S PERFORMANCE OF THE CONTRACT AND THE COMPLETION OF THE PROJECT AS OUTLINED IN THE SCOPE OF WORK AND THE BID TASKS IN THE REQUEST FOR BID AND THE TERMS SHOWN IN SUBSECTION B2. THE HOMEOWNERS ASSOCIATION SHALL PAY THE CONTRACTOR THE CONTRACT SUM OF

ONE HUNDRED FIFTY THOUSAND EIGHT HUNDRED AND 68/100 ------- DOLLARS ($150,800.68 TAX INCLUDED), subject to all additions and deductions as provided in the Contract Documents and all change orders attached as part of this Contract.

C2. ANY ALTERNATIVES TO THE CONTRACT THAT WILL ALTER THE SCOPE OF WORK, THE QUANTITIES, OR THE PROJECT TASKS WILL HAVE TO BE LISTED BELOW. THIS WILL INCLUDE THE BASE BID FROM SECTION C1 AND ANY AND ALL SUBSEQUENT AMOUNTS AS THEY WILL ALTER THE FINAL AMOUNT OF THE CONTRACT.

1. BASE BID $150,800.68 TAX INCLUDED

THIS AREA IS FOR ANY ADDENDUM COSTS ABOVE THE BASE PRICE

C3. UNIT PRICES, IF ANY, ARE LISTED AS FOLLOWS SHOULD THE REQUEST FOR BID NOT BE ATTACHED. FOR UNIT PRICES ASSOCIATED WITH THIS BID, REFER TO THE

ATTACHED REQUEST FOR BID AND THE TASK BREAKDOWN IN THE REQUEST FOR BID.

ITEM #   UNIT PRICE   TOTAL

N/A

SECTION D
PROGRESS PAYMENTS

D1. UNLESS SO AGREED, THE CONTRACTOR SHALL BE PAID UPON COMPLETION OF THE WORK AS STATED IN THE SCOPE OF WORK AND AS OUTLINED IN THE TASKS FOR BID OF THE ATTACHED REQUEST FOR BID. ANY AND ALL PROGRESS PAYMENTS AND TERMS WILL BE LISTED IN SECTION E.

D2. ALL PAYMENTS ON THE CONTRACT SHALL BE PAID TO THE FOLLOWING CONDITIONS AND SCHEDULE:

a. 90 PERCENT PAYMENT WITHIN TWENTY (20) DAYS AFTER SUBSTANTIAL COMPLETION OF THE PROJECT. SUBSTANTIAL COMPLETION IS DETERMINED WHEN THE WORK OUTLINED IN THE SCOPE OF WORK IS COMPLETED, EACH TASK LISTED ON THE BID BY TASK BREAKDOWN OF THE REQUEST FOR BID ARE COMPLETED, ALL CONDITIONAL WAIVERS ARE SUBMITTED, AND ALL SIGNED CHANGE ORDERS HAVE BEEN SUBMITTED BY THE CONTRACTOR.

b. 10 PERCENT PAYMENT WITHIN TWENTY (20) DAYS UPON COMPLETION OF ALL PUNCH LIST ITEMS. COMPLETION OF THE PUNCH LIST ITEMS SHALL BE DETERMINED WHEN ALL ITEMS LISTED ON THE PUNCH LIST PROVIDED FROM A FINAL WALK-THROUGH ATTENDED BY THE CONSULTANT, HOMEOWNERS ASSOCIATION'S REPRESENTATIVE, AND THE CONTRACTOR ARE COMPLETED AND ACCEPTED BY THE HOMEOWNERS ASSOCIATION.

c. ALL SPECIAL CONDITIONS (IF ANY). THESE SPECIAL CONDITIONS WILL BE LISTED IN SECTION H OF THIS CONTRACT.

D3. THE 90-DAY PERIOD FOR THE TIME LIMIT FOR THE RIGHT TO LIEN SHALL BE SET FOR TWO PORTIONS OF THE CONTRACT IDENTIFIED AS THE SUBSTANTIAL COMPLETION PORTION (90 PERCENT AND THE PUNCH LIST PORTION (10 PERCENT OF THE CONTRACT. THE TIME LIMITS FOR ALL LIEN RIGHTS SHALL BE EXECUTED UPON ATTAINING THE SUBSTANTIAL COMPLETION OF THE CONTRACT FOR THE INITIAL CONTRACTED AMOUNT. THE REMAINDER PUNCH LIST ITEMS ARE CONSIDERED A SEPARATE CONTRACT, AND THE TIME LIMITS FOR ALL LIEN RIGHTS SHALL BE EXECUTED UPON COMPLETION OF ALL PUNCH LIST ITEMS, ONCE ALL PUNCH LIST ITEMS ARE COMPLETED AND SIGNED OFF AS COMPLETED BY THE HOMEOWNERS ASSOCIATION'S AGENT AND/OR THE CONSULTANT.

D4 INVOICES, DEMANDS, AND REQUESTS FOR PAYMENT WILL NOT BE SUBMITTED TO THE HOMEOWNERS ASSOCIATION OR MANAGER FOR PAYMENT UNTIL ALL WAIVERS ARE SUBMITTED FROM THE CONTRACTOR, SUBCONTRACTORS, SUB-SUBCONTRACTORS, AND MATERIALMEN. SUBMITTAL OF ALL WAIVERS (CONDITIONAL WAIVER WHICH BECOMES UNCONDITIONAL UPON PAYMENT OF DEMAND BY THE CONTRACTOR) IS REQUIRED AND IS CONSIDERED AN INTEGRAL PART OF THE CONTRACT AND

THE SCOPE OF WORK. FAILURE TO SUBMIT ALL DOCUMENTS BY THE CONTRACTOR WILL DELAY THE TIMING FOR RIGHT TO LIEN SINCE THE SUBMITTAL OF ALL DOCU-MENTS FOR PAYMENT ARE CONSIDERED A PART OF THE REQUIREMENTS OF THE 90 PERCENT PAYMENT OF THE CONTRACT.

D5. PAYMENTS DUE AND UNPAID UNDER THE CONTRACT SHALL BEAR INTEREST FROM THE DATE PAYMENT IS DUE AT THE RATE STATED BELOW. THIS RATE AND ALL TERMS ARE TO BE AGREED TO BY THE HOMEOWNERS ASSOCIATION AND THE CONTRAC-TOR PRIOR TO SIGNING OF THE CONTRACT.
(INSERT RATE OF INTEREST AGREED UPON, IF ANY.)

N/A

(Usury laws and requirement under the Federal Truth in Lending Act, similar state and local consumer credit laws, and other regulation at the Owner/Homeowners Association's and Contractor's principal places of business, the location of the project, and elsewhere may affect the validity of this provision. Legal advice should be obtained with respect to deletions or modifications and also regarding requirements, such as written disclosures or waivers.)

SECTION E
FINAL PAYMENT

E1. FINAL PAYMENT, CONSTITUTING THE ENTIRE UNPAID BALANCE OF THE CONTRACT SUM, SHALL BE MADE BY THE HOMEOWNERS ASSOCIATION OR THE HOMEOWNERS ASSOCIATION'S AGENT TO THE CONTRACTOR WHEN THE WORK HAS BEEN COM-PLETED AND VERIFIED BY THE CONSULTANT, ALL PUNCH LIST ITEMS CORRECTED AND VERIFIED BY THE CONSULTANT, THE CONTRACT FULLY PERFORMED AND VERI-FIED BY THE CONSULTANT, AND ALL CHANGE ORDERS, ALL FINAL CERTIFICATES FOR PAYMENT, AND WAIVERS OF LIENS BY THE CONTRACTOR, SUBCONTRACTOR, SUB-SUBCONTRACTORS, AND MATERIALMEN HAVE BEEN ISSUED. NO FINAL PAYMENT WILL BE SUBMITTED FOR PAYMENT UNTIL ALL THE ABOVE DOCUMENTS HAVE BEEN SUBMITTED TO THE HOMEOWNERS ASSOCIATION, THE HOMEOWNERS ASSOCIA-TION'S AGENT, OR THE CONSULTANT. FAILURE TO SUBMIT ALL DOCUMENTS BY THE CONTRACTOR WILL DELAY THE TIMING FOR RIGHT TO LIEN SINCE THE SUBMIT-TAL OF ALL DOCUMENTS FOR PAYMENT ARE CONSIDERED A PART OF THE REQUIRE-MENTS OF THE 10 PERCENT RETENTION CONTRACT.

SECTION F
LISTING OF CONTRACT DOCUMENTS

F1. ALL CONTRACT DOCUMENTS ARE LISTED IN ORDER IN SECTION G1 AND ATTACHED IN ORDER TO THIS CONTRACT DOCUMENT EXCEPT FOR MODIFICATIONS, CHANGE ORDERS, AND ADDENDA ISSUED AFTER EXECUTION OF THIS AGREEMENT, AND ARE LISTED AS FOLLOWS:

F2. THIS CONTRACT DOCUMENT IS THE COVER DOCUMENT AND ATTACHED AS SEC-TION 1.

F3. ANY SUPPLEMENTARY AND/OR ANY OTHER CONDITIONS OF THIS CONTRACT ARE CONTAINED IN THE REQUEST FOR BID OR ANY APPLICABLE PROJECT MANUALS DATED N/A.

F4. ANY AND ALL SPECIFICATIONS LISTED IN ANY APPLICABLE PROJECT MANUALS DATED N/A.

F5. ANY AND ALL DRAWINGS, BLUEPRINTS, SKETCHES LISTED AND DATED N/A.

F6. THE ADDENDA AND CHANGE ORDERS TO THIS CONTRACT ARE LISTED AS FOLLOWS, WITH AN EXPLANATION OF THE CAUSE OF ANY ADDENDA AND CHANGE ORDERS.

NO ADDENDA WITH THIS CONTRACT

ADDENDUM
CHANGE ORDER
NUMBER   **DATE**   **PAGES**

EXPLANATION:

ADDENDUM
CHANGE ORDER
NUMBER   **DATE**   **PAGES**

EXPLANATION:

ADDENDUM
CHANGE ORDER
NUMBER   **DATE**   **PAGES**

EXPLANATION:

ADDENDUM
CHANGE ORDER
NUMBER   **DATE**   **PAGES**

EXPLANATION:

F7. ALL OTHER DOCUMENTS, IF ANY, FORMING A PART OF THIS CONTRACT ARE LISTED IN ORDER AND AS FOLLOWS:

1. REQUEST FOR BID DATED MARCH 16, 2009
2. INSTRUCTIONS TO BIDDER AS CONTAINED IN THE REQUEST FOR BID DATED 3/16/2009
3. STANDARD SPECIFICATIONS
4. NOTICE OF AWARDED BID AND CONTRACTOR SCHEDULE

SECTION G
GENERAL CONDITIONS OF THIS CONTRACT

The Contractor is responsible for reading and understanding all the general conditions of this Contract. Signature of this Contract indicates the Contractor has read, completely understands, and acknowledges all the terms of Section G, which are the General Conditions of this Contract. There shall be no deviation or alteration of these terms as set forth in this Contract.

G1. Contract Documents

a. The Contract Documents consist of this Agreement with all the conditions of the Contract (General, Supplementary, and any other Conditions), Drawings, Blueprints, Sketches, Specifications, Addenda issued prior to the execution of this Agreement, Other Documents listed in this Agreement, and all Modifications issued after execution of this Agreement. The intent of the Contract Documents is to include all items necessary for the proper execution and completion of the Scope of Work by the Contractor. The Contract Documents are complementary, and what is required by one shall be as binding as if required by all parties. The performance of the Contractor shall be required only to the extent that is consistent with the Contract Documents, Scope of Work, Request for Bid, Specifications, and all other Contract Documents (Addenda, Change Orders, and Modifications) and reasonably inferable from them as being necessary to produce the intended results.

b. The Contract Documents shall not be construed to create a contractual relationship of any kind between the Consultant and the Contractor, the Owner/Homeowners Association and a Subcontractor or Sub-subcontractor, or between any persons or entities other than the <u>Owner/Homeowners Association and the Contractor.</u>

c. Execution of this Contract by the Contractor is a representation that the Contractor has visited the work site and has become completely familiar with the local conditions under which the Scope of Work is to be performed. The Contractor agrees to have examined all aspects of the work site, the Scope of Work, and the Bid Tasks as outlined and agreed to in the Request for Bid for this project. Prior to submitting a completed proposal, the Contractor certifies a complete and careful examination of the Proposal and Bid Documents, visited the work site, and fully informed themselves as to all existing conditions and limitations, and shall include in the Proposal a sum to cover the cost of all items included in the Request for Bid and the Bid package. The Contractor, when awarded this Contract, shall not be allowed any extra compensation by reason of any matter or thing, that would have become known if the Contractor would have fully informed himself/herself prior to submitting a completed bid. Signing this Contract shall indicate the Contractor has agreed to the existing specified Scope of Work and will complete the project as specified. The quantities listed in the Request for Bid documents are the measurements for bidding purposes only, and by signing this Contract, the Contractor has certified he/she has verified all quantities and will complete the project as outlined in the Scope of Work and as specified.

d. The term Work means the construction and services required by the Contract Documents, whether completed or partially completed, and includes all other labor, materials, equipment, and services provided or to be provided by the Contractor to fulfill the Contractor's obligations. The Work may constitute the whole or a part of the project.

G2. THE OWNER/HOMEOWNERS ASSOCIATION AND MANAGEMENT COMPANY

a. The Owner/Homeowners Association may appoint an Agent (Manager, Management Company, or Asset Manager) to represent the Owner/Homeowners Association's best interest. Should the Owner/Homeowners Association appoint an Agent, the Agent shall be the contact person for the Owner/Homeowners Association during the entire project. All correspondence shall be directed to the Owner/Homeowners Association or the appointed Agent.

b. The Owner/Homeowners Association or the Owner/Homeowners Association's Agent (Manager, Management Company, or Asset Manager) shall furnish and provide surveys, blueprints, and legal descriptions of the work site.

c. Except for permits and fees, which are the responsibility of the Contractor under the Contract Documents, the Owner/Homeowners Association or the Owner/Homeowners Association's Agent shall secure and pay for necessary approvals, easements, assessments, and charges required for the construction, use, or occupancy of permanent structures or permanent changes in existing facilities.

d. Should the Contractor fail to correct Work which is not in accordance with the requirements of the Contract Documents or persistently fails to carry out the Work in accordance with the Contract Documents, the Owner/Homeowners Association or the Owner/Homeowners Association's Agent, by a written order, may order the Contractor to stop the Work, or any portion thereof, until the cause for such order has been eliminated; however, the right of the Owner/Homeowners Association to stop the Work shall not give rise to a duty on the part of the Owner/Homeowners Association to exercise this right for the benefit of the Contractor or any other personal entity. The Owner/Homeowners Association or the Owner/Homeowners Association's Agent may hire or appoint a Consultant to represent their interests with the same authority as authorized by the Owner/Homeowners Association.

G3. THE CONTRACTOR

a. The Contractor shall possess all required, proper licenses in the proper category for the type of work performed with the local licensing agency or the Registrar of Contractors. The Contractor shall also have the proper license bonding or reserve on record with the local licensing agency and/or the Registrar of Contractors.

b. The Contractor shall have the capabilities of acquiring all required bonding or letters of credit necessary and as required by the Owner/Homeowners Association in the form of Bid Bonds and Performance Bonds.

c. The Contractor shall supervise and direct the Work, using the Contractor's best skill and attention to ensure the project is being installed to the Scope of Work and Specifications. The Contractor shall be solely responsible for and have control over the construction means, methods, techniques, sequences, and procedures, and coordinating all portions of the Work under the Contract unless the Contract Documents or Scope of Work give other specific instructions concerning these matters. The Owner/Homeowners Association and/or the Owner/Homeowners Association's Agent or Consultant shall only advise and offer direction unless so specified in the Contract Documents.

d. Unless provided in the Contract Documents, the Contractor shall provide and pay for all labor, material, equipment, tools, construction equipment and machinery, water, heat, utilities, transportation, and all other facilities and services necessary for the proper execution and completion of the Work, whether temporary or permanent and whether or not incorporated or to be incorporated into the Work.

e. The Contractor shall enforce strict discipline and good conduct among the Contractor's employees and other persons carrying out the Contract. There shall be no taunting or harassing of any persons employed or not employed by the Contractor that will make any person feel uncomfortable. The employees shall treat all persons in and around the project with complete respect. The Contractor's employees shall wear proper clothing and dress in a manner acceptable to the Owner/Homeowners Association. Any employee of the Contractor who is not dressed appropriately shall be asked to leave the project until he/she can return wearing proper clothing. The Contractor will ensure the employees are participating in a company drug awareness program under the Contractor's guidelines. At no time shall any employee be allowed on the work site who is intoxicated or under the influence of any drug, chemical, or alcohol. The Contractor shall have a competent and skilled leader and employees performing the tasks for this project. The leader and the majority of the skilled employees shall be on the project throughout the entire workday. All unskilled persons and trainees shall be used in a limited fashion and under the supervision of a skilled leader and other skilled employees.

f. The Contractor warrants to the Owner/Homeowners Association and Consultant that all materials and equipment furnished under the Contract will be of high quality and as specified in the Specification section of the Request for Bid unless otherwise required or permitted by the Contract Documents, that the Work will be free from defects not inherent in the quality required or permitted, and the Work will conform to the requirements of the Contract Documents, the Scope of Work, and the Specifications. Work not conforming to the requirements of the Request for Bid, Scope of Work, or the Specifications, including substitutions not properly approved and authorized, may be considered unacceptable. The Contractor's warranty excludes remedy for damage or defect caused by abuse, modifications not executed by the Contractor, improper or insufficient maintenance, improper operation, or normal wear and tear under normal usage. If required by the Owner/Homeowners Association or the Consultant, the Contractor shall furnish satisfactory evidence as to the kind and quality of materials and equipment.

g. Unless otherwise provided in the Contract Documents, the Scope of Work, or the Request for Bid, the Contractor shall pay sales, consumer, use, and other similar taxes which are legally enacted when Bid are received or negotiations concluded, whether or not yet effective or merely scheduled to go into effect, and where applicable and so stated in the Request for Bid, shall secure and pay for all building, engineering, public works, and other permits and governmental fees, licenses, and inspections necessary for proper execution and completion of the Work.

h. The Contractor shall comply with and give notices required by laws, ordinances, rules, regulations, and lawful orders of public authorities bearing on performance of the Work. The Contractor shall promptly notify the Owner/Homeowners Association, and Consultant if the Drawings and Specifications are observed by the Contractor to be at variance therewith.

i. The Contractor shall be responsible to the Owner/Homeowners Association for the acts and omissions of the Contractor's employees, Subcontractors, Sub-subcontractors, and their Agents and employees, and other persons performing portions of the Work under a Contract with the Contractor.

j. The Contractor shall review, approve, and submit to the Owner/Homeowners Association, Owner/Homeowners Association's Agent, or Consultant all Shop Drawings, Product Data, Samples, Materials Safety Data Sheets, and all similar submittals required by the Contract Documents, Scope of Work, or Request for Bid with reasonable promptness. The Work shall be in accordance with approved submittals. When professional certifications of performance criteria of materials, systems, or equipment is required by the Contract Documents or the Request for Bid, the Owner/Homeowners Association, the Owner/Homeowners Association's Agent, or the Consultant shall be entitled to rely upon the accuracy and completeness of such certifications.

k. The Contractor shall keep the premises and surrounding area free from accumulation of waste materials or rubbish caused by operations under the Contract. The Contractor shall, before starting work, remove or otherwise cover and/or protect all landscaping, curbing, equipment, accessories, signs, lighting fixtures, resident/tenant and customer personal property, and similar items or provide ample protection of such items. Upon completion of each work area, replace any of the above items if damaged during the maintenance or construction phase. Protect adjacent surfaces as required or directed. Any damage done shall be repaired by the Contractor at his/her expense. The Contractor shall observe all safety precautions as recommended by OSHA and the governing agency in which the project resides. All debris from the operation shall be removed daily from the job site and disposed of at an approved, suitable disposal site. The Contractor will not dispose of any debris or materials in the dumpsters. Should the Contractor dispose of any material in the dumpsters, the Contractor will pay all charges and fees to have the material hauled off and disposed of. The Contractor shall clean all sidewalks, driveways, building entrances, or any other surfaces adjacent to the work area of all dust, dirt, rocks, or any other debris by sweeping or air blowing immediately after the maintenance material has been applied. When the Contractor destroys, alters, defaces, or in any way damages any property adjacent to the project work area, the Contractor shall repair the damaged property to the condition equal to or better than it was before the Work began. This includes, but is not limited to, structures, walls, lawns, sprinkler systems, plumbing, irrigation, landscaping, all concrete structures, etc. The debris shall be cleaned away from all building entrances.
The Contractor and his/her employees shall not track or walk back over the new Work, or track materials onto adjacent areas such as walks, patios, curbs, outdoor carpeting, etc. Final cleanup of the project shall consist of replacing any property disturbed by the Work; removing all debris, scraps, and containers from the work site; and repairing, cleaning, painting, or replacing defaced or disfigured finishes and surfaces caused by Work performed, including curbs, sidewalks, walls, or any and all surfaces where construction materials have splashed or been oversprayed. The Contractor shall be responsible for cleaning all walks, floors, carpets, etc. should the material being used be tracked onto these surfaces. The Contractor shall be responsible for cleaning all property (autos, walls, windows, or any personal property or building fixtures) of material or residue as a result of the construction on all legitimate claims. These claims will be determined legitimate by the Owner/Homeowners Association, the Owner/Homeowners Association's Agent, or the Consultant.

The Contractor shall NOT dump or put construction materials, waste, debris, or anything into dry wells or dumpsters, or on landscaped areas at any time. The Contractor shall properly dispose of any and all material, either removed from the job site or leftover product, in a proper and environmentally safe manner.

l. The Contractor shall provide the Owner/Homeowners Association, the Owner/Homeowners Association's Agent, or the Consultant access to the Work in preparation and progress wherever located.

m. The Contractor shall pay all royalties and license fees; shall defend suits or claims for infringement of patent rights; and shall hold the Owner/Homeowners Association, the Owner/Homeowners Association's Agent, or the Consultant harmless from loss on account thereof, but shall not be responsible for such defense or loss when a particular design, process, or product of a particular manufacturer or manufacturers is required by the Contract Documents unless the Contractor has reason to believe that there is an infringement of patent.

n. The Contractor agrees to indemnify and hold the Owner/Homeowners Association and its Agents, Consultants, employees, and residents harmless from claims, costs, suits, judgments, expenses, attorney's fees, and any and all professional fees on account of any damages which arise out of the respective acts or omissions of the Contractor, its Subcontractors, its Sub-subcontractors, its Consultants, its Agents, and its employees.

The Owner/Homeowners Association agrees to indemnify and hold the Contractor, its Subcontractors, its Sub-subcontractors, its Consultants, its Agents, and its employees harmless from claims, suits, judgments, expenses, attorney's fees, and any and all other professional fees on account of any damages which arise out of the respective acts or omissions of the Owner/Homeowners Association and its Consultants, Agents, and employees.

Any indemnification or hold harmless obligations of the Contractor shall extend only to the claims relating to bodily injury and property damage (and then only to that part or proportion) of any claims damage, loss, or defect that results from the negligence or intentional act of the indemnifying party or someone for whom they are responsible. The Contractor shall not under any circumstance have a duty to defend any other persons or entities not connected to the Contractor by this Contract.

The Contractor shall not be required to indemnify any other parties from damages, attorney's fees, or personal or property damage for any amount exceeding the proportional amount of the Contractor's direct cause of such damages. The Contractor shall be liable for fines or assessments made against the Owner/Homeowners Association that are directly the fault or negligence of the Contractor or the Owner/Homeowners Association, and the Contractor shall have the right of subrogation for claims to the proportion of their responsibility. The Contractor shall be responsible for all claims against their Subcontractors and Sub-subcontractors, their Agents, their Consultants, and their employees as it pertains to the Work. The Contractor shall be responsible for indemnifying the Owner/Homeowners Association of their actions and the actions of their Subcontractors, Sub-subcontractors, Agents, Consultants, and employees for negligence, personal property, and personal injury.

o. Claims against any persons or entities indemnified under this Paragraph by an employee of the Contractor, a Subcontractor, a Sub-subcontractor, or anyone directly or indirectly employed by them or anyone for whose acts they may be liable, the indemnification obligation under this Paragraph shall not be limited by a limitation on amount or type of damages, compensation, or benefits payable by or for the Contractor, a Subcontractor, and a Sub-subcontractor under workers' or workmen's compensation acts, disability benefit acts, or other employee benefit acts.

p. The obligations of the Contractor under this Paragraph shall not extend to the liability of the Consultant, the Consultant's Consultants, architects, Agents, and employees of any of them arising out of (1) the preparation or approval of maps, drawings, opinions, reports, surveys, Change Orders, Construction Change Directives, designs, or Specifications, or (2) the giving of or the failure to give directions or instructions by the Consultant, the Consultant's Consultants, architects, Agents, and employees of any of them provided such giving or failure to give is the primary cause of the injury or damage.

q. Should the completion of the project be delayed due to the fault of the Contractor by incomplete work, unacceptable work, poor or improper application of construction materials, or insufficient cleanup and corrective Work that will require additional services, testing, mediation, or arbitration of the Consultant, the Contractor, by the decision of the Owner/Homeowners Association, will be responsible for all additional fees charged by the Consultant for these additional services. These fees will be deducted from any retention left unpaid to the Contractor, and any balance left unpaid shall be invoiced by the Consultant to the Contractor.

G4. ADMINISTRATION OF THE CONTRACT

a. The Owner/Homeowners Association's Agent and/or the Consultant will provide administration of the Contract and will be the Owner/Homeowners Association's Representative (1) during construction, (2) until final payment is due, and (3) with the Owner/Homeowners Association's concurrence, from time to time during the correction period described in Section G6.

b. The Owner/Homeowners Association's Agent and/or the Consultant will visit the site at intervals appropriate to the stage of construction to become generally familiar with the progress and quality of the completed Work and to determine in general if the Work is being performed according to the Scope of Work and the Specifications attached to the Request for Bid and attached with this Contract. The Owner/Homeowners Association's Agent and/or the Consultant will not be required to make exhaustive or continuous on-site inspections to check the quality or the quantity of the Work. On the basis of on-site observations, the Owner/Homeowners Association's Agent and/or the Consultant will keep the Owner/Homeowners Association informed of the progress of the Work and will endeavor to guard the Owner/Homeowners Association against defects and deficiencies in the Work.

c. The Owner/Homeowners Association's Agent and/or the Consultant will not have control over or charge of and will not be responsible for construction means, methods, techniques, sequences, or procedures, or for safety precautions and programs in connection with the Work, since these are solely the Contractor's responsibility as provided and stated in Sections G1a and G10a. The Owner/Homeowners Association's Agent and/or the Consultant will not be responsible for the Contractor's failure to carry out the Work in accordance with the Scope of Work, Specifications, Change Orders, or the Contract Documents.

d. Based on the Owner/Homeowners Association's Agent's and/or the Consultant's observations and evaluations of the Contractor's Applications for Payment, the Owner/Homeowners Association's Agent and/or the Consultant will review the amounts due the Contractor upon receipt of the Contractor's invoice, will compare the amounts invoiced to the amount of the Contract and any change orders, and after approval shall submit all invoices in the approved amounts, with appropriate waivers of liens from the Contractor, Subcontractor, Sub-subcontractors, and Materialmen, for payment from the Owner/Homeowners Association.

e. The Owner/Homeowners Association's Agent and/or the Consultant will interpret and will have input to the matters concerning performance under and requirements of the Scope of Work, Specifications, and Contract Documents on written request of either the Owner/Homeowners Association or Contractor. The Owner/Homeowners Association's Agent and/or the Consultant will make initial decisions on all claims, disputes, or other matters in question between the Owner/Homeowners Association and Contractor, but will not be liable for results of any interpretations or decisions rendered in good faith. The Owner/Homeowners Association shall have all final decisions in matters relating to the maintenance of his pavement or related concrete. The Owner/Homeowners Association's decisions in matters relating to any aesthetic effect will be final if consistent with the intent expressed in the Scope of Work, Specifications, or Contract Documents.

f. The Owner/Homeowners Association's Agent and/or the Consultant will have the authority to reject Work which does not conform to the Scope of Work, Specifications, or the Contract Documents, and will do so with the permission of the Owner/Homeowners Association or the Owner/Homeowners Association's Agent.

g. The Owner/Homeowners Association's Agent and/or the Consultant will review and approve, or take whatever appropriate action needed, upon the Contractor's submittals such as Shop Drawings, Product Data and Samples, Change Order requests, Material Data and Specification Sheets, and Laboratory Materials Test and Design Data Records, but only for the limited purpose of checking for conformance with information given and the design concept expressed in the Specifications and the Contract Documents.

h. <u>Arbitration of Disputes:</u> Any dispute or claim in law or equity arising out of this Contract or relating to the Scope of Work described in this Contract, that is not settled through mediation, shall be determined by neutral, binding arbitration. The arbitration shall be conducted in accordance with Rules 72 through 76, Arizona Rules of Civil Procedure or the rules of the American Arbitration Association (AAA). The parties to the arbitration may agree in writing to use different rules and/or arbitrator(s). In all other respects, the arbitration shall be conducted in accordance with the Arizona Rules of Civil Procedure. Judgment upon the award rendered by the arbitrator(s) may be entered in any court having jurisdiction thereof. Any action that is within the jurisdiction of a probate or small claims court is excluded from arbitration under this agreement. The filing of judicial action to enable the recording of a notice of pending action, for order of attachment, receivership, injunction, or other provisional remedy, shall not constitute a waiver of the right to arbitrate under this provision. Venue for any mediation, arbitration, or litigation relating to this Contract shall be in the local county and state.

i. <u>Attorney's Fees:</u> The prevailing party in any dispute (whether or not it is submitted to arbitration) shall be awarded its reasonable attorney's fees, any and all costs, and disbursements of counsel and expert witness fees, in addition to any other relief awarded as set forth and determined by said arbitrator.

SECTION G5. SUBCONTRACTS AND SUBCONTRACTORS

a. A Subcontractor is a person or entity who has a direct Contract with the Contractor to perform a portion of the Work at the work site. The Subcontractor must be a licensed contractor in the work category the Subcontractor performs. A Sub-subcontractor is a person or entity who has a direct Contract with the Contractor through the Subcontractor to perform a portion of the Work at the work site. The Sub-subcontractor must be a licensed contractor in the work category the Sub-subcontractor performs.

b. Unless otherwise stated in the Request for Bid, the Contract Documents, or the bidding requirements, the Contractor, as soon as practicable after award of the Contract, shall furnish in writing to the Owner/Homeowners Association through the Consultant the names of all the Subcontractors, Sub-subcontractors, and Material Suppliers for each of the principal portions of the Work. The Contractor shall not contract with any Subcontractor or Sub-subcontractor about whom the Owner/Homeowners Association, Owner/Homeowners Association's Agent, or Consultant has made reasonable and timely objection. The Contractor shall not be required to contract with anyone about whom the Contractor has made reasonable objection. Contracts between the Contractor and Subcontractors, Subcontractors, and Sub-subcontractors shall (1) require each Subcontractor and Sub-subcontractor, to the extent of the Work to be performed by the Subcontractor or Sub-subcontractor, to be bound to the Contractor by the terms of the Scope of Work, Request for Bid, Specifications, and the Contract Documents, and to assume toward the Contractor all the obligations and responsibilities which the Contractor, by the Scope of Work, Request for Bid, Specifications, and the Contract Documents, assumes toward the Owner/Homeowners Association and Consultant, and (2) allow the Subcontractor and Sub-subcontractor the benefit of all rights, remedies, and redress afforded to the Contractor by the Scope of Work, Request for Bid, Specifications, and the Contract Documents.

SECTION G6. CONSTRUCTION BY THE HOMEOWNERS ASSOCIATION AND ADDITIONAL CONTRACTORS

a. The Owner/Homeowners Association reserves the right to perform construction or operations related to the project with the Owner/Homeowners Association's own forces, and to award separate Contracts in connection with other portions of the project or other construction or operations on the site under conditions of the contract identical or substantially similar to these, including those portions related to insurance and waiver of subrogation. If the Contractor claims that delay or additional cost is involved because of such action by the Owner/Homeowners Association, the Contractor shall make such claim as provided elsewhere in the Contract Documents.

b. The Contractor shall afford the Owner/Homeowners Association and separate contractors reasonable opportunity for the introduction and storage of their materials and equipment and performance of their activities, and shall connect and coordinate the Contractor's construction and operations with theirs as required by the Contract Documents.

c. Costs caused by delays, improperly timed activities, or defective construction shall be the responsibility of the party responsible therefor.

SECTION G7. CHANGES IN THE SCOPE OF WORK AND CHANGE ORDERS TO WORK

a. The Owner/Homeowners Association, without invalidating the Contract, may order changes in the Work consisting of additions, deletions, or modifications, the Contract Sum and Contract Time being adjusted accordingly. Such changes in the Work shall be authorized by written Change Order signed by the Owner/Homeowners Association, Contractor, and Consultant, or by written Construction Change Directive signed by the Owner/Homeowners Association and Consultant.

b. The Contract Sum and Contract Time shall be changed only by a Change Order issued by the Owner/Homeowners Association signed by the Owner/Homeowners Association and the Contractor.

c. The cost or credit to the Owner/Homeowners Association from a change in the Work shall be determined by mutual agreement between the Owner/Homeowners Association and the Contractor.

d. Any additional Work or materials required above and beyond the Contract shall be executed with an agreed change order between the Owner/Homeowners Association, Manager, Consultant, and Contractor. Any additional Work performed or materials used will not be compensated for if not approved prior to application and accompanied by an authorized change order. All labor and materials used for change orders must be verified, and any unauthorized labor and/or materials will not be paid for.

e. Any tasks added to the Contract by change order shall be at the bid price and/or bid unit price as listed on the bid sheet in the Request for Bid. At no time shall the price for change order items be more than the listed price for similar tasks in the Request for Bid. Should the price for a change order task be in excess of the bid task amount and/or the unit amount, the price will be adjusted accordingly.

f. Where a minimum amount is listed for a task and any additional Work is included which will keep the unit amount under the minimum charge, no additional compensation will be added.

SECTION G8. CONTRACT AND WORK TIME, AND COMPLETION OF WORK TIME

a. Time limits stated in the Scope of Work and the Contract Documents are of the essence of the Contract. The Contractor is given the opportunity to set a reasonable schedule on the Bid by Tasks of the Request for Bid by declaring the number of days it will take to complete the project. Also, the Contractor has submitted a schedule in writing which is acceptable to the Owner/Homeowners Association and the Consultant. This schedule has become part of the Contract Documents and will be referred to for any and all liquidated damages caused by over-running the scheduled time to complete the Work. By executing the Agreement, the Contractor confirms that the Contract Time is a reasonable period for performing the Work.

   The Contractor will be allowed to work between the hours set out in the Request for Bid. The Contractor agrees to work only within the hours set out in the Request for Bid including cleanup time at the end of the workday. It is recommended that the Contractor bid according to his/her ability to complete the project in the designated number of days predicted by the Contractor in order to complete the project with diligence and prudence. The Contractor shall not be allowed to work after a set time as called out in the Request for Bid. Therefore, cleanup and shutdown time will need to be allowed in the workday.

   The Owner/Homeowners Association will assess an established penalty per day, as shown in the Request for Bid, that the Contractor works beyond the submitted schedule of days. The exception to this rule is delays due to bad or inclement weather, or other acts of God that are not the fault of the Contractor and further listed in item c below.

b. The date of Substantial Completion shall be designated when all tasks listed in the Bid by Tasks in the Request for Bid are completed, all change orders (if any) are signed by the Contractor and submitted to the Owner/Homeowners Association for approval, and all Conditional Waivers of Liens from the Contractor, Subcontractors, and Sub-subcontractors have been submitted with the Contractor's Invoice. Should the Work be completed and the Contractor has not submitted all signed change orders (if any) to the Contract to the Owner/Homeowners Association, the Owner/Homeowners Association's Agent, or the Consultant, then the date of substantial completion shall be the date the signed change orders are received from the Contractor.

c. If the Contractor is delayed at any time in progress of the Work by changes ordered in the Work, labor disputes, fire, unusual delay in deliveries, abnormal adverse weather conditions not reasonably anticipatable, unavoidable casualties, or any causes beyond the Contractor's control, or other causes which the Consultant determines may justify delay, then the Contract Time shall be extended by Change Order for such reasonable time as the Consultant may determine.

d. Should the Work be delayed for any reason, the Contractor shall continue work on the project the first day following the delay.

SECTION G9. PAYMENTS ON CONTRACTS AND FINAL PAYMENT ON COMPLETION

a. Payments shall be made as provided in Sections D and E of this Contract.

b. Payments may be withheld on account of (1) defective Work not remedied, (2) claims filed by third parties, (3) failure of the Contractor to make payments properly to Subcontractors and Sub-subcontractors or for labor, materials, or equipment, (4) reasonable evidence that the Work cannot be completed for the unpaid balance of the Contract Sum, (5) damage to the Owner/Homeowners Association or another Contractor, (6) reasonable evidence that the Work will not be completed within the Contract Time and that the unpaid balance would not be adequate to cover actual or liquidated damages for the anticipated delay, or (7) persistent failure to carry out the Work in accordance with the Scope of Work, Specifications, or the Contract Documents.

c.   When the Consultant agrees that the Work is substantially complete and notifies the Owner/Homeowners Association, the Consultant shall provide the Owner/Homeowners Association with the invoices and Waivers of Liens from the Contractor for payment as outlined in Section D of this Contract.

d.   Final payment shall not become due until the Contractor has delivered to the Owner/Homeowners Association all completed releases of all liens from the Contractor, Subcontractors, Sub-subcontractors, and Materialmen arising out of this Contract or receipts in full covering all labor, materials, and equipment for which a lien could be filed and a Preliminary Notice of Lien (20-day letter), or a bond satisfactory to the Owner/Homeowners Association to indemnify the Owner/Homeowners Association against any liens. If any liens remain unsatisfied after payments are made, the Contractor shall refund to the Owner/Homeowners Association all money that the Owner/Homeowners Association may be compelled to pay in discharging any liens, including all costs and reasonable attorney's fees.

e.   The making of the final payment shall constitute a waiver of claims by the Owner/Homeowners Association except those arising from:

1.   Liens, claims, security interests, or encumbrances arising out of the Contract and unsettled;
2.   Failure of the Work to comply with the requirements of the Contract Specifications and Contract Documents; or
3.   Terms of special warranties required by the Contract Documents.

Acceptance of final payment by the Contractor, a Subcontractor, a Sub-subcontractor, or Material Supplier shall constitute a waiver of claims by that payee except those previously made in writing and identified by that payee as unsettled at the time of final Application for Payment.

f.   The Contractor shall provide, prior to submitting invoices for final payment, all waivers of liens from all Subcontractors, Sub-subcontractors, Material Suppliers, and any persons or entities that have submitted a Notice of Right to Lien (commonly referred to as a 20-day letter) to the Owner/Homeowners Association, the Owner/Homeowners Association's Agent, or the Consultant.

g.   All invoices must be accompanied with all Conditional Waiver of Liens from the Contractor, Subcontractors, Sub-subcontractors, and Material Suppliers who have served notice with a Preliminary Notice of Lien (which will become an Unconditional Waiver of Lien upon receipt and endorsement of the demand deposit document) for the 90 Percent Payment Portion.

WITHOUT THE ACCOMPANYING CONDITIONAL WAIVERS, THE INVOICES WILL NOT BE SUBMITTED FOR PAYMENT. ONLY INVOICES WITH PROPER WAIVER DOCUMENTS WILL BE SUBMITTED FOR PAYMENT.

h-   Upon completion of the Punch List Items, an invoice for final payment should be submitted and accompanied with a Conditional Waiver of Lien (which will become an Unconditional Waiver of Lien upon receipt and endorsement of the demand deposit document) for the 10 percent retention amount.

WITHOUT THE ACCOMPANYING CONDITIONAL WAIVERS, THE INVOICES WILL NOT BE SUBMITTED FOR PAYMENT. ONLY INVOICES WITH PROPER WAIVER DOCUMENTS WILL BE SUBMITTED FOR PAYMENT.

SECTION G10. PROTECTION FROM BODILY INJURY AND DAMAGE

a.   The Contractor shall be responsible for initiating, maintaining, and supervising all safety precautions and programs in connection with the performance of the Contract. The Contractor shall take reasonable precautions for safety of, and shall provide reasonable protection to prevent damage, injury, or loss to:

  1. The Contractors, Subcontractors, Sub-subcontractor's employees on the Work site, the Owner/ Homeowners Association and his/her Agent and Consultant, the Owner/Homeowners Association's employees, the Agent's employees, the Tenants, the Tenant's employees and customers, the Consultant's employees, and all other persons who may be affected thereby;
  2. The Work and materials and equipment to be incorporated therein; and
  3. All other property of the Owner/Homeowners Association, the Owner/Homeowners Association's Agent and Consultant, Tenants, and customers of the Owner/Homeowners Association and Tenants at the work site or adjacent thereto.

The Contractor shall give notices and comply with applicable laws, ordinances, rules, regulations, and lawful orders of all public authorities bearing on safety of persons and property and their protection from damage, injury, or loss. The Contractor shall promptly remedy damage and loss to property at the site caused in whole or in part by the Contractor, a Subcontractor, a Sub-subcontractor, or anyone directly or indirectly employed by any of them, or by anyone for whose acts they may be liable and for which the Contractor is responsible except for damage or loss attributable to acts or omissions of the Owner/Homeowners Association, the Owner/ Homeowners Association's Agent, or the Consultant or by anyone for whose acts either of them may be liable, and not attributable to the fault or negligence of the Contractor. The foregoing obligations of the Contractor are in addition to the Contractor's obligations under Subsection G3-n (The Contractor).

b.   The Contractor shall advise the Owner/Homeowners Association, Owner/Homeowners Association's Agent, or Consultant of any and all damage to private and personal property or personal injury on the project or the property immediately and in writing. The Contractor shall also, in writing, indicate the action taken, or to be taken, to correct any and all damage or damages and claims. This notification shall be within twelve (12) hours of the occurrence.

c.   The Contractor shall not be required to perform without consent any Work relating to asbestos or poly-chlorinated biphenyl (PCB).

SECTION G11. INSURANCE, ADDITIONAL INSUREDS, INSURANCE RIDER

a.   The Contractor shall purchase and maintain, from a company or companies lawfully authorized to do business in the jurisdiction in which the project is located, insurance for protection from claims under workers' or workmen's compensation acts and other employee benefit acts which are applicable, claims for damages because of bodily injury, including death, and claims for damages, other than to the Work itself, to property which may arise out of or result from the Contractor's operations under the Contract, whether such operations be by the Contractor or by a Subcontractor, Sub-subcontractor, or anyone directly or indirectly employed by any of them. This insurance shall be written for not less than limits of liability specified in the Insurance Rider listed in Subsection G11-g, the Request for Bid, and the Contract Documents or required by law, whichever coverage is greater, and shall include contractual liability insurance applicable to the Contractor's obligations. Certificates of such insurance shall be filed with the Owner/Homeowners Association prior to the commencement of the Work naming the Owner/Homeowners Association, the Owner/Homeowners Association's Agent, and the Consultant as additional insureds.

b. The Owner/Homeowners Association shall be responsible for purchasing and maintaining the Owner/Homeowners Association's usual liability insurance. Optionally, the Owner/Homeowners Association may purchase and maintain other insurance for self-protection against claims which may arise from operations under the Contract. The Contractor shall not be responsible for purchasing and maintaining this optional Owner/Homeowners Association's liability insurance unless specifically required and stipulated by the Contract Documents or in the Request for Bid.

c. Unless otherwise provided, the Owner/Homeowners Association shall purchase and maintain, from a company or companies lawfully authorized to do business in the jurisdiction in which the project is located, property insurance upon the entire Work at the site to the full insurable value thereof. This insurance shall be on an all risk policy form and shall include interests of the Owner/Homeowners Association, the Owner/Homeowners Association's Agent, the Consultant, the Contractor, Subcontractors, and Sub-subcontractors in the Work and shall insure against the perils of fire and extended coverage and physical loss or damage including, without duplication of coverage, theft, vandalism, and malicious mischief.

d. A loss insured under Owner/Homeowners Association's property insurance shall be adjusted with the Owner/Homeowners Association and made payable to the Owner/Homeowners Association as fiduciary for the insureds, as their interests may appear, subject to the requirements of any applicable mortgagee clause.

e. The Owner/Homeowners Association shall file a copy of each policy with the Contractor before an exposure to loss may occur. Each policy shall contain a provision that the policy will not be cancelled or allowed to expire until at least thirty (30) days' prior written notice has been given to the Contractor.

f. The Owner/Homeowners Association and Contractor waive all rights against each other and Owner/Homeowners Association's Agent, the Consultant, Consultants to the Consultant as Associates, separate contractors described in Section H, if any, and any of their subcontractors, Sub-subcontractors, Agents, and employees, for damages caused by fire or other perils to the extent covered by property insurance obtained pursuant to Subsection G11 or any other property insurance applicable to the Work, except such rights as they may have to the proceeds of such insurance held by the Owner/Homeowners Association as fiduciary. The Contractor shall require similar waivers in favor of the Owner/Homeowners Association and the Contractor by Subcontractors and Sub-subcontractors. The Owner/Homeowners Association shall require similar waivers in favor of the Owner/Homeowners Association and Contractor by the Owner/Homeowners Association's Agent, Consultant, Consultants to the Consultant as Associates, separate Contractors described in Section H, if any, and the Subcontractors, Sub-subcontractors, Agents, and employees of any of them.

g. Supplementing the terms of Section G11 of this Agreement, the Contractor shall obtain and keep in full force and effect the following insurance coverage (the "Insurance Coverage"):

 A. Comprehensive General Liability Insurance in the form satisfactory to the industry and the insurance carrier as required by state law and with the following coverage: premises-operations, products/completed operations (the insured shall keep the insurance coverage in force until completion of the project and for ninety [90] days after), personal injury, blanket contractual liability, independent contractors, and broad form property damage (which shall include coverage for explosion and underground hazards). The limit of such coverage shall be in an amount not less than three million dollars ($3,000,000) combined single limit per occurrence for bodily injury and property damage.

 B. Workers' Compensation Insurance in accordance with state and local statutory limits, in the form satisfactory to the industry and the insurance carrier as required by state law, including employer's liability in an amount not less than five hundred thousand dollars ($500,000) per accident.

C. Automobile Liability Insurance (owned, non-owned, and hired) in the form satisfactory to the industry and the insurance carrier as required by state law in the amount not less than one million dollars ($1,000,000) combined single limit per occurrence for bodily injury and property damage.

1. On or before the date of this agreement, the Contractor shall provide to the Owner/Homeowners Association Certificate(s) of Insurance in form and substance satisfactory to the Owner/Homeowners Association, evidencing the Insurance Coverage. In no event shall (a) the Contractor (or Subcontractors, or Sub-subcontractors) enter the property or (b) any Work or Service by the Contractor (or any Subcontractor, or Sub-subcontractor) commence prior to receipt of such certificates.

2. Certificates of Insurance and Insurance Policies set forth herein, regardless of completed Contract price, must include as Additional Insureds: the Owner/Homeowners Association, the Owner/Homeowners Association's Agent (if any), and the Owner/Homeowners Association's Consultant (if any).

3. All Subcontractors, Sub-subcontractors, Consultants, Architects, Materialmen, and Others involved in the project shall provide required Certificates of Insurance set forth herein, regardless of completed Contract price, and must include as Additional Insureds: the Owner/Homeowners Association, and the Owner/Homeowners Association's Agent (if any), or be listed as Additional Insureds of the Contractor.

4. All Certificates and Insurance Policies set forth herein must provide that the Owner/Homeowners Association be given thirty (30) days' notice in the event of cancellation and/or reduction of coverage limits, and ten (10) days' notice in the event of non-renewal of any of the insurance documents and coverage required and stipulated.

SECTION G12. CORRECTION OF WORK AND PUNCH LIST ITEMS

a. The Contractor shall promptly correct all Work rejected and all punch list items identified by the Owner/ Homeowners Association, the Owner/Homeowners Association's Agent, and/or the Consultant or failing to conform to the requirements of the Scope of Work, the Specifications, and/or the Contract Documents, whether observed before or after Substantial Completion and whether or not fabricated, installed, or completed, and shall correct any Work found to be not in accordance with the requirements of the Scope of Work, Specifications, and/or the Contract Documents within a period of one year from the date of Substantial Completion of the Contract or by terms of an applicable special warranty required by the Request for Bid, and/or the Contract Documents. The provisions of this Section apply to Work done by Subcontractors, Sub-subcontractors, as well as to Work done by direct employees of the Contractor. THE WARRANTY PERIOD ON THIS PROJECT SHALL BE IN ACCORDANCE AND AS STATED IN THE REGULATIONS SET OUT BY THE ARIZONA REGISTRAR OF CONTRACTORS. THE STATE LAW REQUIREMENT IS ONE (1) YEAR FROM SUBSTANTIAL COMPLETION AND/ OR ACCEPTANCE OF THE PROJECT.

b. Nothing contained in this Section shall be construed to establish a period of limitation with respect to other obligations which the Contractor might have under the Contract Documents. Establishment of the time period of one year as described in the previous paragraph (a) relates only to the specific obligations of the Contractor to correct the Work and punch list items, and has no relationship to the time within which the obligation to comply with the Contract Documents may be sought to be enforced, nor to the time within which proceedings may be commenced to establish the Contractor's liability with respect to the Contractor's obligations other than specifically to correct the Work and all identified punch list items.

c. Correction of Work and punch list items are considered as a separate Work item, and any monies retained for punch list items is considered a separate portion of the Work, and substantial completion for correction to the Work and the punch list items will be the date the corrections to the Work and/or the punch list items are completed. This paragraph will refer to the 10 percent retention of the contracted payment for lien purposes. The right to lien on the retention or punch list items shall begin upon completion of the punch list items as determined by the Owner/Homeowners Association, the Owner/Homeowners Association's Agent, and/or the Consultant.

All punch list items will be completed within thirty (30) days of notification by the Owner/Homeowners Association, the Owner/Homeowners Association's Agent, and/or the Consultant. Should the Contractor not complete the punch list items within the thirty (30) days of notice, or decline to complete any of the punch list items, the Owner, the Owner's Agent, and/or the Consultant shall hire a Contractor to complete the punch list items, and the cost of completing the punch list items shall be subtracted from the 10 percent retention prior to release of the retention funds to the Contractor.

SECTION G13. OTHER DIRECTIONS AND OTHER ITEMS

a. The Contract shall be governed by the laws incorporated by the property location/homeowners association's state, the local licensing agency, and the local county and city where the project is located.

b. As between the Owner/Homeowners Association and the Contractor, any applicable statute of limitations shall commence to run and any alleged cause of action shall be deemed to have accrued:

1. Not later than the date of Substantial Completion for acts or failures to act occurring prior to the relevant date of Substantial Completion;
2. Not later than the date of issuance of the final Payment for acts or failures to act occurring subsequent to the relevant date of Substantial Completion and prior to issuance of the Payment; and
3. Not later than the date of the relevant act or failure to act or acts by the Contractor for acts or failures to act occurring after the date of the final Payment.

SECTION G14. TERMINATION OF THE CONTRACT

a. If the Consultant fails to recommend payment for a period of thirty (30) days through no fault of the Contractor, or if the Owner/Homeowners Association fails to make payment thereon for a period of thirty (30) days, the Contractor may, upon seven additional days' written notice to the Owner/Homeowners Association and the Consultant, terminate the Contract and recover from the Owner/Homeowners Association payment for Work executed and for proven loss with respect to materials, equipment, tools, and construction equipment and machinery, including reasonable overhead, profit, and damages applicable to the project.

b. If the Contractor defaults or persistently fails or neglects to carry out the Work in accordance with the Request for Bid, the Specifications, and/or the Contract Documents or fails to perform a provision of the Contract, the Owner/Homeowners Association, after seven (7) days' written notice to the Contractor and without prejudice to any other remedy the Owner/Homeowners Association may have, may make good such deficiencies and may deduct the cost thereof, including compensation for the Consultant's services and expenses made necessary thereby, from the payment then or thereafter due the Contractor. Alternatively, at the Owner/Homeowners Association's option, and upon certification by the Consultant that sufficient cause exists to justify such action, the Owner/Homeowners Association may terminate the Contract and take possession of the site and of all materials, equipment, tools, and construction equipment and machinery thereon owned by the Contractor and may finish the Work by whatever method the Owner/Homeowners Association may deem expedient. If the unpaid balance of the Contract Sum exceeds costs of finishing the Work, including compensation for the Consultant's services and expenses made necessary thereby, such excess shall be paid to the Contractor, but if such costs exceed such unpaid balance, the Contractor shall pay the difference to the Owner/Homeowners Association.

SECTION H.
ALL ADDITIONAL ITEMS AND INSTRUCTIONS TO THE CONTRACT

1. SCOTTSDALE CURB AND GUTTER DETAIL

The undersigned have read all portions of this Contract, and the attached Request for Bid and Specifications, and agree to all terms and conditions as set forward and outlined in the Contract Documents. By signature of this document, the Contractor agrees to complete the Work under the terms of the Contract, the Request for Bid, and the Specifications.

This Agreement and Contract is entered into as of the day and year written above and on page 1 of this Contract Document.

HOMEOWNERS ASSOCIATION:

MY HOMEOWNERS ASSOCIATION
(NAME OF HOMEOWNERS ASSOCIATION (OR AGENT FOR THE HOMEOWNERS ASSOCIATION)

(SIGNATURE)

(PRINTED NAME AND TITLE)

CONTRACTOR:

ANY SEAL COATING AND PAVING COMPANY, INC. AN ARIZONA CORPORATION
(NAME OF CONTRACTOR)

(SIGNATURE)

(PRINTED NAME AND TITLE)

After the contract has been issued and the work started, some unforeseen or additional work might or can be added. The only correct instrument to initiate this change is a written change order. Do not cancel an executed contract after the work has begun and issue a new contract with the changes in it. Always issue a change order to the original contract. This will keep the original scope of work in place with the changes added on. Following is a recommended change order form to use. The forms shown can be used for commercial, industrial, or residential properties as well as homeowners associations. Our example uses a homeowners association.

Change Order Form

_____   _____
(Name of Contractor)   (Date of Change Order)

Re: Change Order to Contract to _____

for   NAME OF CLIENT
_____

DESCRIBE WORK

located at the _____
             LOCATION OF WORK (parking lot, association streets, etc.)

of the Contract and Agreement between Owner and Contractor for Pavement and Concrete Repair and Maintenance.

Change Order Number	Date	Pages

Explanation

OWNER: CONTRACTOR:

_____   _____
(Signature)   (Signature)

_____   _____
(Printed name and title)   (Printed name and title)

Figure 7.3: Properly Completed Change Order

Any Seal Coating & Paving, Inc   October 19, 2008
(Name of Contractor)   (Date of Change Order)

Re: Change Order to Contract to Any Homeowners Association

                                        NAME OF CLIENT

for Pavement Crack Sealing, Patching, and Seal Coating
DESCRIBE WORK
located at the ANY Homeowners Association in Chandler, AZ
                        (LOCATION OF WORK: parking lot, association streets, etc.)
of the Contract and Agreement between Association and Contractor for Pavement and Concrete Repair and Maintenance.

Change Order Number	Date	Pages
1	10/22/2008	1

Explanation

Cost for development of certified barricade and traffic plan for closure of bike lane on MY Way and setup of barricading by a licensed barricade and traffic contractor (Barricades Ltd.) as required by the local governing agency. Needed plan for permit issue for parking variance on MY Way. Also, two (2) Notification Bulletin Boards required at each end of the bike path with closure information as required by the local governing agency.

Change Section C, paragraphs C1 and C2 to Read

C1. THE HOMEOWNERS ASSOCIATION SHALL PAY THE CONTRACTOR IN CURRENT FUNDS FOR THE CONTRACTOR'S PERFORMANCE OF THE CONTRACT AND THE COMPLETION OF THE PROJECT AS OUTLINED IN THE SCOPE OF WORK AND THE BID TASKS IN THE REQUEST FOR BID AND THE TERMS SHOWN IN SUBSECTION B2. THE HOMEOWNERS ASSOCIATION SHALL PAY THE CONTRACTOR THE CONTRACT SUM OF FIFTY THOUSAND SEVEN AND 00/100 ----------------- DOLLARS ($50,007.00 TAX INCLUDED), subject to all additions and deductions as provided in the Contract Documents and all change orders attached as part of this Contract.

C2. ANY ALTERNATIVES TO THE CONTRACT THAT WILL ALTER THE SCOPE OF WORK, THE QUANTITIES, OR THE PROJECT TASKS WILL HAVE TO BE LISTED BELOW. THIS WILL INCLUDE THE BASE BID FROM SECTION C1 AND ANY AND ALL SUBSEQUENT AMOUNTS AS THEY WILL ALTER THE FINAL AMOUNT OF THE CONTRACT.

1. BASE BID $47,752.21 TAX INCLUDED
2. CHANGE ORDER #1 $1,523.51 TAX INCLUDED (traffic plan and barricade service)
3. CHANCE ORDER #2 $731.28 TAX INCLUDED (required bulletin boards x 2)

OWNER: CONTRACTOR:

_____   _____
(Signature)   (Signature)

_____   _____
(Printed name and title)   (Printed name and title)

Chapter 8

Project Management, Inspection, and Completion

Now that the contractor has been selected and a contract put in place, it's time to set up the project and implement a project management plan. The project is not complete without project management, inspection, and proper completion with all the required documents. There are nine steps for proper project management:

1. Project scheduling
2. Permitting if needed for traffic control, use of public parking, trucking, excavating, dust emissions, work in the public way (including driveway approaches on the public right of way), etc.
3. Resident or tenant notifications and constant contact
4. Inspection
5. Material verification
6. Substantial completion
7. Punch list items
8. Final completion
9. Waivers of liens to cover all preliminary lien letters

A problem and distress development table (Appendix A) has been created to help identify problems and suggest corrective actions that should be applied (Troubleshooting Table for Pavement Problems).

PROJECT SCHEDULING

Project scheduling is an agreed time between the contractor and the owner or HOA board of directors. According to contract documents, the ways and means of completing the project is the sole responsibility of the contractor; however, the project must be completed with as least inconvenience as possible. The contractor will provide the owner, manager, or HOA with a schedule to do the project. Take the time to examine the schedule and check if it will fit with your tenants' or residents' needs and schedules. With commercial and industrial properties, there are delivery and shipping schedules, customers, and parking arrangements to consider. With homeowners associations and apartment communities, the everyday life of the residents must be considered. There are special needs and requests, unscheduled deliveries, trash pickup, and in some HOAs, streets without sidewalks, which will require special scheduling. With all property types, there are towing arrangements to make and notifications to agencies (post office, fire department, solid waste pickup, etc.). Also, there might be permits required by local agencies, such as parking in the public right of way, construction permits for work in the public right of way (this includes driveways, work permits, dust permits), excavation, and any other local agency permits required. Check with local agencies prior to bidding the project (especially overlays, removal and repaving, and reconstruction) to make sure you are in compliance when the project begins.

When any excavation is included in the project, make sure all underground utilities have been located. Most municipalities have a service that will locate utilities in homeowners associations with local utility companies. In some cases, a private locator will need to be employed to locate utilities. Any utility damaged during the project will be the responsibility of the owner and contractor if no effort is made to locate them.

Working with on-site security will also aid in a successful project, so include all security companies in the scheduling process.

Using a site plan (or map), section out the pavement to accommodate the issues and unique problems of your property. Referring to the contractor's schedule, check to see if the sectioning matches and is workable for both parties and all tenants or residents. Tenants and residents are going to have all kinds of reasons for special sectioning or needs. Only concern yourself with the availability of commercial/industrial suites or homes and access for the contractor's crews. If anyone has a disability, he/she will come forward when notices are handed out, and adjustments can be made at that time. Never ask for special problems or dispositions, as everyone will have one. When sectioning out a project, it is important to include the contractor's input, as he/she knows what his/her production capabilities are (e.g., how many square yards they can seal in a day, how many square feet they can pave, how many linear feet of cracking they can seal). With homeowners association streets, try to stay away from half streets as much as possible because creating two-way traffic in one lane creates liability situations.

TENANT AND RESIDENT NOTIFICATION

It is very important to keep the tenants of commercial properties and residents of apartment communities and homeowners associations notified of the project schedule and all tasks. After these groups are notified, they can make arrangements with vendors and service providers, make adjustments to any appointments they may have, and also notify the manager, contractor, or consultant of any scheduling problems they may have. Providing a section map with a notification letter will help these individuals visualize the instructions in the letter. In the letter, you also need to include a point of contact (usually the manager or consultant) and a phone number to reach them.

In the notification letter, include the map of the sections with corresponding days and dates, hours of restricted traffic or street closure, and any alternate routes and parking for commercial and industrial tenants or apartment and HOA residents. Figure 8.1 is a sample resident notification letter. Figure 8.2 is a sample commercial property notification letter, and Figure 8-3 is a sample of a question and answer form to hand out with the notification letter for apartment or HOA residents, which is handy and recommended and does reduce some confusion. It is recommended to personally hand out the letters to make sure everyone receives one. With commercial properties, have the recipient sign for it so you have a record the letter was delivered. With commercial properties, it is also helpful and recommended to personally stop in on each tenant daily and talk to him/her. This extra effort will avoid many hours of aggravation later on in the project. Also, if there are any special circumstances or needs, this is the time they all come out and can be dealt with. In the letter include a delay cause, which is a short paragraph stating how future notifications and scheduling will be handled due to weather conditions, equipment breakdowns, or many other problems that can develop with a project. Commercial and industrial tenants can be notified by the daily drop-in visits. With apartment and HOA residents, a bulletin board can be designated where changes can be posted (e.g., when they see their pavement wasn't serviced on the scheduled day).

Keeping constant and open communication with tenants and residents is vital and critical in making a project run smoothly.

Figure 8.1 Resident Notification Example

IMPORTANT
PAVEMENT MAINTENANCE
<u>NOTICE</u>

Dear Resident   October 20, 2008

The board of directors has contracted to do pavement maintenance to our streets. Please refer to the attached map for locations, sections, and dates. The sections are identified as section 1 and section 2. Please refer to the attached map for section identification.

- Section 1 consists of Mark Lane from east of 104th Way to 108th Place Gate, 106th St., Running Deer Trail, Blue Sky Drive, and the street west of 108th Place.

- Section 2 consists of 103rd Street, White Feather Lane, Mark Lane, the street between White Feather Lane and Mark Lane to east of 104th Way, and 104th Way.

Both sections will be crack sealed Monday, October 29. The streets will not be closed to traffic during this application, as the Contractor moves very rapidly. Please drive carefully and avoid the Contractor's equipment and crews.

On Tuesday, October 30, section 1 will be sealed. This section will be closed for 24 hours (7:00 AM Oct. 30 to 7:00 AM Oct. 31). Please use the 103rd Street Gate and park your vehicles in section 2 or other locations within the association.

On Wednesday, October 31, section 2 will be sealed. This section will be closed for 24 hours (7:00 AM Oct. 31 to 7:00 AM Nov. 1). Please use the 108th Street Gate and park your vehicles in section 1 or other locations within the association.

Vehicles must be moved out by 6 AM the day **YOUR** section is being sealed or it cannot be moved until the following morning.

Should a section be cancelled due to weather or unforeseen circumstances, a notice will be placed at the mailbox center to explain any changes.

<u>**RESIDENT ACCESS ONLY DURING THE ROAD SEAL!!!**</u>

NO WORKERS (maids, landscapers, pest or pool contractors, etc.) OR DELIVERIES.

PLEASE RESCHEDULE SERVICES <u>NOW</u>!!!

TRASH WILL BE PICKED UP ON SCHEDULED DATES; HOWEVER, IT WILL BE EARLY IN THE MORNING. PLEASE HAVE TRASH OUT BY 5:00 AM

SHOULD YOU PARK YOUR VEHICLE IN THE WORK AREA OR YOUR VEHICLE INTERFERES WITH THE CONTRACTOR'S WORK, IT WILL BE TOWED AT YOUR EXPENSE.

The streets will be closed for 24 hours in each section to allow for curing. Should you breach the barricades in anyway and cause any damage, you will be financially responsible for all the damage caused.

Please be sure to see the attached map and also read the "most frequently asked questions" sheets.

Thank you. Your patience is appreciated.

Any questions or problems please call:

Consultant _____

Consultant Name _____

Office Phone No. _____

Cell Phone No. _____

Figure 8.2 Commercial Tenant Notification Example

COMMERCIAL PROPERTY NAME

IMPORTANT
PAVEMENT MAINTENANCE
<u>NOTICE</u>

Dear Tenant   May 9, 2008

The Owner and Manager have contracted to do pavement maintenance May 16 to May 23. The parking lot will be divided into 3 sections and the rear delivery drive will be divided into 2 sections. Sections 1, 2, and 3 are the parking lot sections, and sections 4 and 5 are the delivery drive sections. Section 4 will be completed with section 1 and section 5 will be completed with section 3. The schedule for the maintenance project is listed below.

On Friday, May 16, the Contractor will be patching the parking lot without closing a section. However, the Contractor will be blocking off the areas to patched with red cones and barricades. Please avoid the areas being patched.

On Monday, May 19, the Contractor will be crack sealing and seal coating section 1 and section 4. These sections will be closed at 7:00 AM Monday and opened at 7:00 AM Tuesday after restriping. Please do not park in these sections and please notify your delivery companies and vendors of this schedule.

On Wednesday, May 21, the Contractor will be crack sealing and seal coating section 2. This section will be closed at 7:00 AM Wednesday and opened at 7:00 AM Thursday after restriping. Please do not park in these sections and please notify your delivery companies and vendors of this schedule.

On Friday, May 23, the Contractor will be crack sealing and seal coating section 3 and section 5. These sections will be closed at 7:00 AM Friday and opened at 7:00 AM Saturday after restriping. Please do not park in these sections and please notify your delivery companies and vendors of this schedule.

There will be some inconveniences during this process; however, the Contractor will do everything possible to keep delivery and shipping flow moving according to the schedule. Please be patient with us during this process. Thank you for your cooperation during this maintenance phase.

Please do not breach any of the barricades or cones set out to mark the sections. Should you move and/or drive through the barricades or cones and cause damage to the material or equipment, you will be financially responsible for all damages. Should you get any material on your vehicle, contact the management company immediately. Please drive carefully to avoid the equipment and workers.

If you have questions or need assistance, please contact one of us immediately. The phone numbers are:

Consultant Name_____
Office Phone No._____
Cell Phone No._____

Contractor Name_____
Office Phone No._____

Contractor Project Manager Name_____
Cell Phone No._____

Owner or Manager Contact Name_____
Office Phone No._____
Cell Phone No._____

Figure 8.3 Frequently Asked Questions
FREQUENTLY ASKED QUESTIONS

What if I have services, workers, or contractors scheduled ?
Please re-schedule them to another week, not during the sealing, since it will be RESIDENT ACCESS ONLY!

How will UPS, FED EX, DHL, etc. reach me ?
Deliveries will not be allowed in. The package will have to be delivered on another day.

When will mail be delivered ?
During the sealing of your street, the post office will be notified. Delivery is the option of the individual carrier.

What about parking ?
When your section is being sealed, you will have to park the vehicles you want to use in the section not being sealed. PARK SMART: You will have to walk to and from your home to your vehicle.

What if I didn't move my car and I need to get out ?
YOU CANNOT DRIVE ON NEW ROAD SEAL FOR 24 HOURS!!! PLAN AHEAD!!! Should you park in the parking stalls being sealed, your vehicle will be towed and it will cost you one hundred dollars ($100).

What if I drive over the new road seal ?
Driving over the wet sealer will scar the surface of the drives and parking areas and they will have to be re-sealed. The HOA will be charged for this and will pass the cost on to you. You could also damage your sealer and/or your car. All those costs will be your responsibility.

When can I walk on the new road seal ?
The sealer will be dry enough to walk on after three (3) hours. Remember, if it looks wet it probably is!!! Rule of thumb if the pavement is shinny, it is wet.

Will trash and recycling be collected ?
YES! The drivers have been given notice and maps and will come in at 6:00 AM to do a pickup in the scheduled areas for sealing before the work starts. Have your trash out by 5:00 AM so you don't miss the pickup !!

What if I have a 911 emergency ?
Emergency vehicles will have full access to drive wherever they need to go.

I just bought or sold my home, how will the movers reach me ?
THEY CANNOT! Please change the date.

What happens if it rains on my scheduled seal coat day ?
The work that is scheduled for that day will be moved to the end of the schedule. Change notice will be posted at the mail boxes and community bulletin board.

What communications will we receive?
You are receiving this communication with the information and the map, and you will receive a reminder letter.

Don't wait—*please* do your re-scheduling now!

Along with proper resident or tenant notification, keep in touch with everyone on a consistent basis. Some residents or tenants do not want to cause waves or trouble and will not mention anything they feel is important. Remember, every question, comment, or input they may have is important. If it is important to them, it should be important to you, so respond in a timely manner. Make a tour (especially commercial and industrial properties) daily. This makes the project easier and more tolerable for all parties. Any problems residents or tenants have will surface with notices and personal contacts. At this time, schedules can be adjusted or special needs can be addressed prior to starting the project.

Now that a final schedule is set, permits (if any) are approved and issued, and notifications are distributed, the project can start. At this time, inspection and material verification and testing is the next important phase of the project.

During each phase or task of the project, inspection is essential to ensure that the proper quantities of materials are used and the material is installed to the specifications included in the RFB. It is also important for the contractor to keep daily records of the materials being used on the project and notify the owner, manager, or consultant of any material overruns during the project. This prevents "after the fact" change orders that create hard feelings between both the owner and the contractor and gives the Owner a better understanding of why the change orders are necessary. Material overruns that require change orders can occur due to various circumstances on the project. The following are some common causes for overruns:

- Cracks are wider after being cleaned out, which will require more crack seal material.
- The surface is rougher than anticipated. This will cause the use of more crack seal for banding the edges of the crack and applying seal coat or slurry seal material.
- The pavement is old and oxidized, which will increase the absorption of seal coat material, which will increase the amount of seal coat used.
- The existing pavement is thicker than anticipated for remove and replace patching and removal and repaving.

These items should have been observed prior to bidding, but in some instances they do come up after the contract is awarded. If the contractor and the owner's representative are keeping track of the material being used on the project, overruns and material change orders can be dealt with in a professional and expedient manner. This chapter contains tables showing various types of materials and expected amounts of use. Using measurements, apply the factors given for each material type (crack seal, hot-mixed asphalt, seal coat, and type II slurry seal).

During the inspection process, project preparation and applications can be observed and monitored. It is in the best interests of the contractor to monitor their material usage and quantities (especially amount of materials used) to determine if the amount of materials ordered for the project are correct. It is a benefit to the owner, manager, or HOA board of directors to monitor the materials used in order to ensure that the correct amount is the amount bid. By monitoring the materials, any overruns can be determined and projected before the project is completed or the material quantities get beyond the amount bid to complete the project. It is much easier to correct any overruns or overcharges before the material is placed. By monitoring materials, it is also easier to adjust the applications if the overruns are a result of overapplication or obtain a change order if the existing pavement condition is causing the overrun. Monitoring materials eliminates any surprises for the Contractor and customer at the end of the project.

Monitoring Crack Sealing

When monitoring crack sealing, there are several problems that can cause a material overrun. Crack seal material is applied either with a pour pot or a mechanical hot pot machine with a pump and hose. The pour pot can have a material overrun if the pot leaks or the pavement surface is rough, or both. When a pavement surface is rough, it takes more material to band each side of the crack when filling in the gaps and voids in the rough surface. Also, when the pour pot is filled and it leaks while getting it to the crack, material is wasted. With the larger mechanical hot pot machines, the surface texture is a concern the same as with the pour pot. A rough surface will use more material when banding the crack. Another problem is not placing the tip of the wand in or just on top of the crack. If the wand operator projects or shoots material at the cracks and lets the squeegee operator spread the material, a larger band of more than 1 inch on each side of the crack is placed, and at the end of the crack, a large amount of material is left over. This material is spread on the pavement that does not have cracks. Filling cracks less than ¼ inch in opening is a waste of material because the material will not penetrate the crack and provides no benefit. The contractor bid to seal cracks with an opening of ¼ inch; filling smaller cracks will require material the contractor did not account for. Another problem area with crack sealing is caused when cleaning the crack. When the estimator bids the crack sealing, it is assumed the crack opening is ¼ to ½ inch wide. Sometimes the crack width is actually larger after cleaning. All these problems cause overruns in crack seal material. Inspecting the cracks as they are cleaned and applying the material properly will reduce the amount of overruns of crack seal material that can occur.

Table 8.1 explains how to monitor crack seal material. During the application process, make sure to maintain an accurate count of the boxes of material used so you can compare that number to the amount of boxes bid. Knowing the amount or linear feet of crack seal and the condition of the existing pavement surface (smooth to rough) will help you determine the amount of crack seal material needed to seal the linear feet of cracking in the bid. In all cases, bid cracking by the linear foot of ¼ inch or more in width, never gallons or pounds. Contracting crack sealing by gallons or pounds of material greatly increases chances of overruns and future change orders.

Application rates using a 50-lb. block (9.4 lb./gallon and 5.32 gallons/50 lb.-block or 3.19 gallons/30-lb. block):

Table 8.1: Estimating Crack Sealing Material Quantities

Low-Grade Rubber (Factor .44-Normal Surface Condition*)

Crack Dimension Depth × Width (In. × In.)	Material for Crack Fill (lb./ft.)	Material for Banding (lb./ft.)	Total Material to Seal Crack (lb./ft.)	Cracks per lb. Material (ft./lb.)
¼" × ¼"	0.0275	0.11	0.14	7.15
¼" × ½"	0.055	0.11	0.17	5.89
½" × ½"	0.11	0.11	0.22	4.55
½" × 1"	0.22	0.11	0.33	3.04
¾" × 1"	0.33	0.11	0.44	2.28
1" × 1"	0.44	0.11	0.55	1.82
1" × 2"	0.88	0.11	0.99	1.02
2" × 2"	1.76	0.11	1.87	0.54

*Rougher pavement surfaces will require more material for banding the crack

Medium-Grade Rubber (Factor .49-Normal Surface Condition*)

Crack Dimension Depth × Width (In. × In.)	Material for Crack Fill (lb./ft.)	Material for Banding (lb./ft.)	Total Material to Seal Crack (lb./ft.)	Cracks per lb. Material (ft./lb.)
¼" × ¼"	0.030625	0.1225	0.16	6.25
¼" × ½"	0.06125	0.1225	0.19	5.27
½" × ½"	0.1225	0.1225	0.25	4
½" × 1"	0.245	0.1225	0.37	2.71
¾" × 1"	0.3675	0.1225	0.49	2.05
1" × 1"	0.49	0.1225	0.62	1.62
1" × 2"	0.98	0.1225	1.11	0.91
2" × 2"	1.96	0.1225	2.09	0.48

*Rougher pavement surfaces will require more material for banding the crack

High-Grade Rubber (Factor .52-Normal Surface Condition*)

Crack Dimension Depth × Width (In. × In.)	Material for Crack Fill (lb./ft.)	Material for Banding (lb./ft.)	Total Material to Seal Crack (lb./ft.)	Cracks per lb. Material (ft./lb.)
¼" × ¼"	0.0325	0.13	0.17	5.89
¼" × ½"	0.065	0.13	0.2	5
½" × ½"	0.13	0.13	0.26	3.85
½" × 1"	0.26	0.13	0.39	2.57
¾" × 1"	0.39	0.13	0.52	1.93
1" × 1"	0.52	0.13	0.65	1.54
1" × 2"	1.04	0.13	1.17	0.86
2" × 2"	2.08	0.13	2.21	0.46

*Rougher pavement surfaces will require more material for banding on each side of the crack

Pavement temperature plays a large role in the type of material selected for the project. Because crack seal softens and becomes more plastic as temperatures increase, and hardens (becoming less plastic) as temperatures decrease, knowing the environment of the project site is critical to the selection of the material. When deciding which grade to use, evaluate the high and low pavement temperatures of the project site.

For example, Phoenix, Arizona, pavement temperatures are extremely high in the summer months and moderately low in the winter months (pavement temperatures range from 180°F to 35°F). You would use high-grade rubber and a factor of 0.52. This will reduce the potential for tracking the material as it becomes too soft with the high temperatures.

Fargo, North Dakota, pavement temperatures are moderately high in the summer months and very low in the winter months (pavement temperatures range from 80°F to -5°F). You would use low-grade rubber and a factor of 0.44. This allows the material to flex while temperatures are low in order to prevent rupturing during expansion and contraction associated with extreme temperature changes.

Pavements in climates between high and low temperatures require a medium grade. Contractors, consultants, and material distributors can contact the material manufacturers for the proper crack seal rubber grade to use in their climate.

Some contractors will bid crack sealing by gallons or pounds needed to complete the project. This quantity will need to be converted to linear feet sealed in order to properly compare the other bids where the contractors bid by linear foot. The following example shows how to calculate the linear feet of cracking (¼" x ½") using 1,500 gallons or 12,000 pounds of high-grade rubber crack seal in 50 pound boxes.

Example 8.1:   Calculating total linear feet of crack sealing using 1,500 gallons of crack sealant

1.   1,500 gallons ÷ 5.32 gallons (50 pound box) = 281.95 (50-pound boxes)

2.   281.95 (50-pound boxes) x 50 (lbs/50-lb box)= 14,098 lbs of crack sealer

3.   14,098 lbs x 5 ft/lb (from table 8.1) = 70,490 linear feet of cracks sealed

The total linear of feet of cracking (1/4" or wider with a depth of ½") that can be sealed using 1,500 gallons of crack sealer is 70,490.

Example 8.2:   Calculating total linear feet of crack sealing using 12,000 pounds of crack sealant

1.   12,000 lbs of crack sealer x 5 (lf. of crack seal material per lb.) = 60,000 Linear feet of crack

The total linear of feet of cracking (1/4" or wider with a depth of ½") that can be sealed using 12,000 pounds of crack sealer is 60,000.

Crack widths greater than ¾ inch will require backer rod to be placed in the crack prior to sealing. It also may be recommended to R&R patch cracks with an opening width greater than 1 inch. Cracks with an opening of ¾ to 1 inch will require a double application, which will require doubling the quantities (e.g., 1" × 1" = 75 LF/2 blocks or 14 LF/2 gallons). This can also cause an overrun if the contractor only bid for a single application.

Use the following formula to figure boxes of material needed or used to seal 18,500 LF of cracks ½ inch wide by ½ inch deep and banding 1 inch on each side of the crack, using high softening point material (with a crack seal factor .52) and an average smooth surface:

18,500 ÷ 3.85 (lbs. of crack seal / lf.) = 4,805 total pounds of crack sealant

Using 50-lb. boxes, the amount of boxes should be 96 boxes (4,805 ÷ 50), which would take 511 gallons (96 × 5.32).

Using 30-lb. boxes, the amount of boxes should be 160.17 boxes (4,805÷ 30), which would also take 511 gallons (160.17 × 3.19).

Application rates will increase if the surface is rough or rocky. This is caused by the additional material required when banding on both sides of the sealed crack.

Monitoring Patching

When monitoring patching, it is easy for the patch crew to patch more than what is actually bid. Marking the areas completely (using white spray paint) to be skin or surface patched, or saw cut for remove and replace patching, aids in reducing overruns or deficiencies of patching materials. Marking patch areas provides the bidding contractors with the same areas for measurements and consistencies in bids.

When monitoring R&R patching, check the thickness of the pavement as it is being removed. If possible, core the pavement to check for thickness prior to bidding R&R patching. Most R&R patches are bid for 2 to 2½ inches thick if cores are not previously done or no prior pavement history is available. In some cases, there are isolated areas that can be thicker than the rest of the streets or parking lots, which can create overrun problems. These areas are an anomaly to the total project area and should be accounted for separately.

When monitoring skin or surface patching, the material used should not be installed too thick; it not only creates overruns of material, but it also creates a bump or hump in the pavement.

Increases in R&R and surface patch areas, increases or decreases in pavement thickness in areas to be R&R patched, and skin patch areas installed too thick will cause overruns or deficiencies in asphalt pavement materials. Any changes that occur should be verified at the time of patching. Also, make sure the R&R patch is installed in the actual marked area. A patch installed in a smaller area than what is marked will create a deficiency of material and should be adjusted with a credit. Close observation during the application time can reduce overruns and deficiencies of material while patching.

Patching is the easiest of the maintenance applications to estimate; however, with R&R patching, it is the most expensive maintenance application that can be done to a pavement.

Skin or surface patching is applying an asphalt patch to the asphalt surface. There is no saw cutting or excavation involved. To calculate the material needed (either a fine aggregate hot mix or type II slurry seal), measure the dimensions of the patch area in square feet. Installing a pavement fabric or geotextile between the skin or surface patch and pavement is a slight variation to the surface patch and is used to repair an alligator cracked area or areas until funds can be reserved for total pavement removal. This type of patching is used in place of the more expensive R&R patching where the pavement is going to be removed in three to five years. The process involved is where a fabric inner layer is tacked or attached to the original pavement and the hot asphalt skin patch is applied. This type of patch helps preserve the pavement and reduce any reflective cracking. This type of patch is also an excellent temporary patch for pavements with a large amount of structural damage, which requires R&R patching until funding can be budgeted for total removal and repaving. This will increase the use of the pavement for a short period of time at a fraction of the cost of R&R patching, especially if the pavement is going to be removed or replaced in the near future (three to five years). When inspecting skin patching, the total area is determined for all patches (use the same estimating formulas in Table 8.4). Where a fabric inner layer is used, add a per-square-foot cost for the fabric to each surface patch area. Also, a cost should be added as a patch surcharge (usually $1 to $2 per patch) for an extra labor requirement.

When monitoring remove and replace patching, the depth of the existing pavement should be known prior to starting; measure the dimensions in square feet and calculate the amount of material to be removed and how much material must be installed and compacted. Use the calculations in Table 8.5 to determine the tonnage of material to use for patching and calculate the tonnage removed. Also, add the cost per linear foot of measurement for saw cutting all sides of the patch area. Using the scale tickets collected from the trucking company or the contractor, the costs of hot asphalt material can be verified.

Monitoring Seal Coating and Slurry Seal

When monitoring seal coats and slurry seals, the existing surface texture is a major concern. A rough surface will require more material than a smooth surface, and an old, oxidized surface will require more material than a newer flexible surface. If the seal coat project was bid using 0.40 gallon of sealer per square yard and the surface texture is rough, the application rate could be as much as 0.50 to 0.60 gal./yd.² The same applies for type II slurry seal. Instead of 16 lb. of material per square yard, 20 to 25 lb. could be used. Keep track of material being used and make adjustments or change orders as the project goes along, not after the project is completed. If a surface is old and oxidized, the surface will absorb some of the seal coat material, causing a material overrun.

The specifications for the project have a stated application rate for seal coats and type II slurry seals. These application rates are a function of the existing pavement surface condition or texture. Materials for seal coats and type II slurry seals are delivered or applied from an application machine with a known volume. Verify the amount of material and measure the surface area sealed to obtain the application rate or coverage rate. Use Table 8.3 to determine the coverage area in square feet and compare the amount of material delivered and later used.

Seal Coats

Seal coats are applied using equipment with tanks that hold the material during application. Knowing the volume of material (gallons) and the desired application rate (gallons/square yard), the total area to be covered by one tank of material can be determined. Most tanks have a storage capacity of 250, 500, 1,000, 2,000, or 3,000 gallons (the larger tanks are used for spray applications). In Table 8.2, total areas covered are given for specific volumes at certain application rates. The coverage areas are for smooth to medium pavements only. Table 8.2 can be used for rougher or rockier surfaces only if a spray application is being used. If the material is applied using a squeegee method, then the total yards covered will decrease as the roughness of the pavement increases.

Application Rate Gallon/yd.^{2}	Coverage (yd.^{2}/gallon)	Coverage (yd.^{2}) 250 Gallons	Coverage (yd.^{2}) 500 Gallons	Coverage (yd.^{2}) 1,000 Gallons	Coverage (yd.^{2}) 2,000 Gallons
0.1	10	2,500	5,000	10,000	20,000
0.15	6.7	1,675	3,350	6,700	13,400
0.2	5	1,250	2,500	5,000	10,000
0.25	4	1,000	2,000	4,000	8,000
0.3	3.4	850	1,700	3,400	6,800
0.35	2.9	725	1,450	2,900	5,800
0.4	2.5	625	1,250	2,500	5,000
0.45	2.3	575	1,150	2,300	4,600
0.5	2	500	1,000	2,000	4,000

Table 8.2: Seal Coat Coverage

When applying two coats, the average material used on a smooth existing pavement is 0.25 to 0.30 gal./yd.^{2} per application. As the surface becomes rougher or older, more seal coat material will be required per sq/yd.

Dilution Rate
Monitoring the dilution rate is another critical step to having the correct consistency and solids in the final applied material. Dilution rates are figured as a percentage of the concentrate amount delivered or picked up. An example of a proper dilution rate is diluting 800 gallons of concentrate 25 percent.

800 gallons of concentrate x 0.25 = 200 gallons of water

This gives the total amount of water that needs to be added to the concentrate for a 25 percent dilution rate. Adding the volume of water to the volume of concentrate will give the total volume of the sealer.

200 gallons of water + 800 gallons of concentrate = 1000 gallons of diluted sealer.

It is important to remember when diluting the concetrate that the water being added is a percentage of the existing concentrate. So if a dilution is 25 percent, then the water added is 25 percent of the orginal concetrate volume. This is a 4-parts concentrate to 1 part water mixture. If the specifications call for a 20 percent dilution rate, the mixture would be 1 part water to 5 parts concentrate.

Because sealer is typically transported in tankers (2000, 1000, and 500 gal) or placed using a squigee tank (250 gal), it is necessary to know how to calculate the dilution of the sealer based on the volume of the tank and not the amount of concentrate. This is accomplished by dividing the total volume of the tank by one plus the dilution rate. Remember, if the dilution rate is specified as 25 percent, the actual number used in the equation will be 0.25.

The following example calculates the amount of water and concentrate in gallons for a 1000 gallons with a specified dilution rate of 25 percent.

concentrate in gallons = volume of tank / 1+ dilution rate
concentrate in gallons = 1000/1.25 = 800 gallons of concentrate
water in gallons (dilution) = 1000 − 800 = 200 gallons of water

An example of the <u>incorrect</u> method of calculating would be taking 25 percent of the 1000 gallon tanker resulting in 250 gallons of water. That figures to be 250 gallons of water and 750 gallons of concentrate. When the calculations for dilution of concentrate are applied (250 ÷ 750 = 0.333), the result is a 33.3 percent dilution of concentrate. This can cause a reduction of solids and reduce the life expectancy of the seal coat and can promote premature wearing. Always calculate the dilution rate as a percent of concentrate.

Table 8.3 shows the amount of water and concentrate (in gallons) needed for 2000, 1000, 500, and 250 gallon tankers for given dilution rates as a percentage of concentrate using the tank volume method.

A Volume of Tank Gallons	B Specified Dilution of Concentrate %	C Sealer Concentrate Gallons A/(1+B)	D Water Gallons A-C
2000	40	1429	571
2000	30	1538	462
2000	25	1600	400
1000	40	714	286
1000	30	769	231
1000	25	800	200
500	40	357	143
500	30	385	115
500	25	400	100
250	40	179	71
250	30	192	58
250	25	200	50

Table 8.3: Dilution Rates of Seal Coat Concentrate and Water

The normal dilutioin rate for asphalt emulsion is 25 percent except when the ambient and the pavement surface is hotter than 100°F. Then the dilution rate should be increased to 30 percent and the application rate increased to compensate for the increase in dilution rate. The normal dilution rate for coal tar and coal tar/asphalt emulsion blends is 30-40 percent as specified by the manufacturer.

If the material is delivered diluted from the manufacturer to the project, a certificate of compliance should be included to show the dilution rate. If the material is delivered in concentrate form, the dilution rate will have to be determined by the volume of the distribuion tank. Most tanks, (except the mobile squeegee machine, which is square) will be cylindrical. The only way to determine the amount of concentrate and water is using a calibrated rod to check the depth of concentrate in the tank prior to diluting.

The dilution rate is important for controlling the percent of solids in the total sealer mixture. Solids are important for wearability, the higher the dilution rate the lower the solids count. Therefore do not exceed the specified dilution rate. Remember, dilution rate is calculated as a percent of concentrate, not percent of tank volume. Calculating the dilution as a percentage of the tank volume will yield too much dilution and lower solids than specified in the dried product.

Solids are determined by evaporation of the water from the material. Any additional sand or minerals added to the concentrate will increase the solids count. Any additional sand or minerals added will have to be adjusted by the percent of additon.

Measuring the unit weight by gallon of diluted material is an indication of the blend of water and concentrate. A gallon of seal coat concentrate should weigh 9 to 10 pounds. Once the material is diluted and arrives on the project, it should be within the parameters of 9.4 to 10 pounds per gallon. Using the unit weight of the seal coat product and the unit weight of water, combined with the proper dilution percentages, the unit weight of the diluted material can be determined. Following is an example of determining the unit weight of a diluted sample.

The unit weight of seal coat is 10.0 lbs per gallon and water is 8.33 lbs per gallon. The percentage of concentrate is 80 and water is 20. The calculations are as follows:

10.0 lb/gallon x .8 = 8.0 lbs per gallon concentrate
8.33 lb/gallon x .2 = 1.66 lbs per gallon water

Therefore the total unit weight of the diluted seal coat should be 9.66 lbs per gallon.

8.0 lbs/gallon (concentrate) + 1.66 lbs/gallon (water) = 9.66 lbs/gallon (total)

Type II Slurry Seal

Type II slurry seal is a mixture of slurry seal sand, asphalt emulsion, water, and cement (depending on daily high temperatures). One truckload of material with 10,000 lb. of slurry seal sand will cover an average surface of 500 yd.^{2} at an application rate or yield of 20 lb./yd.^{2}. The rougher or rockier the surface, the higher the application rate or yield, which will reduce the surface area covered. Conversely, the smoother the surface, the less the application rate or yield, which will increase the surface area covered. Table 8.4 shows the expected coverage per 10,000-lb. load of sand with a given yield on an average surface to rough surface texture.

Application Rate or Yield (lb./yd.^{2})	Total Coverage (yd.^{2} per 10,000 lb. of material)
15	670
20	500
25	400
30	335

Table 8.4: Type II Slurry Seal Yield

Monitoring Asphalt Paving and Overlays

When monitoring overlays or paving projects, check the depth of the hot-mixed asphalt as it is being installed. This can be accomplished by using a device known as a poker, which is usually a thin metal rod or a wooden dowel with the desired depth clearly marked on the lower tip. The contractor should have a poker that is with the paving crew. Check the measurement of the end of the poker to make sure it is set at the proper setting or thickness. Should an owner, manager, or consultant check the depth, a wooden dowel works best after it has been marked for the proper depth. Make sure the poke hole is filled after checking the depth to prevent a small hole in the finished surface. Checking the depth of the pavement is very important because this will allow for adjustments during paving in order to maintain the proper depth. The depth of hot-mixed, uncompacted asphalt pavement should be ¼ inch for each inch installed. For example, when 1 inch compacted is specified, then 1¼ inch uncompacted should be installed. Likewise, when 2 inches of compacted asphalt is specified, the depth of the uncompacted asphalt should be 2½ inches. When 3 inches compacted is specified, then the uncompacted asphalt should be placed 3¾ inches thick. Hot-mixed asphalt pavement is very expensive, and overruns can be costly to both the customer and the contractor. Deficiencies can be harmful to the customer by reducing the life expectancy of the street or parking lot and can also harm the integrity of the contractor.

Paving and Overlay Materials

The hot-mixed asphalt batch plants have a mix design, which is known as a Marshall, Hveem, or Superpave mix design. The most commonly used control for pavement maintenance in parking lots and HOA streets is the Marshall test. This provides a field compaction control used to determine how well the asphalt pavement is compacted. When monitoring pavement, it is important to know what this Marshall value is. It is measured in pounds per cubic feet (PCF), and the tons of asphalt used on your project can be calculated using this figure. Typical Marshall densities have a parameter of 138 PCF to 155 PCF, depending on local rock deposits and geographical locations. Table 8.5 shows the total tons required to pave 1 ft.^{2} from 1 inch to 4 inches thick using a Marshall value from 140 to 154 PCF. The formula to calculate tons of asphalt required is easy to use and is set out below in Example 1. Table 8.5 is set up with the factors used to figure how many tons of asphalt will be needed for various depths per square foot of total surface area.

Marshall Density (PCF)

Pavement Thickness (in.)	140	142	144	146	148	150	154
1	0.0059	0.006	0.006	0.0061	0.0062	0.0063	0.0065
2	0.0117	0.0119	0.012	0.0122	0.0124	0.0125	0.0129
3	0.0175	0.0178	0.018	0.0183	0.0185	0.0188	0.0193
4	0.0234	0.0237	0.024	0.0244	0.0247	0.025	0.0257

Table 8.5: Total Weight of 1 ft.^{2} of Asphalt (tons) at a Given Depth (in.) with Various Marshall Values

To figure out the factors for other PCF figures, find the volume of the asphalt (in this case, it's the depth in feet multiplied by 1 ft.^{2}) and multiply it by the Marshall value to obtain the weight of the asphalt (for 1 ft.^{2} area) in pounds. Divide the weight by 2,000 lb. per ton to obtain the total tons for 1 ft.^{2}.

Example 1:

The Marshall value provided by a testing lab, state highway lab, or the manufacturer is 143 PCF (lb./ft.^{3}). How many tons are needed for a 1 ft.^{2} area at 2 inches thick?

Step 1: Find the volume of the 1 ft.^{2} area.

Volume = Area × Depth (in feet)
= 1 ft.^{2} × (2 in. ÷ 12 in./ft.)
= 1 ft.^{2} × 0.166 ft.
= 0.166 ft.^{3}

Step 2: Find the weight of the asphalt using the Marshall value.

Weight = Volume (ft.^{3})   ×   Marshall value (lb./ft.^{3})
= 0.166 ft.^{3} ×   143 (lb./ft.^{3})
= 23.74 lb.

Step 3: Convert pounds to tons.

Tons	=	pounds ÷		pounds per ton (lb./ton)
	=	23.74 lb.	÷	2,000 lb./ton
	=	**0.0119 tons**		

0.119 tons per square foot of area of asphalt will be needed to pave an area of 1 ft.² at 2 inches in depth.

Once the per-square-foot factors have been calculated, the tonnage for a project can be calculated by multiplying the factor by the total area of the project.

Example 2:

With a square-foot factor of 0.0119 tons/ft.² at 2 inches in depth (Example 1), find the total tonnage needed for a parking lot 320 feet wide by 420 feet long.

Step 1: Find the total area of the parking lot.

Area	=	320 ft. ×	420 ft.
	=	134,400 ft.²	

Step 2: Multiply the area by the square-foot factor to find the total tonnage.

Tonnage	=	134,400 ft.²	×	0.0119 tons/ft.²
	=	1,599 tons		

Striping and Painting

Verifying the material used for striping and painting is done based on the gallons of each color used, compared to the area applied. Check with local paint suppliers (DOT-approved paints) for application rate and costs. Calculate the area to be covered and multiply it by the manufacturer's application rate, which will give the number of gallons needed for the project. Stripes are normally 3 to 4 inches wide and 16 to 20 feet in length. Curbing is painted the length of the curb and is usually 4 inches in width and is painted on one side and the top. Stencils are priced per letter or number, and the best way to monitor stencils is to inventory all of the letters or numbers used. All other shapes and dimensions are measured by square foot and application of the paint color desired. Colors other than white or yellow will be higher in cost, and adding glass beads for higher reflection will add about 43 to 72 percent to the cost of striping.

Concrete Curbs, Gutters, and Sidewalks

When verifying the cubic yards for concrete structures like curbs, combination curb and gutters, valley gutters (concrete swales), dumpster pads, and sidewalks, use the measurement of length multiplied by width and then multiply by the thickness in feet. The volume will then need to be converted from cubic feet to cubic yards. *For details of various concrete structures, check with the local municipalities. It is important to remember that there is no uniform detail, so check with the prevailing specification in the area.*

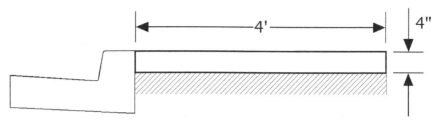

Figure 8.4: Concrete Sidewalk

Example 3: Calculating cubic yards of concrete for sidewalk

Calculate the cubic yards needed to construct a concrete sidewalk with the cross sectional dimensions shown in Figure 8.4, which is 200 feet long.

Step 1: Convert depth from inches to feet.

Depth (in.)	=	4 in.	÷	12 in./ft.
	=	0.33 ft.		

Step 2: Calculate the cross sectional area in square feet (ft.^{2}).

Area (ft.^{2})	=	Width	×	Depth
	=	0.33 ft.	×	4 ft.
	=	1.32 ft.^{2}		

Step 3: Calculate the volume of the entire sidewalk in cubic feet (ft.^{3}).

Volume (ft.^{3})	=	Length	×	Area
	=	200 ft.	×	1.32 ft.^{2}
	=	264 ft.^{3}		

Step 4: Convert the volume from cubic feet (ft.^{3}) to cubic yards (yd.^{3}).

Volume (ft.^{3})	=	264 ft.^{3} ÷ 27 ft.^{3}/yd.^{3}
	=	9.78 yd.^{3}

In this example, the Contractor will probably order 10 yd.^{3} of concrete.

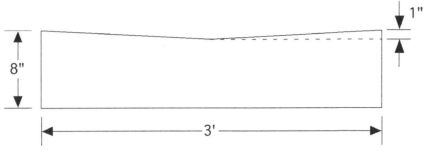

Figure 8.5: Valley Gutter (Concrete Swale)

Example 4: Calculating cubic yards of concrete for valley gutter (swale)

Calculate the cubic yards needed to construct a concrete valley gutter (swale) shown in Figure 8.5, 3 feet wide, 300 feet long, and 8 inches in depth.

Step 1: Convert depths from inches to feet.

Depth (in.)	=	8 in.	÷	12 in./ft.
	=	0.67 ft.		
Swale depth	=	1 in.	÷ 12 in./ft.	
	=	0.0833 ft.		

Step 2: Calculate the cross sectional area of the gutter.

The cross sectional area of the valley gutter (swale) is width multiplied by height minus the area of the 1-inch swale. The swale is a triangle so its area is base multiplied by height divided by 2. Remember, quantities in inches need to be converted to feet.

Area	=	(width × height) – ((swale height × swale base)/2)
	=	(0.66 ft. × 3 ft.) – (0.0833 ft. × 3 ft.)/2
	=	2 ft.^{2} – 0.125 ft.^{2}
	=	1.875 ft.^{2}

Step 3: Find the volume of entire the valley gutter (concrete swale).

Volume =	1.875 ft.^{2} × 300 ft.	
=	563 ft.^{3}	

Step 4: Convert the volume from square feet (ft.^{2}) to cubic feet (ft.^{3}).

Volume (cu ft)	=	563 ft.^{3} ÷ 27 ft.^{3}/yd.^{3}
	=	20.85 yd.^{3}

In this example, the contractor will probably order 21 yd.^{3} of concrete.

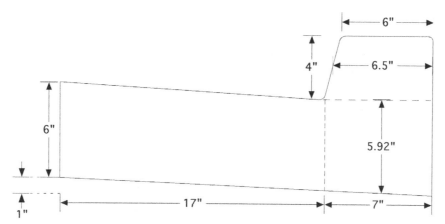

Figure 8.6: Concrete Curb and Gutter

Example 5: Calculating cubic yards of concrete for curb and gutter

Calculate the total volume in cubic yards of concrete needed to construct 300 linear feet of the curb and gutter shown in Figure 8.6.

Step 1: Calculate the cross sectional area of the curb.

In this example, the cross section of the curb can be broken into three pieces: the top of the curb, the section under the top of the curb, and the gutter section. The top of curb and the section under it are trapezoids, and the gutter portion is simply length multiplied by width. Note: the area of a trapezoid is $((b_1 + b_2)/2) \times h$. For this example, the mean base has been included in the drawing, so those values will be used in place of $((b_1 + b_2)/2)$.

Cross sectional area =
Area of top of curb + Area below top of curb + Area of gutter

$$= (6.5 \text{ in.} \times 4 \text{ in.}) + (5.92 \text{ in.} \times 7 \text{ in.}) + (17 \text{ in.} \times 6 \text{ in.})$$
$$= 26 \text{ in.}^2 + 41.44 \text{ in.}^2 + 102 \text{ in.}^2$$
$$= 169.44 \text{ in.}^2$$

Step 2: Convert the area from square inches (in.^{2}) to square feet (ft.^{2}).

Area in square feet = $\quad$ 169.44 in.^{2} ÷ 144 in.^{2}/ft.^{2} (12 in. × 12 in.)
$\qquad\qquad\qquad$ = $\quad$ 1.18 ft.^{2}

Step 3: Calculate the volume of the entire curb.

Volume = $\quad$ Area $\quad\times\quad$ Length
$\qquad$ = $\quad$ 1.18 ft.^{2} × $\quad$ 300 ft.
$\qquad$ = $\quad$ 354 ft.^{3}

Step 4: Convert the volume from cubic feet (ft.^{3}) to cubic yards (yd.^{3}).

Volume (yd.^{3}) $\quad$ = $\quad\quad$ Volume (ft.^{3}) $\quad$ ÷ $\quad$ 27 (ft.^{3})/(yd.^{3})
$\qquad\qquad\qquad$ = $\quad\quad$ 354 ft.^{3} ÷ $\quad$ 27 (ft.^{3})/(yd.^{3})

$$= \qquad \textbf{13.11 yd.}^3$$

In this example, the contractor will probably order 13.5 yd.³ of concrete.

By requesting load tickets from the contractor or the supplier, you can check to make sure the calculated quantities were delivered to your project. Curb and gutter is the most complicated to calculate. All other structures (sidewalks, valley gutters or (swales), dumpster pads, and other pads) are calculated the same as the sidewalk example.

Materials Testing and Acceptance

While verifying quantities of materials used is an essential part of project inspection, management, quality control, and quality assurance, testing the construction materials is also an important part of project management.

Pavement and concrete designs are the first phase of project testing. The information obtained from field testing and laboratory testing will determine how the actual pavement section (thickness of the aggregate base course and asphalt pavement) of a new street or parking lot will be constructed. The project plans list a pavement section with a detail (diagram or drawing) of the pavement section. This diagram shows and lists the designed thickness of the ABC and pavement. In order to develop this section design, a testing laboratory (geotechnical engineering firm) obtains soil samples at the project site during the preliminary site surveying or investigation. Then, in the laboratory, the testing lab will perform tests to determine the resistance of loads (or "R-value") for the soils, Atterberg limits (liquid limit, plastic limit, and plasticity index [PI]), particle analysis (sieve analysis), and Proctors (moisture and density curve). The PI with the sieve analysis and Proctor also indicates how clayey the subgrade material is and what will be the moisture and dry density requirement for compaction purposes. Also, these tests will be used to develop the base preparation for initial scarifying and adding water to obtain required density and moisture content to subgrade (existing) materials.

While the project is being developed, testing will be performed on each phase (subgrade, ABC, and pavement) to determine if each application complies with project specifications. Testing labs are used to monitor compaction by conducting compaction tests using either a nuclear density gauge or sand cone. Either method can be used for both subgrade and ABC testing. However, the sand cone is the most accurate test method, and the nuclear gauge should be calibrated to the sand cone. Hot-mixed asphalt pavement can be tested for compaction and density either by using a nuclear density gauge or by extracting core samples. The control for the density testing of asphalt can be Marshall, Hveem, or Rice test methods (maximum theoretical density test). The most common test control used for paving and patching parking lots and low-density traffic streets is the Marshall test. Like the Proctor, the Marshall test (as well as the Hveem and Rice tests) is used to determine the density of the asphalt after compaction. Core sample testing involves extracting a core from the compacted pavement and performing a specific gravity test to determine the density. The specific gravity test is the primary or absolute density value. Therefore, the nuclear gauge should be calibrated to a core test. Asphalt pavement is difficult to test with a nuclear gauge because it tends to pick up the density in the base under the pavement.

During the paving process, temperatures should be taken to monitor the actual temperature of the asphalt pavement. Temperature of the asphalt pavement is a very important factor to patching and paving, and an infrared thermometer should be used to monitor temperatures in the dump trucks, at the delivery point, and after the pavement has been laid or installed. Most asphalt should be delivered to the project site between 290°F and 320°F. The initial rolling with a steel wheel roller should occur at or about 280°F (or as soon as the roller can get on the hot asphalt without picking it up), which should achieve close to the required density of 95 percent of the Marshall density. However, if the pavement is too hot, the roller will pick up the material on the steel drum, leaving marks in the pavement. Remaining compaction should be done by a second rolling, followed by a finish rolling. Testing labs should be used to develop a roller pattern for all phases of rolling and compaction. This information is used to control the compaction process to ensure that density will be met and overcompaction is avoided. Overcompaction will cause the pavement to pull apart or check, resulting in the degradation of the pavement structure. Extreme care should be taken to prevent over-rolling or overcompaction, which will create future pavement failure or premature deterioration.

Testing is also used as a project management tool for pavement maintenance. When removing and repaving a street or parking lot, the exposed base will need to be tested for density. This will require a Proctor test in order to obtain the values needed to compare the density tests. Should the exposed base have soft (oversaturated) areas, then additional tasks will need to be performed by the contractor such as chemically treating the areas, scarifying and air drying, or excavating the areas and replacing them with new materials imported from off-site sources. Should chemical treatment be required (lime treatment), then pH tests should be performed to monitor the lime content. New designs and Proctors will be needed because the chemical and physical properties of the material have changed. Where remove and replace patching is a part of the scope of work, density testing of the base will be required if the base material is disturbed when the pavement is removed, prior to the installation of the new hot-mixed asphalt. Wherever density tests are performed, Proctors are required. When a new pavement overlay is installed, density test of the patched areas should be required.

Seal coating tests are used to determine the dilution rate of the concentrated material and the amount of solid materials in the seal coat after evaporation of the water from the material. There are several tests performed in the laboratory for designing and quality assurance checks of seal coat. These test results are provided to the project manager by the seal coat manufacturer and are listed on a manufacturer's product and specification sheet. It is important to monitor the amount of water that is added to the manufacturer's concentrate at the job site. Most manufacturers specify 20 to 25 percent dilution by volume to the concentrated material of asphalt emulsions and 45 to 55 percent dilution by volume for coal tars and resurfacer blends (asphalt and coal tar blends). The volume of the storage tank of the material to be used on your pavements should be known at the start of the project. For 100 percent volume of a tank, the tank should hold 75 to 80 percent of concentrate and 25 to 20 percent of water (for coal tars or blends, it should hold 55 to 45 percent of concentrate and 45 to 55 percent water). Sealers can be monitored by viewing the tank before and during the dilution process (a 1,000-gallon tank will take 750 to 800 gallons of concentrate and 250 to 200 gallons of water for asphalt emulsion and 550 to 450 gallons of concentrate and 450 to 550 gallons of water for coal tars or blends). Again, concentrate to dilution is listed by the manufacturer and is shown on the manufacturer's spec sheet.

Upon completion of dilution, the material should be agitated to mix the water and concentrate. Most concentrate is delivered from the manufacturer and contains sand or aggregate required in the finished product. However, if sand is added on-site, then the volumes of sealer and water will have to be adjusted by the percentage of sand added to the material. Adding sand will increase the solid count, which will increase the volume of concentrate. Material application rates will also have to be monitored during the application. This is accomplished by dividing the gallons of diluted sealer by the surface area covered, as explained in chapter 4. Field tests for seal coats to maintain quality control include testing gradation of the fine aggregates, residue by evaporation, dilution rate, unit weight, scrub test, and wet track abrasion test, which are also explained in chapter 4.

All sealers should be prequalified by a series of tests (unit weight, residue by evaporation, scrub test, and wet track abrasion on the concentrate material). The other controlling factor to achieving a good seal coat application is ground temperature. Should the existing pavement be over 140°F during the cooler part of the twenty-four-hour curing period, then the dilution and application rates should be adjusted upward (e.g., 25 percent dilution should be adjusted upward to 30 to 35 percent, and the application rate should be adjusted from 0.16 gal./yd.^{2} upward to 0.18 gal./yd.^{2}). Seal coat application guidelines are shown in chapter 4. The existing pavement surface condition will also affect the seal coat application. Should the existing paved surface be rough, oxidized, and brittle, it will use more sealer material. Rough, rocky surfaces can only have the material application rate adjusted upward. Should the pavement be dry and oxidized, the application rate can be adjusted for absorption; however, the surface should be dampened by a water mist to reduce the full absorption rate. Always require a certificate of compliance (COC) from the contractors, issued by the material manufacturer.

Quality control for type I, II, and III slurry seals includes testing the gradation of the aggregate used and the amount of emulsion, water, and cement (if specified) that is added to the mix prior to blending and applying the slurry seal on the paved surface. The most common slurry seal material used on parking lots and HOA streets is type II, and if specified, it has a polymer additive to produce a more stable material. The blending of materials is controlled on the project site by dials and gauges on the application truck for pounds or gallons of the various materials added. The material suppliers will perform the tests for the gradation of the aggregate and the asphalt tests for the emulsion. The results of these tests from the manufacturers can be obtained from the contractor. Another test for slurry seal is the wet track abrasion test. This test determines the amount of wear, by percentage, or material lost. All of this information is available and supplied to you by the material manufacturer's specification sheets and tests completed by the contractor or the material manufacturers (emulsion and aggregate). Prior to accepting the bids, you can request the specification sheets or test results. Also, the yield (in pounds per square yard on the pavement) should be checked during the application. This should be done by an experienced materials person or a materials testing laboratory. The yield is specified in the scope of work and is a function of how rough the existing pavement surface was prior to installation of the type II slurry seal materials.

Hard rock chip seals are monitored by the yield (the amount of rock applied in pounds per square yard) and the amount of asphalt binder applied (shot rate in gallons per square yard) directly on the existing pavement. The size of aggregate used is determined by a gradation or sieve analysis that is specified in the Scope of Work (e.g., maximum size aggregate of ⅜-inch or ½-inch rock). Also, other tests that should be performed on the aggregate (LA abrasion or hardness of the rock, freeze and thaw of the rock, etc.) and on the asphalt binder (penetration, viscosity, hardening, etc.) are provided by the aggregate producer and refinery (liquid asphalt producer) through the contractor. The shot rate (or amount) of tack oil is a function of chip rock size and the surface texture of the existing pavement. Typically, chip seals call for a shot rate for the tack coat of 0.10 to 0.30 gal/yd.[2]

All testing procedures as well as test results are controlled or governed by the American Society for Testing Materials (ASTM) or American Association of State Highway and Transportation Officials (AASHTO), project-specific specifications and procedures, and local government agencies where the work is being performed. A project manager (owner, consultant, or contractor) should be familiar with these tests and be knowledgeable enough to review the test results in order to determine why tests failed (if any) and if there are any inconsistencies in tests results. For example, some test results do not represent the material being tested (material changed, testing equipment failed, material is difficult to test, and testing equipment operator failure).

Substantial Completion

When all tasks listed in the request for bid (RFB) and the contract have been completed, the contractor has achieved substantial completion. At this time, a certificate of substantial completion should be issued, stating that the contractor has fulfilled the obligation of the contract per the contract scope of work and specifications, and should be paid for substantial completion of the contract. The purpose of a substantial completion certificate is to establish a target to issue an agreed payment to the contractor for a percentage of his/her contracted sum. Most contracts are written where the contractor will receive 85 to 90 percent of the contract amount upon substantial completion, less progress payments already paid. Since the project was inspected and all materials were tested during each construction phase, authorization of payment for substantial completion is warranted. The remainder of the contract (15 to 10 percent) is a retainer to ensure the contractor will complete any punch list items developed from a final walk-through. The following is an example of a letter to the owner or manager stating the contractor has achieved substantial completion and should be paid for the amount agreed.

CERTIFICATE OF SUBSTANTIAL COMPLETION

PROJECT:

DATE OF ISSUANCE:

OWNER:

CONTRACTOR:

This Certificate of Substantial Completion applies to all Work as outlined, the Scope of Work, and all Specifications contained in the Request for Bid, Contract Documents, Scope of Work, Specifications, or to the following specified parts thereof.

The Work to which this Certificate applies has been inspected by an authorized representative of _____

_____,

(Owner/Association)

and that Work is hereby declared to be substantially complete in accordance with the Request for Bid, Contract Documents, Scope of Work, and Specifications on

_____

(Date of Substantial Completion)

A tentative (punch) list of items to be completed or corrected is attached hereto. This list may not be all-inclusive, and the failure to include an item in it does not alter the responsibility of the CONTRACTOR to complete all the Work in accordance with the Request for Bid, Contract Documents, Scope of Work, and Specifications. The items in the tentative (punch) list shall be completed or corrected by the CONTRACTOR within sixty (60) days of the date of issuance of Substantial Completion.

The following documents are attached to and made a part of this Certificate:

This certificate does not constitute an acceptance of Work not in accordance with the Contract Documents, Scope of Work, or Specifications, nor is it a release of CONTRACTOR's obligation to complete the Work in accordance with the Contract Documents, Scope of Work, or Specifications.

The OWNER/ASSOCIATION will release 90 percent of the Contract and change order items in accordance with the Contract for Substantial Completion.

Executed on _____.

(Date of Certificate)

OWNER/ASSOCIATION (or Agent for the Owner/Association):

By: _____. for the Owner/Association

_____

(Authorized Signature)

Completion of this certificate notifies the owner or homeowners association and the contractor that substantial completion has been achieved, and the owner or homeowners association agrees to pay the agreed percentage of the contract amount and change orders. Once the contractor has completed the punch list items and has satisfactorily completed the project, a final certificate of completion is issued, which notifies the owner, HOA, and manager that the project is completed according to the contract. Below is an example of a form used for punch list items generated from a final walk-through and inspection, followed by an example of a certificate of final completion. At this point, the retention funds are released to the contractor in exchange for all proper waivers of liens.

Contractor
Address   Date

Re: Punch list for the pavement maintenance project located at

_____

Dear _____ :

The following list contains items that need attention at the _____ project. A walk-through was conducted and the following items need attention.

1-

2-

3-

Please contact us with your schedule to complete this punch list so we can close out the project and get all funds released for any retention the association may be holding.

Thank you for your attention to this matter.

Sincerely,

_____

CERTIFICATE OF FINAL COMPLETION

PROJECT:

DATE OF ISSUANCE:

OWNER:

CONTRACTOR:

This Certificate of Final Completion applies to all Work as outlined, the Scope of Work, and all Specifications contained in the Contract Documents, Scope of Work, or Specifications, or to the following specified parts thereof.

The final Work and tentative (punch) list items to which this Certificate applies have been inspected by an authorized representative of OWNER/ASSOCIATION, and that Work and tentative (punch) list is hereby declared to be completed in accordance with the Contract Documents, Scope of Work, and Specifications on:

_____

(Date of Final Completion and Acceptance of Project)

A tentative (punch) list of items has been completed or corrected as required and stated in the Contract and Scope of Work.

The following documents are attached to and made a part of this Certificate:

This certificate does not constitute an acceptance of Work not in accordance with the Contract Documents, Scope of Work, or Specifications, nor is it a release of CONTRACTOR's obligation of the warranty of any deficiencies that may occur in the state statute of one (1) year from issue of this certificate for the Work in accordance with the Contract Documents, Scope of Work, or Specifications.

The OWNER/ASSOCIATION will release $ _____ in accordance with the Contract for Final Completion, which was retained until the completion of the tentative (punch) list items.

Executed on _____

(Date of Certificate)

OWNER/ASSOCIATION (or Agent for the Owner/Association):

By: _____ for the Owner/Association

_____

(Authorized Signature)

Typically, the contractor, his/her subcontractors, sub-subcontractors, and material suppliers have issued a preliminary notice of lien, which is commonly referred to as a 20-day notice or 90-day letter. These notices appear like an official lien on the property; however, they are only a notice to the owner, HOA board of directors, or the manager of the right to lien the property for nonpayment as stated in the contract documents. Since a preliminary notice of right to lien has been issued, a release of lien for payment must be acquired from all persons or companies that provided a preliminary notice of lien document. Prior to the owner or HOA releasing any payment, a conditional release (or waiver) of lien must accompany the contractor's invoice. Also, the contractor must submit releases (waivers) of lien from any subcontractors, Sub-subcontractors, and suppliers in order to obtain any payment. Waivers of liens from subcontractors, sub-subcontractors, and material suppliers are substantial proof that the contractor has paid the subcontractors, sub-subcontractors, and suppliers. If the contract allows for scheduled or progress payments, waivers must be submitted for the amount of the progress payment and any materials used during that contract period. Remember, just because the general contractor has provided a waiver of lien, this does not mean the subcontractors, sub-subcontractors, or suppliers have been paid. Should the general contractor have not paid the subcontractors, sub-subcontractors, or suppliers, they still have the right to lien your property for nonpayment. It is wise to have the contractor list all subcontractors, sub-subcontractors, and material suppliers that will be used on the project prior to signing the main contract. It is also recommended that a general contractor (or contractor of record on the project) receive a list of any sub-subcontractors that the subcontractor intends to use. With this information, all proper paperwork and releases can be obtained and the project kept in clean order.

There are two types of waivers of lien: unconditional and conditional. An unconditional waiver of lien is a document that releases any right that the contractor has to lien the property regardless of payment. A conditional waiver is a document that releases lien rights upon transfer of funds from the payer to the payee. At that time, the conditional waiver becomes an unconditional waiver. In many cases, a contractor will supply an unconditional waiver of lien upon receipt of payment.

Sample Waiver of Lien Form

CONDITIONAL WAIVER AND RELEASE

Upon verifiable exchange of funds for payment, this Waiver and Release becomes Unconditional. This is the most commonly used form for progress and final payments.

Project _____Job Number_____

On receipt by the undersigned of a check from _____
(Maker of Check)

in the sum of $_____ payable to _____
(Amount of check)   (Payee or Payee of Check)

and when the check has been properly endorsed and has been paid by the bank on which it is drawn, this document becomes effective to release any mechanic's lien, any state or federal statutory bond right, any private bond right, any bonded stop notice, any claim for payment, and any rights under any similar ordinance, rule, or statute related to claim or payment rights for persons in the undersigned's position, that the undersigned has on the job of _____
(Owner)

located at_____
(Job Description and/or Location)

to the following extent. This release covers the final payment to the undersigned for all labor, services, equipment, or materials furnished to the job site or to

_____
(Person to whom undersigned contracted)
except for disputed claims in the amount of

_____
(Written Amount)

$_____. Before any recipient of this document relies on it, (Dollar Amount) the person should verify evidence of payment to the undersigned.

The undersigned warrants that he/she either has already paid or will use the monies received from this final payment to promptly pay in full all of his/her laborers, subcontractors, materialmen, and suppliers for all work, material, equipment, or services provided for or to the above referenced project up to the date of this waiver.

Upon receipt of payment, deposit to Contractor's bank account and release, the amount remaining due from the Owner to the Contractor is $ _____
(Balance due to Contractor)

Date _____

_____
(Company/Contractor Name)

By_____
(Signature)

_____
(Printed Name and Title)

Sample Unconditional Waiver of Lien

UNCONDITIONAL WAIVER AND RELEASE UPON PROGRESS PAYMENT

The undersigned has been paid and has received a progress payment in the sum of $_____ for labor,
services, equipment, or material furnished to _____   (Amount of Check)
_____ on the project of _____
      (Your Customer)   (Owner)
located at _____ and does hereby release
                    (Project Description)
any mechanic's lien, stop notice, or bond right that the undersigned has on the above referenced project to
the following extent.

This release covers a progress payment for labor, services, equipment, or material furnished to
_____ through _____
                    (Your Customer)   (Date)
only and does not cover any retentions retained before or after the release date; extras furnished before the
release date for which payment has not been received; extras or items furnished after the release date. Rights
based upon work performed or items furnished under a written change order which has been fully executed
by the parties prior to the release date are covered by this release unless specifically reserved by the claimant
in this release. This release of any mechanic's lien, stop notice, or bond right shall not otherwise affect the
contract rights, including rights between parties to the Contract based upon a rescission, abandonment, or
breach of the Contract, or the right of the undersigned to recover compensation for furnished labor, services,
equipment, or material covered by this release if that furnished labor, services, equipment, or material was not
compensated by the progress payment.

_____
      (Dated)   (Company/Contractor Name)

By: _____
      (Title)

NOTICE: THIS DOCUMENT WAIVES RIGHTS UNCONDITIONALLY AND STATES THAT
THE CONTRACTOR HAS BEEN PAID FOR GIVING UP THOSE RIGHTS. THIS DOCUMENT
IS ENFORCEABLE AGAINST THE CONTRACTOR IF THE CONTRACTOR SIGNED IT, EVEN
IF THE CONTRACTOR HAS NOT BEEN PAID. IF THE CONTRACTOR HAS NOT BEEN PAID,
USE A CONDITIONAL RELEASE FORM.

   NOTE: This form is to be used to release claims to the extent that a progress payment has actually been
received by the releasing party.

Sample of Conditional Waiver of Lien Upon Payment

CONDITIONAL WAIVER AND RELEASE UPON FINAL PAYMENT

Upon receipt by the undersigned of a check from _____
(Maker of check)

In the sum of $_____payable to _____
(Amount of Check)   (Payee or Payees of Check)

and when the check has been properly endorsed and has been paid by the bank upon which it is drawn, this document shall become effective to release any mechanic's lien, stop notice, or bond right the undersigned has on the job of _____
(Owner)

located at _____
(Job Description)

This release covers the final payment to the undersigned for all labor, services, equipment, or material furnished on the job, except for disputed claims for additional work in the amount of $_____. Before any recipient of this document relies on it, the party should verify evidence of payment to the undersigned.

_____ _____
(Dated)   (Company/Contractor Name)

By: _____
(Title)

NOTE: This release is not effective until the check that constitutes final payment has been properly endorsed and has cleared the bank.

Sample of Unconditional Waiver of Lien Upon Payment

UNCONDITIONAL WAIVER AND RELEASE UPON FINAL PAYMENT

The undersigned has been paid in full for all labor, services, equipment or material furnished to _____ on the project of _____

       (Your Customer)   (Owner)

located at _____

                         (Project Description)

and does hereby waive and release any right to a mechanic's lien, stop notice, or any right against a labor and material bond on the project, except for disputed claims for extra work in the amount of $_____.

_____   _____

      (Dated)   (Company Name)

By: _____

                      (Title)

NOTICE: THIS DOCUMENT WAIVES RIGHTS UNCONDI-TIONALLY AND STATES THAT THE CONTRACTOR HAS BEEN PAID FOR GIVING UP THOSE RIGHTS. THIS DOCUMENT IS ENFORCEABLE AGAINST THE CONTRACTOR IF THE CONTRACTOR SIGNED IT, EVEN IF THE CONTRACTOR HAS NOT BEEN PAID. IF THE CONTRACTOR HAS NOT BEEN PAID, USE A CONDITIONAL RELEASE FORM.

NOTE: This form is to be used to release claims to the extent that a progress payment has actually been received by the releasing party.

These waivers and release forms are only example forms. For the exact form, you must contact your local licensing agency or bonding agency. Following is a brief description of the use of each form.

Conditional Waiver and Release Upon Progress Payment

Use this form when the claimant is required to execute a waiver and release in exchange for, or in order to induce the payment of, a progress payment and the claimant has not been paid. This form is useful when the claimant has not been paid yet, but will be paid out of a progress payment that is not the final payment. This conditional waiver and release is only effective if the claimant is *actually paid*. This release does not cover all items.

Unconditional Waiver and Release Upon Progress Payment

Use this form when the claimant is required to execute a waiver and release in exchange for, or in order to induce payment of, a progress payment and the claimant asserts in the waiver that he/she has in fact been paid the progress payment. This release does not cover all items.

Conditional Waiver and Release Upon Final Payment

Use this form when the claimant is required to execute a waiver and release in exchange for, or in order to induce the payment of, a final payment and the claimant has not been paid. This release is only binding if there is evidence of payment to the claimant. Evidence of payment may be demonstrated by the following:

- The claimant's endorsement on a single check or a joint payee
- Check that has been paid by the bank upon which it was drawn
- Written acknowledgment of payment given by the claimant

Unconditional Waiver and Release Upon Final Payment

Use this form when the claimant is required to execute a waiver and release in exchange for, or in order to induce payment of, a final payment *and* the claimant asserts in the waiver he/she has in fact been paid the final payment.

Caution: in the case of a conditional release, the release is only binding if there is evidence of payment to the claimant. Evidence of payment may be demonstrated by the following:

- The claimant's endorsement on a single check or a joint payee check that has been paid by the bank upon which it was drawn
- Written acknowledgment of payment given by the claimant

Upon receipt of all final waivers of liens for final payment and the issuance of final certificates of completion, the project is completed and can be closed. The project now falls under any stated warranty period. The warranty period is usually governed by the state where the work is performed by statute. The warranty period can also be stated in the contract documents. The normal warranty period is one (1) year, or it can be extended to two (2) years depending on how the conditions of the warranty period are stated.

About the Authors

Thomas and Patrick McDonald are owners and principals of PMIS, Inc. (dba Pavement Maintenance Information Source). PMIS, Inc. is a pavement maintenance consulting firm specializing in pavement distress surveys and development of maintenance plans and budgets. The company also provides project management; they develop project-specific scopes of work, specifications, bid documents, and contract documents. PMIS also manages project monitoring materials and material application rates, maintains resident and tenant communications, develops final needed punch lists, and collects all closing documents needed from the contractors. PMIS is constantly researching pavement maintenance applications for new technology, better use of materials, new agency regulations, and new application equipment advances in order to provide up-to-date pavement maintenance applications for clients. For more information about PMIS, please visit our Web site: www.pavementmaint.com.

Thomas McDonald

Thomas McDonald graduated from the University of Colorado with a BS degree in business with an emphasis in finance and economics; he did postgraduate work at the University of Denver. He is a partner and owner of PMIS, Inc. (dba Pavement Maintenance Information Source). Thomas founded PMIS in 1992 to provide pavement maintenance consulting services to property owners and managers and homeowners associations. He has been a member and director of the Building Owners and Managers Association (BOMA), International Facilities Managers Association (IFMA), National Apartment Association (NAA), Community Association Institute (CAI), American Society of Civil Engineers (ASCE), and Association of Asphalt Paving Technologist (AAPT), among others. He has been an instructor in pavement maintenance to property owners and managers and has developed training guides and classes for general contractors. Thomas has been involved in pavement construction and maintenance since 1964. He was lead project materials technician for the Colorado Department of Transportation (DOT) on several highway projects and a member of the materials testing team at the Eisenhower Tunnel. He was a construction and street maintenance inspector for the City of Lakewood, Colorado. He worked for geotechnical and construction materials consulting engineering firms as a field inspector and pavement maintenance specialist. He was the quality control supervisor for materials manufacturing, testing, and production for a major heavy highway contractor, as well as an estimator for a national pavement maintenance contractor. Thomas was an instructor for the annual Federal Highway Administration's materials school in Denver and is trained in the Corps of Engineers' PAVER® pavement maintenance program. He developed a computerized version of the PAVER program (ECOPAVE™) for a major geotechnical engineering firm.

Patrick McDonald

Patrick is a graduate of the Del Webb School of Construction, College of Engineering at Arizona State University. He has been a pavement maintenance consultant and principal with PMIS, Inc. since 1992. He has completed pavement surveys and distress analyses for several commercial, industrial, and multihousing properties and homeowners associations. Patrick has been project manager on large pavement construction and maintenance projects in Phoenix, San Diego, Los Angeles, Denver, and Las Vegas. Patrick has authored a manual on developing proper request for bids (RFB) and contract documents. Patrick is an associate member of the American Society of Civil Engineers. He has developed training materials about pavement maintenance for property owners and managers and training classes and guides for general contractors.

Appendix A. Troubleshooting Guide

Each asphalt application is different and as result, there are different problems that can arise depending on the application used. Here is an easy to use guide that breaks down some problems commonly seen with various applications along with the cause and a solution.

CRACK SEALING

Problem	Cause	Solution
CRACK SEAL MATERIAL OPENS (RUPTURES)	This problem occurs when not enough material was installed in the crack.	Reseal the crack/s with the SAME crack seal material if hot pour material was used to initially seal the cracks. If cold pour material was used, hot or cold pour material should be used to reseal the cracks.
CRACK SEAL MATERIAL EASILY PULLS OUT OF THE CRACK	This problem is caused from inadequate cleaning of the crack prior to installing the crack seal material or the crack had moisture in it prior to installing the crack seal material. In either case, the material cannot adhere to the edges of the crack to create a bond.	The material will have to be removed and the crack resealed properly according to local specifications and manufactures recommended application.

CRACK SEAL (CONT.)

Problem	Cause	Solution
MATERIAL STICKS TO VEHICLE TIRES AND PULLS OUT OF CRACK	This problem exists when the crack seal material cannot cool down, solidify and set up before exposure to traffic. Crack seal material should be allowed to cool and set up prior to letting traffic drive over it. Vehicles pull material out of crack with tires because it has not set up. This will occur when the temperature of the material stays high due to high ground or ambient temperatures or when vehicles enter the work area before the material has had a chance to cool.	Apply a light coat of sand, silica sand, or cement over the surface to prevent the tacky pickup of the material by vehicle tires. Also, a water spray can be used to cool down the material rapidly. If the problem is caused from high pavement temperatures, a light coating of sand, silica sand, or cement is required.
CRACK SEAL MATERIAL WILL NOT SET UP OR STAYS SOFT AND PLIABLE OVER TIME	This will occur when the wrong material (too low of a softening point for the climate conditions) is used for the climate where it is installed or a cut back (gasoline, diesel fuel, paint thinner, etc) is used in the material as an extender.	Keep the material coated with sand, silica sand, or cement dust until the material will sets up and does not adhere to vehicle tires or foot traffic. The material will have to be examined and tested several times to check on tacky condition. DO NOT ALLOW SEALING to occur until the material has set and if a cut back was used in the crack seal material or suspected, until the cutback material has thoroughly evaporated out of the material.

SURFACE (SKIN) PATCHING

Problem	Cause	Solution
RAVELING	Raveling is a result of applying the patch material after it has cooled (is cold or under 200° F) and can not be compacted, or has cooled below the 200° F temperature due to cold ground, ambient temperatures or both. Ground and ambient temperatures should be 55° F and rising before installing a thin surface patch consisting of hot-mixed asphalt pavement. A skin or surface patch will cool very rapidly and should be compacted with a vibratory plate compactor, or a 3 to 5 ton static roller as soon as it is applied.	Re-apply a new patch over the original patch. The problem with this method is that it will make the patch area thick and create a bump in the surface that will be very noticeable to drivers and can become a trip hazard. Heat the patch area with exposed flame or infrared heater and reheat the patch area. Compact with vibratory plate or 3 to 5 ton roller as soon as the heat exposure is removed. The problem with this method is overheating which can cause the asphalt binder to "coke" and become brittle. This will reduce the life of the patch. An experienced applicator is a must when using this method. Apply a thin amount of asphalt emulsion (not coal tar) to the patch area to bind the surface together. When the warmer or hotter temperatures expose the patch, it will continue to bond itself. This method is recommended as the skin or surface patch is being initially installed where the ground and ambient temperatures are below 70° F for a high daily temperature. Heat the patch area, completely remove the patch area and install a new patch in accordance to the skin or surface patch specifications.

SURFACE (SKIN) PATCHING (CONT.)

Problem	Cause	Solution
BREAKS OFF OR SEPERATES FROM THE SURFACE	When a skin or surface patch is installed or applied without enough tack material on the older surface, the patch will not adhere to the older pavement and will separate or lift off. Should the patch area not be properly tack coated (e.g., tack coating the edges of the patch area) the patch will not properly adhere to the older pavement and separate or lift off.	The remaining patch will need to be removed and re-applied.
EDGES RAVEL AND SEPARATE OFF THE PATCH	When a skin or surface patch is applied with our enough tack coat on the edges of the patch area, or the hot-mixed asphalt has cooled below 200°F the edges will not adhere to the older pavement. When this happens, the edges will separate from the older pavement and the new skin or surface patch and lift off.	Heat the edges of the original skin or surface patch area apply new tack coat on the edges of the existing patch and approximately 6 inches of the older pavement and apply new hot-mixed asphalt pavement. This strip of pavement will need to be compacted as soon as it is applied for adherence purposes.

SURFACE (SKIN) PATCHING (CONT.)

Problem	Cause	Solution
REFLECTIVE CRACKING	When a skin or surface patch is applied to an area where cracking has developed in the older pavement, it will reflect through the new patch in a short period of time.	There is no remedy to cure reflective cracking in a skin or surface patch, but initial preparation or the older surface can be accomplished to reduce the effects of reflective cracking. Prior to installing the skin or surface patch, apply hot pour crack seal into the cracks, then apply the hot-mixed asphalt over the original pavement and crack seal. The cracks will reflect through the patch, however, the crack has been sealed which will reduce water penetration into the base material. Another method is to apply a fabric inner layer on the older pavement over the entire patch area. Apply the patch material to the area and compact using a vibratory plate or 3 to 5 ton roller. Should the original area be too cracked up or the cracks are structural (alligator type) then the area will have to be removed and patched in lieu of skin or surface patching.

REMOVE AND REPLACE (R&R) PATCHING

Problem	Cause	Solution
RAVELING	Raveling is a result of applying the patch material after it has cooled (is cold or under 200° F) and can not be compacted, or has cooled below the 200° F temperature due to cold ground, ambient temperatures or both. Ground and ambient temperatures should be 50° F and rising before installing a remove and replace (referred to as an R&R) patch consisting of hot-mixed asphalt pavement. The patch material should be monitored for temperature and compacted with static roller before the temperature reaches 260° F for best results.	Heat the patch surface area with exposed flame burner or infrared heater and compact with a vibratory plate or 3 to 5 ton roller. The problem that exists with this method is overheating which can cause the asphalt binder to "coke" and become brittle. This will reduce the life of the patch. An experienced applicator is a must when using this method Install a skin or surface patch over the R&R patch area. The problem with this method is that it will cause a bump or raised area that will be noticeable to drivers and can be a trip hazard. Apply a thin amount of asphalt emulsion (not coal tar) to the patch area to bind the surface together. When the warmer or hotter temperatures expose the patch surface, it will continue to bond itself. This method is recommended as the skin or surface patch is being initially installed where the ground and ambient temperatures are below 60° F for a high daily temperature. Remove the patch area and replace

REMOVE AND REPLACE (R&R) PATCHING (CONT.)

Problem	Cause	Solution
HOLDS WATER (PUDDLES)	This problem occurs when the base below the patch is not compacted prior to installing the patch material, the patch material is not compacted properly, or the patch material was over compacted. The material is compacted with a vibratory plate compactor or minimum 3 to 5 ton roller, usually with vibrating capabilities. Should the roller stop with the vibrator on, a depression forms.	Using string lines in lateral and vertical directions, install a skin or surface patch to reduce or possibly eliminate the puddle area. In most cases puddles will only be reduced to an acceptable depth (1/4 inch). It is very difficult if not impossible to eliminate a puddle totally.
PATCH SETTLES	This happens when the base is not compacted properly prior to installing the patch material. All patches will settle after installation normally. Most specifications will require the patch material be installed to yield a final compacted height ¼ inch above the patch areas. This will allow for settling while still remaining above the patch area. Should the base under the patch material not be compacted properly, the patch will settle below the original surface.	Apply a skin or surface patch to level the patch with the original surface.

REMOVE AND REPLACE (R&R) PATCHING (CONT.)

Problem	Cause	Solution
SAW JOINTS HAVE OPENED WIDE	This problem is a result of improper or poor compaction of the base prior to installing the hot-mixed asphalt pavement patch.	Remove patch material, compact the base material and repatch
PATCH FAILS AND CRACKS UP and/or DEPRESSES	This problem is caused from improper base compaction. The base was not compacted or compacted to acceptable standards prior to installing the patch material.	Remove the patch, compact the base to required standards and repatch.

SEAL (SLURRY) COAT

Problem	Cause	Solution
TRAFFIC IS PICKING UP MATERIAL AFTER THE CURING PERIOD	This happens when the material is not blended properly or a soft asphalt material is used. Also, should the ground temperatures and ambient temperatures be excessive, the asphalt sealer will take longer to set up and cure. Should the surface be too wet at the time of application will also cause the material to not adhere properly to the existing pavement and will also retard curing. Coal tar and Coal tar blends are harder materials and do not pick up easily after curing.	Allow the material to cure out usually 2 to 3 months until it has had the opportunity to cure and some air dust has settled on the surface. Apply sealer to the areas where the new sealer has picked up after the extended curing period. These seal coat patches will blend together with the sealer on the surface in a short period of time. Where the situation exists that a lengthy curing period is not acceptable, a lime water solution can be used to accelerate the curing. However, the surface will develop a white appearance, which will go away over a short period of time. ALWAYS INVOLVE THE SEAL COAT MANUFACTURER when resolving this problem as he/she can recommend the proper solution for your particular situation.

SEAL (SLURRY) COAT (CONT.)

MATERIAL IS PEELING OFF OLDER OR EXISTING PAVEMENT	This occurs when the original pavement was not cleaned properly and there is still some dirt, dust, or other fine debris residue left on the pavement. When the existing pavement is not cleaned properly or thoroughly, the seal coat material can not adhere or bind to the original surface and will peel off. Another cause for peeling is freezing. Seal coat material is 40% or more water and will freeze when exposed to freezing temperatures. This will cause the material to peel off.	The seal coat material will have to be removed by aggressive cleaning and sweeping. A new application can be applied once the original seal coat has been removed and the surface thoroughly cleaned.
SEAL COAT MATERIAL TRACKED OR SPRAYED ON CONCRETE SIDEWALKS, DRIVEWAYS, CURBS, ETC.	This problem occurs when the contractor over sprays or trims sealer material on adjacent concrete while trimming or applying the material. pedestrian and vehicle traffic will breach the barriers and track the material on adjacent concrete.	Power wash with high-pressure water. The sooner the material is power washed off the concrete, the easier it is to remove and the less power wash marks will be left.

SEAL (SLURRY) COAT (CONT.)

Problem	Cause	Solution
SEALER IS WEARING OFF THE ORIGINAL OR EXISTING PAVEMENT MORE RAPIDLY THAN USUAL	This occurs when the sealer is over diluted or applied thinly (thin application rate). When sealer is diluted from a concentrate form to the recommended mixture it should wear 3 to 4 years for a single coating and 4 to 5 years for a double application under normal traffic and weather conditions. The manufacturer has a specified application rate (usually 0.14 to 0.20 gallons per square yard of surface) as well as a specified dilution rate (usually 25% of concentrate) to attain the desired life cycle of the seal coat application. When the material is over diluted or applied too thin, it will wear off the original pavement more rapidly than designed.	A new seal coat application will need to be applied. Based on the amount of wear, the original contracted amount will need to be reapplied.

SEAL (SLURRY) COAT (CONT.)

Problem	Cause	Solution
SEALER HAS LIFTED OFF PETROLEUM SPOT AREAS WHERE VEHICLES PARK	Petroleum products (oil, anti-freeze, brake fluid, transmission fluid, gasoline, etc.) will cause pavement to soften and deteriorate. The petroleum product will penetrate pavement and when sealer is used, it will not adhere properly or at all. This will cause the sealer to lift off the pavement and break away.	Petroleum spots should be heated and scraped until the spot is removed. A bonding agent should be used to cover the area and sealer reapplied. In many cases, the spot is too saturated with petroleum products and the area will need to removed and patched.
THE SEAL COAT MATERIAL HAS THE ODOR OF ROTTEN EGGS	Asphalt material will grow bacteria and cause the material to increase (swell) in volume. This is caused by gas created by the bacteria.	An additive must be added to eliminate the bacteria and neutralize the effects of the bacteria on the material or a new material should be brought in.
SEAL COAT MATERIAL ON VEHICLES, CARPETS, ETC.	This will occur when pedestrian or vehicle traffic goes through the material before it has a chance to set up or cure.	Use a tar or mechanics grease remover on a cloth towel and dab lightly. Do not rub, just dab until the material is removed. A good product to use is cream "GO-JO™ " or similar without pumice additive. Most seal coat manufactures also have a product available. When using a tar or grease remover on vehicles or waxed surfaces, it will remove the wax on the area treated. A new wax coating will be required on the area treated.

SEAL (SLURRY) COAT (CONT.)

Problem	Cause	Solution
SEAL COAT MATERIAL DEVELOPS BLACK POWDER	This is referred to as "roll out" and is a result of fillers in the seal coat material separating from the sealer blend. This is generally a normal characteristic from oxidation, traffic, water exposure, etc. Should the black powder be excessive or occurs within a short period of time after application, it is usually caused from over dilution or too much filler material added.	It is best to allow the sealer to wear away and over time the problem will slow down. The pavement will have to be swept more frequently to reduce the build up of the powder in curbs and on the surface. The frequency of sweeping (with a broom street sweeper) will depend on the severity of the roll out or powdering. Adding more sealer to the surface will not solve the problem. It may reduce the problem for a short period. However, the black powder will begin to reappear once the top coating wears away exposing the problem material. There are several different brands of seal coat in the industry, and a top quality sealer with other non-mineral fillers will reduce (in some cases eliminate) medium and heavy roll out or powdering. Always request a manufacturers material specification sheet and check out other projects where the material has been used.

SEAL (SLURRY) COAT (CONT.)

Problem	Cause	Solution
WHITE STRIPES BECOME GRAY OR BOTH WHITE AND YELLOW STRIPES TURN DULL SHORTLY AFTER BEING APPLIED OR PAINTED	Graying of stripes will occur due to the amount of fresh asphalt in the slurry seal material. Dulling will occur when fresh asphalt residue and dirt are tracked onto the stripes. This is a normal occurrence and is not the use of poor quality materials.	The best course of action is to let the stripes go until the surface is seal coated, then repaint or restripe. Another course of action is power washing the stripes to remove the dull appearance. The gray appearance is permanent until the stripes are repainted.
NEW STRIPES TURNED YELLOW OR BROWN	When sealer has not had the proper time to cure and is still wet or damp, the stripes will turn color. In areas where the dew point (humidity) is high at the time of application, the sealer will take longer to cure.	Remedy – Let the slurry seal cure for a longer period (@ 1 week or until the dew point drops) and restripe.

TYPE I, II and III SLURRY SEAL

Slurry seal is a wet application of aggregate, asphalt emulsion, an emulsifying agent, Portland type II cement, and water blended or mixed to a design while being applied. The color of the material is brown at the time of application and will turn black as it sets up (referred to as breaking). Slurry seal is a new wearing surface and will remain tender and subject to power steering sheer until it completely cures in about 6 to 8 months. Type II slurry seal is most commonly used on parking lots and streets.

Problem	Cause	Solution
DEVELOPS SHEAR OR TWIST MARKS	This occurs when vehicle traffic turns wheels while in a static mode (standing still). This action causes the tires to twist the material off the older pavement.	Depending on the severity of the twist mark determines the action required. A small or slight twist mark can be left to cure out and settle out. A large or heavy mark will have to be patched at a later date with type II slurry seal or a hot sand seal patch material.
SAND AND SMALLER ROCKS START SEPERATING FROM THE MATERIAL AND SET ON THE SURFACE.	This is a common curing characteristic of slurry seals. The material will only hold the sand and small rock that is required to fill the voids in the original paved surface. The remainder to the sand and rock will be released and migrate to the surface.	Sweeping the pavement with a street sweeper at 3 and 9 months will remove the majority of the sand and rocks. Once the slurry seal material has cured properly (usually 12 to 18 months) a seal coat application will stop this process.

TYPE I, II and III SLURRY SEAL (CONT.)

Problem	Cause	Solution
MATERIAL DOES NOT CHANGE COLOR TO BLACK OVER THE ENTIRE SURFACE OR IN ISOLATED LOCATIONS.	As slurry seal material cures (breaks as common terminology), it will change color from brown to black. If the entire area sealed is brown, it is still wet and should not be open to any type of traffic. Should there be isolated spots after the surface has cured or turned black, the spots are still wet or have a less amount of one of the mixing materials. Also, should the material be applied in a constant shaded area, it will take longer to cure than the areas exposed to direct sun light. Factors affecting the curing of slurry seal material is ambient and ground temperatures. Ground temperature is the most important factor in curing time of slurry seal material.	Keep the area closed to traffic and allow more time for curing. If there are isolated spots, allow additional time for curing by barricading the spots or shaded area. Should the material be solid and not wet, the brown spots are caused from slow curing in area where ground temperatures were low and the material did not initial break properly. There is nothing wrong with the slurry seal materials, and is acceptable. Once the material oxidizes, the lighter spots will be less noticeable and after an initial seal coat is applied the spots will not be noticeable.

TYPE I, II and III SLURRY SEAL (CONT.)

Problem	Cause	Solution
SAND AND SMALLER ROCKS START SEPERATING FROM THE MATERIAL AND SET ON THE SURFACE.	This is a common curing characteristic of slurry seals. The material will only hold the sand and small rock that is required to fill the voids in the original paved surface. The remainder to the sand and rock will be released and migrate to the surface.	Sweeping the pavement with a street sweeper at 3 and 9 months will remove the majority of the sand and rocks. Once the slurry seal material has cured properly (usually 12 to 18 months) a seal coat application will stop this process.
CRACKING ON THE SURFACE HAS DEVELOPED IN A SHORT PERIOD (@ 2 TO 6 WEEKS)	This is referred to a reflective cracking. The cracks in the underlying surface will reflect through the new slurry seal surface is a short period of time due to thermal reactions (expansion and contraction) in the original paved surface. This is one of the reasons the cracks should be sealed prior to applying a slurry seal.	There is no remedy for this process. Should the cracking widen to a width to allow for water percolation, the surface will have to be crack sealed at the time of seal coating.

TYPE I, II and III SLURRY SEAL (CONT.)

Problem	Cause	Solution
WHITE AND YELLOW STAINS OR WHITE BUBBLES APPEAR AFTER THE SLURRY SEAL HAS CURED.	This usually occurs after the slurry seal material has cured and the surface is exposed to water (sprinklers or rain). A part of the slurry seal process is the use of an emulsifying agent. It is used in the initial blending of the material and after it has cured, the emulsifying agent is no longer needed. It stays suspended until it is exposed to water. That's when the stains and bubbles appear.	There is no remedy for this process. The stains will fade and the bubbles will disappear and lessen over time. This is a part of the curing process of slurry seals.
MATERIAL PEELS (FLAKES) OFF THE UNDERLYING (ORIGINAL) PAVEMENT	This happens when the underlying (original) pavement was not cleaned thoroughly or there is oil, gasoline, or other petroleum-based material on the original surface.	Remove the slurry seal material and clean the area of all dirt and debris. Where there is a petroleum spill, this area will have to be treated with open flame or a bonding agent and patched with the same slurry seal material. In some extreme cases, the area will need to be saw cut, removed and patched, then patched with the same slurry seal material.

TYPE I, II and III SLURRY SEAL (CONT.)

Problem	Cause	Solution
MATERIAL TRACKED OR PLACED ON CONCRETE SIDEWALKS, DRIVEWAYS, CURBS, ETC.	This problem occurs when the contractor places material on adjacent concrete while trimming or applying the slurry material. Also, when pedestrian and vehicle traffic breach the barriers and track the uncured material on adjacent concrete.	Remove all excess material and/or power wash with high pressure water. The sooner the material is power washed off the concrete, the easier it is to remove and the less power wash marks will left.
TIRE TRACKS IN THE SLURRY SEAL MATERIAL	This is caused when someone breached the barriers with their vehicle and drove through the uncured (wet) material.	This is a very difficult problem to correct. The tracking damage is a deep tire path and can not be completely covered or patched. The only corrective action is to patch the damaged area with the same slurry seal material as soon as possible.
SLURRY SEAL MATERIAL IS ON VEHICLES, CARPETS, ETC.	This will occur when pedestrian or vehicle traffic goes through the material before it has a chance to set up or cure.	Use a tar or mechanics grease remover on a cloth towel and dab lightly. Do not rub, just dab until the material is removed. A good product to use is cream "GO-JO™" or equivalent mechanics degreaser hand cleaner or similar without pumice additive. Most seal coat manufactures also have a product available. When using a tar or grease remover on vehicles or waxed surfaces, it will remove the wax on the area treated. A new wax coating will be required on the area treated.

TYPE I, II and III SLURRY SEAL (CONT.)

Problem	Cause	Solution
WHITE STRIPES BECOME GRAY OR BOTH WHITE AND YELLOW STRIPES TURN DULL SHORTLY AFTER BEING APPLIED OR PAINTED	Graying of stripes will occur due to the amount of fresh asphalt in the slurry seal material. Dulling will occur when fresh asphalt residue and dirt are tracked onto the stripes. This is a normal occurrence and is not the use of poor quality materials.	The best course of action is to let the stripes go until the surface is seal coated, then repaint or restripe. Another course of action is power washing the stripes to remove the dull appearance. The gray appearance is permanent until the stripes are repainted.
NEW STRIPES TURNED YELLOW OR BROWN	When sealer has not had the proper time to cure and is still wet or damp, the stripes will turn color. In areas where the dew point (humidity) is high at the time of application, the sealer will take longer to cure.	Remedy – Let the slurry seal cure for a longer period (@ 1 week or until the dew point drops) and restripe.

PAVING (OVERLAYS AND/OR NEW PAVING)

Overlays are where a new paved surface is placed on top of or over an existing pavement. There are 3 types of overlays which are conventional, conventional with a fabric innerlayer, and milling off a portion of the existing pavement and installing a conventional or conventional with fabric inner layer overlay. New paving is where the existing paved area is removed by excavating, milling, or pulverizing; and a new paved surface is installed. Also new paving is where the area to be paved is constructed from the subgrade (natural ground) with the installation of aggregate base course (ABC or gravel), and asphaltic concrete pavement surface installed.

Problem	Cause	Solution
NEW PAVED SURFACE HAS TWIST OR SHEAR MARKS	This occurs when vehicle traffic turns wheels while in a static mode (standing still). This action causes the tires to twist the material. In some very severe cases, the pavement will be twisted or buckled exposing the overlaid surface or the aggregate base course.	Depending on the severity of the twist mark determines the action required. A small or slight twist mark can be left to cure out and settle out. A large or heavy mark will have to be patched at a later date with a hot sand seal or fine hot patch material. The area can be heated with infrared heaters or an open flame (propane weed burner). Then, using a lute (asphalt rake), the area can be repaired and compacted.

PAVING (OVERLAYS AND/OR NEW PAVING) (CONT.)

Problem	Cause	Solution
NEW PAVEMENT HAS SLIPPED EXPOSING THE ORIGINAL PAVED SURFACE OF AGGREGATE BASE COURSE	When this occurs exposing the aggregate base course (gravel base), the base has shifted from moisture in the base or lack of adhesion of the pavement to the base. When this occurs on overlays (over existing pavement), either the older pavement was not cleaned properly prior to installation or there was not enough tack coat (or an insufficient amount of tack coat) applied to adhere the overlay properly to the original paved surface.	The pavement should be cored to determine if a lack of tack coat exists. Should there be a lack of tack coat, require an extended warranty and deduct the cost of the tack coat from the contract amount. The areas where the pavement slipped or shifted should be removed and a new patch installed.
ROCKS ARE SEPERATING FROM THE EDGE ALONG CURBS AND VALLEY GUTTERS (CONCRETE SWAIL)	This will occur when the edge of the pavement is higher in elevation than the edge of the curb or valley gutter. Most specifications require the pavement to be at least ¼ inch higher for settlement along the curb in later years. Should the pavement not settle, the edge will break off from traffic exposure with rocks and sand appearing in the curbs and/or the valley gutters.	There is no remedy for this problem. It is designed to occur. The sand and small rocks will have to be swept up. The shedding of rocks and sand will subside as the pavement cures and oxidizes.

PAVING (OVERLAYS AND/OR NEW PAVING) (CONT.)

Problem	Cause	Solution
STRETCH MARKS APPEAR IN THE PAVEMENT AFTER IT HAS BEEN COMPACTED	This occurs when the pavement is over rolled or there was moisture in the mix when it was installed creating a tender asphalt mix. This is referred to as "checking."	Should over rolling or over compacting be the problem, the cracking will have to be monitored for future widening. Should the cracks continue to widen, the pavement will have to removed and repaved under warranty. Should the cracking stay the same, seal coating will protect the areas affected. . Should the cracking be a result of moisture or a tender mix and the cracks, seal coating will protect the areas affected. Cracking from moisture or tender mixes usually does not increase in width and will remain stable. However, should the cracking increase in width in the first 12 months, it should be removed and repaved under warranty, otherwise seal coating will protect the areas affected.

PAVING (OVERLAYS AND/OR NEW PAVING) (CONT.)

Problem	Cause	Solution
CRACKING ON THE SURFACE HAS DEVELOPED IN A SHORT PERIOD (@ 2 TO 6 WEEKS)	This is referred to a reflective cracking. The cracks in the underlying surface will reflect through the new overlaid surface is a short period of time due to thermal reactions (expansion and contraction) in the original paved surface. This is one of the reasons the cracks should be sealed prior to applying an overlay. Should a fabric inner layer be installed, the reflective cracking will be retarded, however a certain percentage of reflective cracking will appear. The fabric inner layer will be a barrier between the overlay to retard water percolation into the base materials. Cracking in standard or conventional overlays will allow water to infiltrate or percolate to the base material.	There is no remedy for this process. Should the cracking widen to a width to allow for water penetration, the surface will have to be crack sealed at the time of seal coating.

PAVING (OVERLAYS AND/OR NEW PAVING) (CONT.)

Problem	Cause	Solution
WITH A FABRIC INNERLAYER OVERLAY, THE FABRIC IS SHOWING (STICKING OUT) THROUGH OR OUT OF THE OVERLAY	This occurs when the fabric is not adhered to the original pavement either by not enough tack coat or no tack coat. Also, should the fabric be exposed to traffic in warm to hot temperatures, vehicles will pick up the fabric and overlap it. These overlaps will come through the pavement overlay. In some isolated incidents, the thickness of the overlay is less than the minimum recommended or specified. The minimum thickness of an overlay on fabric is 1-1/2 inch in the majority of specifications.	The areas damaged by slippage will have to be removed and a patch installed. Where the fabric is folded and sticking out of the overlay, the pavement should be heated, the pavement removed, the fabric repaired and a patch installed.
NEW PAVEMENT HAS TURNED GRAY OR LIGHTENED IN 12 MONTHS OR LESS AND SMALL ROCKS OR SAND IS RAVELING OUT OF THE OVERLAY MATERIAL	This occurs when the liquid binder (liquid asphalt) is on the low side of the allowable limits. The pavement will oxidize at a more rapid rate and will lighten up or turn gray. Also, smaller rocks and sand will ravel out of the surface.	The pavement should be scheduled for seal coating. Since the pavement is oxidizing at a rapid rate, it is recommended to apply 2 coats of seal coat.

SQUARE	RECTANGLE
$A = s^2$	$A = h \times b$
PARALLELOGRAM	TRAPEZOID
$A = h \times b$	$A = \dfrac{(b_1 + b_2) \times h}{2}$
TRIANGLE	CIRCLE
$A = 1/2\ h \times b$	$A = \pi \times r^2$ or $A = \dfrac{\pi \times d^2}{4}$
HEXAGON	OCTAGON
$A = p \times s \times 3$	$A = p \times s \times 2$

Formulas for an Ellipse

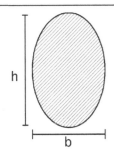

$$A = \frac{\pi \times h \times b}{4}$$

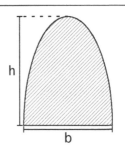

$$A = \frac{\pi \times h \times b}{4}$$

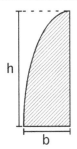

$$A = \frac{\pi \times h \times b}{4}$$

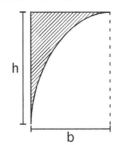

$$A = a \times b \times \left(1 - \frac{\pi}{4}\right)$$

Basic Trigonometry

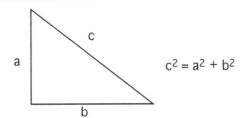

$$c^2 = a^2 + b^2$$

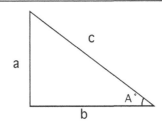

$$A° = \sin \frac{a}{c}$$

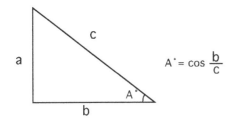

$$A° = \cos \frac{b}{c}$$

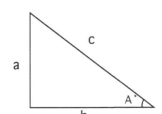

$$A° = \tan \frac{a}{b}$$

Glossary

ADDENDA/ADDENDUM—Written or graphic additions or changes issued prior to the opening of the bids that will clarify, correct, or change the bidding requirements.

ADDITIONAL INSURED—Names of persons, entities, companies, Homeowners Associations (HOA), owners, or HOA board of directors who shall be listed as additional insureds on the contractor's insurance certificates for liability and employees' or workers' compensation insurance.

AGREEMENT—The written contract between the owner, or owner's agent, and the contractor covering the work to be performed. Other contract documents attached to the original agreement shall be considered as part of the original agreement.

AGGREGATE BASE COURSE (ABC)—Typically made of a specific design (recipe) of mixing different sizes of crushed rock together, forming the aggregate, which has certain desirable properties. ¾-*inch Aggregate Base, Class 2*, is used in roadways and parking lots and is an aggregate made of a specific combination or blending of different sizes and quality of rock inclusive of ¾ inch (19.05 mm) to fine dust (-200 or 200 openings per square inch of the sieve). An aggregate is normally made from newly quarried rock, and it is sometimes allowed to be made from recycled asphalt concrete or Portland cement concrete.

AGGREGATE GRADATION—Usually measured by a sieve analysis. In a sieve analysis, a sample of dry aggregate of known weight is separated through a series of sieves with progressively smaller openings. Once separated, the weight of particles retained on each sieve is measured and compared to the total sample weight. Particle size distribution is then expressed as a percentage retained by weight on each sieve size. Results are usually expressed in tabular or graphical format. The typical graph uses the percentage of aggregate by weight passing a certain sieve size.

ALLIGATOR CRACKING—A series of interconnecting cracks caused by fatigue failure of asphalt concrete surface under repeated traffic loading. Cracking begins at the bottom on the asphalt surface (base) where tensile stress and strain are highest under a wheel load. The cracks propagate to the surface initially as a series of parallel longitudinal cracks. After repeated traffic loading, the cracks connect, forming many-sided, sharp-angled pieces that develop a pattern resembling chicken wire or the skin of an alligator. The pieces are less than 2 feet on the longest side. Alligator cracking occurs only in areas subjected to repeated traffic loading, such as wheel paths. Alligator cracking is considered a major structural distress and is often accompanied by rutting and depressions.

APPLICATION RATE—The rate material is applied to a surface in gallons per square yard. Especially critical for seal coat materials to ensure proper application to manufacturer's specifications and job-specific specifications. Wearability and longevity of the material is determined by the application rate on the existing pavement.

ASPHALT/ASPHALT BINDER—Functions as an inexpensive waterproof thermoplastic adhesive. In other words, it acts as the glue that holds the road together. In its most common form, asphalt binder is simply the residue from petroleum refining. To achieve the necessary properties for paving purposes, binder must be produced from a carefully chosen crude oil blend and processed to an appropriate grade. For a few applications, additives (usually polymers) are blended or reacted with the binder to enhance its properties.

ASPHALT CONCRETE FRICTION COURSE (ACFC)—A hot blend of open-graded (or gap-graded) aggregates and liquid binder asphalts (or asphalt emulsions) or special modified bonder to form a friction course to be applied as an overlay on streets, roads, and highways. ACFC is a combination of small aggregate and asphalt binder, batched in an asphalt hot mix plant and applied as a hot paved surface. It has all the same characteristics of a conventional pavement, but is not as dense or thick. An ACFC does require a tack coat application prior to installing, and if too much tack coat is applied, it will have a tendency to bleed later on. ACFC is applied using the same laydown or paving machine, rollers, and dump trucks as overlays and paving.

ASPHALT MIX DESIGN—Two basic ingredients: aggregate and asphalt binder. Hot mix asphalt (HMA) mix design is the process of determining what aggregate to use, what asphalt binder to use, and what the optimum combination of these two ingredients ought to be. There are several different methods used in this process, of which the Hveem, Marshall, and Superpave methods are the most common.

ASPHALT PAVEMENT—A hard wearing surface, often referred to as flexible pavement, used for highways, streets, and parking lots. Asphalt pavement is generally less expensive than Portland cement concrete (PCC) pavement and is less expensive to maintain than concrete in the long-term life cycle.

ASPHALT PAVEMENT WITH FIBER ADDITIVE—A hard wearing surface, often referred to as flexible pavement, used for highways, streets, and parking lots and containing a percentage of fibers in place of or in addition to fine aggregate materials. This blend adds longevity to the asphalt and, through research, reduces the effects of reflective cracking and initial crack expansion. This type of mix design also reduces rutting and shoving caused by heavy loaded vehicles. Asphalt pavement is generally less expensive than Portland cement concrete pavement and is less expensive to maintain than concrete in the long-term life cycle.

ATTERBERG LIMITS—A basic measure of the nature of a fine-grained soil. Depending on the water content of the soil, it may appear in four states: solid, semi-solid, plastic, and liquid. In each state, the consistency and behavior of a soil is different, and thus so are its engineering properties. Thus, the boundary between each state can be defined based on a change in the soil's behavior. The Atterberg limits can be used to distinguish between silt and clay and between different types of silts and clays.

BLEEDING—A film of asphalt binder on the pavement surface. It usually creates a shiny, glass-like reflecting surface that can become quite sticky. Bleeding occurs when asphalt binder fills the aggregate voids during hot weather and then expands onto the pavement surface. Since bleeding is not reversible during cold weather, asphalt binder will accumulate on the pavement surface over time. This can be caused by one or a combination of the following: excessive asphalt binder in the HMA (either due to mix design or manufacturing), excessive application of asphalt binder during bituminous surface treatment application, or low HMA air void content (i.e., not enough room for the asphalt to expand into during hot weather).

BLOCK (THERMAL) CRACKING—The cracking is in the shape of blocks, which are interconnected cracks that divide the pavement into approximately rectangular pieces. The blocks may range in size from approximately one by one feet to ten by ten feet. Block cracking usually indicates that the asphalt has hardened significantly. Block cracking normally occurs over a large proportion of pavement area, but sometimes will occur only in nontraffic areas. Block cracking is also referred to as "thermal" cracking since it is a result of environmental exposures and aging, which makes it hard and oxidized.

BOND, BID—A security instrument issued and delivered with the bid package to guarantee the contractor's bid price.

BOND, PERFORMANCE—A security instrument issued to guarantee the performance of the contractor of the work. Should the Contractor default, then the performance bond will be used to cover any expenses incurred in contracting the second bidder, including the bid price difference.

CAST IN PLACE (VERTICAL) CURBING—This type of structure is designed and placed on the sides of streets and parking lots to confine the pavement, creating a barrier between the street or parking lot and the adjacent landscape. This is a permanent structure and is installed with a major portion buried between the pavement and the adjacent landscape.

CERTIFICATE OF FINAL COMPLETION—Issued upon final completion of all tasks in the contract and scope of work, completion of all cleanup, and completion of any punch list items from the final inspection. Issuing a final certificate of completion indicates final acceptance of the project and all retained funds will paid to the contractor. Once the certificate is issued, the warranty period begins and any defects that develop will be covered under the warranty.

CERTIFICATE OF SUBSTANTIAL COMPLETION—Issued upon completion of all tasks in the contract and scope of work and prior to final inspection and completion of punch list items. Issuing a substantial certificate of completion is not an acceptance of the project, just a document stating the contractor has completed all tasks of the contract. Usually issued for release of payment less any retention for punch list items.

CHANGE ORDER—A document issued by the owner, or the owner's agent with permission and authority of the owner, by or through the consultant, which is signed by the contractor and the owner, or the owner's agent, and authorizes an addition, deletion, or revision in the work, the scope of work, the specifications, or any adjustment in the contract price or the contract times. Change orders are issued on or after the effective date of the agreement.

CHIP SEAL—Constructed by evenly distributing a thin base of hot bitumen or asphalt onto an existing pavement and then embedding finely graded aggregate into it. The aggregate is evenly distributed over the seal spray, then rolled into a smooth pavement surface. A chip seal-surfaced pavement can optionally be sealed with a top layer, which is referred to as a fog seal or cape seal.

CLAYEY MATERIAL—A fine material that has a swelling potential. Clayey materials have a high plasticity index and require more effort to add water for compaction and for compacting to a required density.

COALESCE—In seal coats and slurry seals, the trapped water in the blended material diffuses out, causing the asphalt emulsion to set up (break). At this time, the material will turn from a gray or brown color to a jet black color, and the material will become a solid surface. The droplets then fuse together (coalesce); however, some water will still be trapped in the blended material.

COLD JOINT—A longitudinal joint that occurs in an asphalt pavement when a fresh batch of hot mix asphalt (HMA) is laid adjacent to an existing lane. It is the interface between the two HMA mats. Most often, differences in the temperature and mat plasticity cause an improper bonding of the fresh HMA with the older asphalt lane, and this subsequently causes the longitudinal joint to possess a significantly lower density than the rest of the pavement. Over time, a longitudinal crack usually occurs between the asphalt mats, permitting the intrusion of water, increasing roughness, and potentially limiting the life of the pavement. This type of distress is very common in inverted crowns (drainways in the center of a drive or street and in drain swales of parking lots). Asphalt pavement laid next to a concrete structure (valley gutter or swale, dumpster pad, cast-in-place curb, or combination curb and gutter) will create a cold joint after the asphalt has oxidized and shrunk.

COLD PATCH—Also known as "TEMPORARY PATCH" is a pavement mixture where a cutback (usually diesel fuel, alcohol, paint thinner, to mention a few), is added to prevent setup. Once exposed to air temperature, the cutback vaporizes and releases leaving a solid surface. Mainly used for temporary patching during colder temperatures or until a permanent patch can be installed.

COMBINATION CURB AND GUTTER—Designed to combine a curb and swale or valley gutter. This type of structure is placed on the sides of streets to confine the street, create a barrier between the street and adjacent landscape or sidewalks, and provide a flow or method of movement for water. Combination curb and gutter is a monolithic poured structure.

COMPACTION—Increasing the density of soil, aggregate bases, and hot-mixed asphalt pavement by consolidation through mechanical force, increasing the strength while decreasing permeability, and usually utilized in earthwork, pavement construction, and below-building foundations. Compaction is accomplished by use of rollers (static, vibratory, rubber tired) or heavy equipment. In sands and gravels as well as hot-mixed asphalt pavement, the equipment usually vibrates, to cause reorientation of the particles into a denser configuration. In silts and clays, a sheep's foot roller is frequently used to create small zones of intense shearing, which drives air out of the soil. The result of soil compaction is measured by determining the bulk density of the compacted soil and comparing it to a maximum density (obtained in a laboratory from a Proctor, Marshall, or other compaction test) to determine the relative compaction.

CONCRETE DUMPSTER PAD—A rectangular pad in front of dumpster enclosures constructed of reinforced concrete or concrete with fibers. Used to support the trash truck when picking up dumpsters on a regular pickup schedule. Dumpster pads are usually eight inches thick.

CONCRETE PAD—A flat concrete structure constructed for storage buildings, motorcycle kickstands, outflow of roof drains to truck delivery areas, and bus stops. Usually not reinforced for smaller pads; however, if the structure is large, it will contain fibers, wire reinforcing mesh, or various-sized rebar. Concrete pads can vary in thickness depending on the use and weight it needs to support.

CONCRETE SIDEWALK—A structure formed and poured to provide a solid path or pedestrian walkway.

CONCRETE SWALE—Constructed in the flow line of parking lots and the center of streets to prevent damage to asphalt pavement by water and aid in water flow and direction. A.k.a. concrete valley gutter.

CONCRETE VALLEY GUTTER—Constructed in the flow line of parking lots and the center of streets to prevent damage to asphalt pavement by water and aid in water flow and direction. A.k.a. concrete swale.

CONDITIONAL WAIVER OF LIEN—An instrument issued by the contractor, subcontractor, sub-subcontractor, or materialman that will release the owner or homeowners association of all payment liabilities for the amount paid to the contractor. The term "conditional" means the instrument does not release the owner or homeowners association from the contractors, subcontractor, sub-subcontractor, supplier, or materialman right to lien the property until all funds are transferred. This instrument shall become an unconditional waiver of lien upon transfer of funds to the contractor, subcontractor, sub-subcontractor, supplier, or materialman for the agreed and paid amount.

CONSULTANT—A person, firm, or corporation having a contract with the owner, or the owner's agent on behalf of the owner to provide professional advise and services.

CONTRACT DOCUMENTS—The agreement, addenda (which pertain to the contract documents), and completed and signed original request for bid (RFB) by the contractor (including any documentation accompanying the request for bid (RFB) and any other bid documentation submitted prior to the award of the work) when attached as an exhibit to the agreement. Other documents considered part of the contract documents shall be the notice of award, any bonds if required, all general conditions, all specifications issued with the request for bid (RFB) or addenda or change orders, all drawings, all written amendments, all change orders, all work change directives, field orders, and all written interpretations and clarifications of and by the consultant or the consultant's engineer on or after the effective date of the agreement.

CONTRACTION (TOOLED) JOINT—Joints constructed in concrete to control cracking. These joints are made with grooving tools and saw cutting, and the locations are designed and planned. The typical spacing in feet is two to three times the slab thickness in inches. So, for example, contraction joints in a 6-inch-thick slab would be twelve to eighteen feet apart.

CONTRACTOR—The person, firm, or corporation with whom the owner, or the owner's agent, has entered into agreement to perform and complete the work as stated in the contract documents, the scope of work, and the specifications. Also considered the general contractor. The contractor is responsible for all subcontractors and sub-subcontractors.

CONTRACT SUM—The monies payable by the owner to the contractor for completion of the work in accordance with the contract documents and the request for bid (RFB) and as stated in the agreement.

CORRUGATION AND SHOVING—A form of plastic movement typified by ripples (corrugation) or an abrupt wave (shoving) across the pavement surface. The distortion is perpendicular to the traffic direction. This usually occurs at points where traffic starts and stops (corrugation) or areas where HMA abuts a rigid object (shoving).

CRACK SEALANT (COLD)—Crack seal material that is applied directly from the storage container into a crack with an opening of ¼ inch or more. This is a more temporary application and has less longevity than hot pour crack sealant.

CRACK SEALANT (HOT)—Crack seal material that is applied hot is a more permanent application. The material is heated to 250°F to 400°F and applied by pour pots or a portable hot pot. Hot pour crack sealant is used to seal cracks with an opening of ¼ inch or more to prevent moisture from entering and causing further structural damage

CRACK SEALING—The process of applying hot pour or cold pour crack seal material into cracks with an opening of ¼ inch or more. All cracks must be thoroughly cleaned of dirt, debris, and vegetation and dry of all moisture. The material is applied directly into the crack by pour pots or portable pumping applicators and applied through a hose to the crack. Crack sealing is the first maintenance method to perform on pavement to prevent water infiltration, which can lead to more expensive structural damage.

CURB DRAIN OPENING (WEEP HOLE)—A designed opening in the curb for drainage of water off the street or parking lot. Usually drains into a retention area or wash.

CUT BACK—A chemical (usually diesel fuel, alcohol, paint thinner, naptha, to mention a few) added to an asphalt product (hot-mixed asphalt, crack sealer, and other asphalt products) to prevent setup.

DATE OF COMMENCEMENT OF WORK—The date agreed to by all parties of the agreement as the date the project is to begin.

DATE OF COMPLETION OF WORK—The date agreed to by all parties that the project shall be substantially completed as defined under substantial completion.

DENSITY TEST: CORE/SPECIFIC GRAVITY—The ratio of the density of a given solid or liquid substance to the density of water at a specific temperature and pressure, typically at 4°C (39°F) and 1 atm (760.00 mm Hg), making it a dimensionless quantity. A 6-inch core is extracted from asphalt pavement or concrete, and a specific gravity test is performed on the core. The result of the test is the compacted density of the asphalt or concrete core: ASTM D-2041.

DENSITY TEST: NUCLEAR—The passing of gamma and beta rays through a compacted soil, aggregate base, or hot-mixed asphalt pavement to determine how dense the material is by how fast the particles travel; also, moisture content is determined by particle collisions with hydrogen atoms. Nuclear density testing is a rapid result test; however, the gauge should be calibrated with a sand cone for soils and aggregate bases and a core (specific gravity test) for asphalt pavements: ASTM D-2922.

DENSITY TEST: RUBBER BALLOON METHOD—Determines the in-place soil density of compacted or firmly bonded soil using a rubber-balloon apparatus: ASTM D-2167.

DENSITY TEST: SAND CONE METHOD—A sand replacement method for determining the field unit weight or the in-situ density of natural or compacted soil. The main application of this test is in cases like embankment and pavement construction; this is basically a quality control test where a certain degree of compaction is required: ASTM D-1556.

DEPRESSIONS—Localized pavement surface areas with slightly lower elevations than the surrounding pavement. Depressions are very noticeable after a rain, when they fill with water. Depressions are very common in parking lot construction and overlays.

DRY WELL—A vault built into the pavement or retention area with a slotted opening on the surface. Has the same characteristics as a manhole; however, the bottom is designed to percolate water into the ground and water table. Dry wells can be constructed with a holding vault under the pavement, where water is stored until it can move into the dry well and percolate.

EDGE CRACKING—Generally on the edges of unconfined asphalt pavement. Unconfined (no curbs or other forms of edge barriers) pavement will yield during the compaction process or will develop as the pavement ages, oxidizes, and becomes brittle. Edge cracking is generally in the shape of the letter "C" along the edge of the street, road, or parking lot.

ENGINEER (PE)—An engineer who has been licensed by the state. The license symbolizes that the engineer has completed certain training and has demonstrated a specific degree of competency in a field of engineering.

EXPANSION JOINT—A joint that permits volume change movement of a concrete structure or member. These are usually constructed by installing preformed or premolded elastic/resilient material of approximately ¼-inch to ½-inch thickness as wide as the concrete is thick, before the concrete is placed. Expansion joints should never be less than ¼ inch wide. Premolded expansion joints for installation in residential, commercial, or industrial slabs may be of fiber, sponge rubber, plastic, or cork composition. Such materials must be highly resilient, nonextruding in hot weather, and not brittle in cold weather. An expansion joint should always be utilized where a concrete member will join or abut an existing structure of any type. This would include a junction of sidewalks; a sidewalk joining with a driveway, building, curb, or other similar members; as well as where a floor slab joins a column, staircase, and so on. The square formed by the intersection of two sidewalks should have premolded expansion material enclosing the perimeter. Normally, expansion joints are not provided in sidewalks other than where the walk abuts an existing structure.

EXTRUDED CURBING—A curb that is placed on top of asphalt or concrete and is affixed with an epoxy cement. Extruded curbing is also referred to as a "garden curb." Extruded curbing in parking lots and some low-traffic-density streets is used to border gardens and islands where the area behind the curb is backfilled. This type of curb will not resist the force of vehicles striking it and will break easily.

FINAL PAYMENT—The amount paid to the contractor by the owner, the homeowners association, or their agent on behalf of the owner for any outstanding sums, agreed retention, or additional work which pays the contractor in full for all work performed under the contract and for the exact contract price including any change orders or addenda.

FLOCCULATION—In seal coats and slurry seals, this causes close approach (forms or causes to form into clumps or masses) of droplets, which leads to adhesion between emulsion droplets. Water is then squeezed out.

GENERAL CONTRACTOR—The prime contractor with whom the owner is contracted to complete all tasks of the work listed in the contract documents, the scope of work, and the specifications. Also referred to as the contractor. The general contractor is responsible for all subcontractors and sub-subcontractors.

GEOGRID—manufactured from a punched polypropylene sheet, which is then oriented in three substantially equilateral directions so that the resulting ribs shall have a high degree of molecular orientation, which continues at least in part through the mass of the integral node. Installed between sub-grade material and aggregate base course to stabilize deep saturated or highly unstable soils.

GEOTECHNICAL ENGINEER—The branch of civil engineering concerned with the engineering behavior of earth materials. Geotechnical engineering includes investigating existing subsurface conditions and materials; determining their physical/mechanical and chemical properties that are relevant to the project considered; assessing risks posed by site conditions; designing earthworks and structure foundations; and monitoring site conditions, earthworks, and foundation construction.

GEOTECHNICAL LABORATORY—Contains standard soil mechanics laboratory testing equipment, including triaxial testing devices and equipment for testing soils, aggregate base courses and asphalt pavements, soil classification, permeability, consolidation, and direct shear.

HOT-MIXED ASPHALT—Asphalt concrete pavement, or hot-mixed asphalt pavement, as it is more commonly called, refers to the bound layers of a flexible pavement structure. For most applications, asphalt concrete is placed as HMA, which is a mixture of coarse and fine aggregate and asphalt binder.

IN-PLACE RECYCLED ASPHALT—Hot or cold on-site processing of milled or pulverized asphalt, where asphalt emulsion or liquid asphalt is added to recycle into a new asphalt surface.

INSTRUCTIONS TO BIDDER—Those specific instructions given to the bidder or contractor outlining the requirements to work on the project site, the proper insurance and bond requirements, work hours and days, and all required forms and documents to be provided prior to bidding and starting the project.

JOB MIX FORMULA—A hot-mixed asphalt pavement designed from the existing and available aggregates and liquid binder asphalt. Job mix formulas are unique and specific to the location of the project and the geological deposits. The job mix formula combines course aggregates and fine aggregates in a homogeneous mixture of the two. Liquid asphalt binder is then added to make a solid pavement mass.

JOINT REFLECTION CRACKING—Cracks in a flexible overlay of a rigid pavement. The cracks occur directly over the underlying rigid pavement joints. Joint reflection cracking does not include reflection cracks, which occur away from an underlying joint or from any other type of base (e.g., stabilized cement or lime).

LIABILITY INSURANCE—Insurance required by the owner to cover any losses caused by the contractor, the contractor's workers, and equipment, or from the negligence of the contractor, the contractor's workers, subcontractors, and sub-subcontractors.

LICENSED CONTRACTOR—Contractors who possess the proper category license required to perform the work as stated in the scope of work and the bid by task of the request for bid (RFB). A contractor who is current in fees with the licensing agency and whose license is current and active.

LIEN—Charges, security interests, or encumbrances upon real property or personal property.

LIQUID LIMIT (LL)—The water content where a soil changes from plastic to liquid behavior.

LONGITUDINAL CRACKING—Cracks parallel to the pavement's centerline or laydown direction. Usually a type of fatigue cracking. Cracking is in the direction of flow of traffic, usually at the edge of wheel paths.

MATERIALMAN—Manufacturer or supplier of materials to be used on the project.

MAXIMUM DRY DENSITY—Point where clays, silts, and granular materials will reach a maximum cohesiveness at an optimum moisture content. Maximum dry density is used by geotechnical engineers to certify compaction of clays, silts, and granular materials used to fill low areas.

NOTICE OF RIGHT TO LIEN (20-DAY LETTER)—A document delivered to the owner to disclose the contractor's right to lien the property for nonpayment of the agreed contract amount within ninety (90) days of substantial completion. This letter is required by local government Lien Laws to allow for a Contractor's right to Lien.

NOTICE TO PROCEED—A written notice given by the owner to the contractor fixing the date on which the contract times will commence to run and on which the contractor shall start to perform the contractor's obligations under the contract documents.

Nova Chip®—A paving process that places an ultra-thin, coarse aggregate hot mix over a special asphalt membrane. The NovaChip process is quick, completed in one pass by a specially designed machine, which means your road can be opened to traffic sooner. NovaChip surfacing is applied only by specially trained contractors with a history of quality construction practices. NovaChip surfaces can be used as a preventive maintenance or surface treatment and combines the strength of hot mix with the flexibility of thin maintenance treatments.

OVERLAY—A layer of hot-mixed asphalt placed over an existing pavement to provide a new structural layer and smooth, homogeneous surface. An overlay can range from ½ inch to three inches in thickness.

OWNER—The public body or authority, corporation, association, firm, or person with whom the contractor has entered into the agreement and for whom the work is to be provided.

OWNER'S AGENT OR REPRESENTATIVE—The person, firm, or corporation with whom the owner has contracted to represent them and their best interest.

OWNER'S ASSOCIATION—Whether home, townhome, or condominium, this group shall represent a body of home or property owners who are collectively responsible for their community and hold residency or ownership or both in the same.

PATCH—An area of pavement that has been replaced with new material to repair the existing pavement. There are two types of patches: remove and replace (R&R) and skin (or surface). A patch is considered a defect, no matter how well it performs.

PAVEMENT SECTION DESIGN—The calculated thickness of aggregate bases and hot-mixed asphalt pavement based on the type and classification of subgrade soils, R-value, and traffic count. The pavement section design is developed by a licensed geotechnical engineer and is part of the civil section of the approved project plans and blueprints.

PENALTY CLAUSE—The clause that enables the owner/homeowners association to charge a contractor for failure to complete the work as set out and defined in the agreement, the contract documents, and the request for bid (RFB).

PLANS—The approved drawings, with attached specifications and details, to construct a new street or parking lot. Plans must be approved by the governing agency, and developed and stamped by a licensed architect.

PLASTICITY INDEX (PI)—A measure of the plasticity of a soil. The plasticity index is the size of the range of water content where the soil exhibits plastic properties. The PI is the difference between the liquid limit (LL) and the plastic limit (PI = LL − PL). Soils with a high PI tend to be clay, those with a lower PI tend to be silt, and those with a PI of 0 tend to have little or no silt or clay.

PLASTIC LIMIT (PL)—The water content where soil starts to exhibit plastic behavior. A thread of soil is at its plastic limit when it is rolled to a diameter of 1/8 inch and crumbles.

POLISHED AGGREGATE—Areas of hot-mixed asphalt pavement where either the portion of aggregate extending above the asphalt binder is very small or there are no rough or angular aggregate particles. This occurs after repeated traffic applications. Generally, as a pavement ages, the protruding rough, angular particles become polished and can reduce skid resistance. This can occur quicker if the aggregate is susceptible to abrasion or subject to excessive studded tire wear.

POTHOLE—Sometimes called a kettle, and known in parts of the western United States as a chuckhole, this is a type of disruption in the surface of a roadway where a portion of the road material has broken away, leaving a hole. Most potholes are formed due to fatigue of the pavement surface. As fatigue cracks develop, they typically interlock in a pattern known as "alligator cracking." The chunks of pavement between fatigue cracks are worked loose and may eventually be picked out of the surface by continued wheel loads, thus forming a pothole. The formation of potholes is exacerbated by cold temperatures as water expands when it freezes and puts more stress on cracked pavement. Once a pothole forms, it grows through continued removal of broken chunks of pavement. If a pothole fills with water, the growth may be accelerated as the water washes away loose particles of road surface as vehicles pass. In temperate climates, potholes tend to form most often during spring months when the subgrade is weak due to high moisture content. However, potholes are a frequent occurrence everywhere in the world, including in the tropics. Potholes can grow to several feet in width, though they usually only become a few inches deep, at most. If they become large enough, damage to tires and vehicle suspensions can occur.

PROCTOR—The Proctor compaction test and the related modified Proctor compaction test are tests to determine the maximum practically achievable density of soils and aggregates, frequently used in geotechnical engineering. The test consists of compacting the soil or aggregate to be tested into a standard mold using a standardized compactive energy at several different levels of moisture content. The maximum dry density and optimum moisture content is determined from the results of the test. There are two types of Proctor tests, known as standard Proctor (ASTM D-698/AASTHO T99) and modified Proctor (ASTM D-1557/AASHTO T180). Soil in place is tested for in-place dry bulk density, and the result is divided by the maximum dry density of the designated Proctor to obtain a relative compaction for the soil in place.

PROFESSIONAL LIABILITY (E&O) INSURANCE—Commonly referred to as errors and omissions (E&O) insurance. Professional liability insurance is used to protect the end user from an incorrect analysis or scope of work prescribed for repairs of the streets or parking lots, where financial damage is caused to the property owner or HOA board of directors (e.g., if the parking lot is completely deteriorated and should be replaced, and a contractor or consultant recommends seal coating, this is an error and causes financial damage to the end user).

PROGRESS PAYMENT—A partial payment allowed to the contractor for work completed as agreed between the owner and the contractor.

PROJECT—The actual location and description of the work and referenced by physical name or number.

PROPOSAL—A document submitted by a contractor to perform a requested maintenance application. Most proposals for pavement maintenance are two to three pages, with a condensed scope of work , stated limitations, and stated restrictions; when signed by both parties, it becomes a contract. Most proposals have limited specifications, if any.

PUNCH LIST—The items identified as needing attention, repairs, or corrections upon completion of the work.

RAPID PAY CLAUSE—Most states have a rapid pay clause and a subcontractor pay clause, which states a definite time line for payment to a contractor for services rendered.

RAVELING—The progressive disintegration of a hot-mixed asphalt layer from the surface downward, as a result of the dislodgement of aggregate particles, which is the loss of bond between aggregate particles and the asphalt binder as a result of a dust coating on the aggregate particles that forces the asphalt binder to bond with the dust rather than the aggregate, aggregate segregation, inadequate compaction during construction (poor compaction, cold asphalt pavement, cold weather paving), or exposure to water (from sprinklers or puddles), creating a stripping process.

REGISTRAR OF CONTRACTORS—The government agency responsible for controlling and licensing contractors in their specific trades.

REMOVE AND REPLACE (R&R) PATCHING—Actually removing the badly deteriorated area by saw cutting or jackhammering around the distressed area, removing the existing pavement, compacting or treating the base, and placing a new hot asphalt mix patch back in place. The patch will then be compacted with a static roller or a vibratory plate compactor.

REMOVE AND REPAVE—Removing the existing pavement and repaving to the existing thickness or new design thickness.

REQUEST FOR BID (RFB)—A formal document that contains a scope of work, bidding instructions, insurance requirements, submittal forms, and specifications to properly outline the desired maintenance work desired. Also, all legal document instructions are contained in this document. Also referred to as a request for quotes (RFQ).

REQUEST FOR QUOTE (RFQ)—A formal document that contains a scope of work, bidding instructions, insurance requirements, submittal forms, and specifications to properly outline the desired maintenance work desired. Also, all legal document instructions are contained in this document. Also referred to as a request for bid (RFB).

RESIDUE BY EVAPORATION—A test to determine the amount of solid particle remaining after all water has been evaporated off or out of the material. This is a common test for slurry seals and seal coats. Any additional material (rock, sand, cement) added to the job site mix design will have to be deducted to give a true amount of solids in the mix at the time of manufacturing. The dilution rate can be determined by the results of this test to ensure proper dilution of the seal coat concentrate prior to application.

RETENTION—The amount of monies agreed to by the owner and the contractor to be held from the full contract amount until the punch list items and damaged areas and property are corrected.

RIPRAP—Larger rocks used to reduce erosion on slope drainage. It is normally made from a variety of rock types, commonly granite, limestone, or occasionally concrete rubble from building and paving demolition.

RUBBER ASPHALT CONCRETE (RAC)—A road material made with recycled tires. It has been in use since the late 1970s. RAC is a proven product, one that is a cost-efficient and environmentally friendly alternative to traditional road paving.

RUTTING—Surface depression in the wheel path. Pavement uplift (shearing) may occur along the sides of the rut. Ruts are particularly evident after a rain, when they are filled with water. There are two basic types of rutting: mix rutting and subgrade rutting. Mix rutting occurs when the subgrade does not rut, yet the pavement surface exhibits wheel-path depressions as a result of compaction/mix design problems.

R-VALUE—A measurement of the response of a compacted sample of soil or aggregate to a vertically applied pressure under specific conditions. This test is used by Caltrans for pavement design, replacing the California bearing ratio test. Many other agencies have adopted the California pavement design method and specify R-value testing for subgrade soils and road aggregates. The test method states the R-value of a material is determined when the material is in a state of saturation such that water will be exuded from the compacted test specimen when a 16.8 kN load (2.07 MPa) is applied. Since it is not always possible to prepare a test specimen that will exude water at the specified load, it is necessary to test a series of specimens prepared at different moisture contents. R-value is used in pavement design, with the thickness of each layer dependent on the R-value of the layer below and the expected level of traffic loading, expressed as a Traffic Index.

SANDY MATERIAL—A naturally occurring granular material composed of finely divided rock and mineral particles.

SCOPE OF WORK—The portion of the contract documents and the request for bid (RFB) that explains to the contractor the specific type of work that is expected to occur on the project and the type of materials that are required to accomplish the work.

SCUPPER—A concrete drainway built to allow water to exit a street or parking lot. Scuppers are constructed through curbs, under sidewalks, or on the outfall of a drainway (weep hole) in a curb.

SEAL COAT—A combination of asphalt or coal tar with water, various small mineral particles, polymers, latex, liquid rubber, finely ground tire rubber, carbon black, and other additives for hot or cold weather climates. Used to seal asphalt pavement surfaces to protect them from exposure to and adverse effects of ultraviolet rays, water (from sprinklers or rain), snow, ice, and deicers.

SHOVING—A longitudinal displacement of a localized area of the pavement surface. It is generally caused by braking or accelerating vehicles and is usually located on hills or curves, at intersections, and where the pavement abuts a rigid object (curb, concrete swale or valley gutter, concrete pad, etc.). Shoving is also present with fabric overlays where traffic has moved the pavement on the fabric inner layer. It also may have associated vertical displacement.

SIEVE ANALYSIS—A practice or procedure used (commonly used in civil engineering) to assess the particle-size distribution (also called *gradation*) of a granular material (also known as a gradation test). The size distribution is often of critical importance to the way the material performs in use. A sieve analysis can be performed on any type of nonorganic or organic granular materials including sands, crushed rock, clays, granite, feldspars, coal, soil, a wide range of manufactured powders, grain, and seeds, down to a minimum size depending on the exact method. Being such a simple technique of particle sizing, it is probably the most common test method to determine the particle size composition of particular material.

SILTY MATERIAL—Soil- or rock-derived granular material of a grain size between sand and clay. Silt may occur as a soil or as suspended sediment in a surface water body. It may also exist as soil deposited at the bottom of a water body. Silt is sometimes known as "rock flour" or "stone dust," especially when produced by glacial action. Mineralogically, silt is composed mainly of quartz and feldspar. Sedimentary rock composed mainly of silt is known as siltstone.

SINKHOLE—A hollow area (void) under the pavement or aggregate base. Usually created by water that has washed out material and left a hollowed-out area or void. These areas are exposed during excavation or, in worse cases, by collapse, exposing the sinkhole. Common locations are around or by utilities, manholes, and drywells, where water can accumulate.

SLIPPAGE CRACKING—Crescent- or half-moon-shaped cracks generally having two ends pointed in the direction of traffic. This can also be referred to as edge cracking.

SLURRY SEAL—A wearing surface installed over existing asphalt pavement. It is a specifically designed material consisting of crushed aggregate (type I, ⅛-inch maximum size aggregate; type II ¼-inch maximum size aggregate; and type III, ⅜-inch maximum size aggregate), asphalt emulsion and fillers (usually type II cement), and water, which are mixed together according to the laboratory mix design.

SOIL CLASSIFICATION—Soils native to a particular work site are classified and identified by their characteristics and makeup. The two most commonly used soil classifications are the Unified Soil Classification System (USCS) and the American Association of State Highway and Transportation Officials (AASHTO) Soil Classification System. The Unified Soil Classification System and AASHTO Soil Classification System are used in engineering and geology disciplines to describe the characteristics, texture and grain size of a soil. The classification system can be applied to most unconsolidated materials.

SOLIDS TEST (SEAL COAT)—An evaporation test of a sample of seal coat material to determine how many solids are in the mix. The solids count is determined by a mix design standard. All seal coat materials (asphalt or coal tar) have a solids requirement during manufacturing and at the application site (ASTM D-2939 for asphalts and RP-355e for coal tars).

SPECIAL CONDITIONS—Any special instructions, materials, specif-ications, required documents, or additional work that is not listed or performed in the normal routine of pavement or concrete repair or maintenance. These conditions are unique to the project or work.

SPECIFICATIONS—The portion of the contract documents and the request for bid (RFB) that consists of written technical descriptions of materials, equipment, construction systems, standards, and workmanship as they apply to the project, work, and certain administrative details.

STATE COMPENSATORY INSURANCE—Insurance required by the owner to cover any work-related injuries to employees caused by the contractor, the contractor's workers, and equipment, or from the negligence of the contractor, the contractor's workers, subcontractors, and sub-subcontractors.

STOP WORK—A notice given to the contractor by the owner's agent or the consultant, at the direction of the owner, to stop the work in process due to material, workmanship, or equipment problems, or a violation of specifications.

STRIPPING—The loss of bond between aggregates and asphalt binder that typically begins at the bottom of the hot-mixed asphalt layer and progresses upward. When stripping begins at the surface and progresses downward, it is usually called raveling. This is a very common distress when asphalt pavement is exposed to water (from sprinklers, puddles, ice, snow, etc.).

SUBCONTRACTOR—The person, firm, or corporation with whom the contractor has entered into agreement to perform and complete a portion of the work as stated in the contract documents, the scope of work, and the specifications.

SUBGRADE—The native layer (soil) of an existing project site. This is the foundation that pavements sections are constructed on. Commonly referred to as native soils.

SUBSTANTIAL COMPLETION—The completion of all tasks and the scope of work as defined in the contract document.

SUB-SUBCONTRACTOR—The person, firm, or corporation with whom the subcontractor has entered into agreement to perform and complete a portion of the work which the subcontractor has contracted to perform as stated in the contract documents, the scope of work, and the specifications.

SUPPLIER—A manufacturer, fabricator, supplier, distributor, materialman, or vendor having a direct contract with the contractor, subcontractor, or sub-subcontractor to furnish materials or equipment to be incorporated in the work by the contractor, subcontractor, or sub-subcontractor.

TASK—A single item of the work as listed in the bid by task of the request for bid (RFB) and the contract.

TEMPORARY PATCH—Also known as "COLD PATCH" is a pavement mixture where a cutback (usually diesel fuel, alcohol, paint thinner, to mention a few), is added to prevent setup. Once exposed to air temperature, the cutback vaporizes and releases leaving a solid surface. Mainly used for temporary patching during colder temperatures or until a permanent patch can be installed.

TEMPERATURE—There are two temperatures that affect the application of asphalt maintenance which are ambient (air) and ground. Both have a minimum requirement in order to prevent deterioration of the product by freezing or setting up too rapidly.

TERMINATION OF CONTRACT—The contract shall be cancelled, stopped, discontinued, or ended by decision of the owner or the contractor for reasons of breach of contract or inability to complete the contract.

TERMS—The rules and regulations of performing the work. The agreed items as listed in the contract documents and the request for bid (RFB).

TESTING TECHNICIAN—A trained field technician who conducts the physical testing and reporting.

THERMAL (BLOCK) CRACKING—Interconnected cracks that divide the pavement up into rectangular pieces. Blocks range in size from approximately 1 ft² to 100 ft². Larger blocks are generally classified as longitudinal and transverse cracks. Block cracking normally occurs over a large portion of pavement area but sometimes will occur only in nontraffic areas. Can be from oxidation, aging of the asphalt binder, or poor choice of asphalt binder in the mix design.

TIME SCHEDULE—The time line to complete the project from the start date to the date of substantial completion.

TRANSVERSE CRACKING—Cracks perpendicular to the pavement's centerline, longitudinal joint. Usually a type of thermal cracking.

UNCONDITIONAL WAIVER—An instrument issued by the contractor, subcontractor, sub-subcontractor, supplier, or materialman which releases the owner or homeowners association of all payment liabilities for the amount paid to the contractor, subcontractor, sub-subcontractor, supplier, or materialman.

UNIT PRICE—Price per unit (square foot, square yard, linear foot, etc.) for a listed task.

UNIT WEIGHT (TEST FOR)—A physical test performed to determine the weight of material per cubic yard or cubic foot at the point of application to ensure all the right properties are added or exist.

WAIVER OF LIEN—An official document signed by a contractor, subcontractor, sub-subcontractor, or material supplier to show they have been paid for services and products. A waiver of lien also protects the consumer from recourse for payment. However, a waiver of lien should be provided by all parties involved in the project (contractor, subcontractors, sub-subcontractors, and material suppliers). There are three types of waivers of lien: unconditional upon progress payment, conditional upon progress payment, and conditional upon final payment, as shown below:

Unconditional Waiver and Release upon Progress Payment is when the claimant is required to execute a waiver and release in exchange for or in order to induce payment of a progress payment and the claimant asserts in the waiver that he/she has in fact been paid the progress payment. This release does not cover all items.

Conditional Waiver and Release upon Progress Payment is used when the claimant is required to execute a waiver and release in exchange for or in order to induce the payment of a progress payment and the claimant has not been paid. This form is useful when the claimant has not been paid yet, but will be paid out of a progress payment that is not the final payment. This conditional waiver and release is only effective if the claimant is *actually paid*. This release does not cover all items.

Conditional Waiver and Release upon Final Payment is when the claimant is required to execute a waiver and release in exchange for or in order to induce the payment of a final payment and the claimant has not been paid. This release is only binding if there is evidence of payment to the claimant. Evidence of payment may be demonstrated by the claimant's endorsement on a single check or a joint payee check that has been paid by the bank upon which it was drawn or written acknowledgment of payment given by the claimant.

WARRANTY—A guarantee given by a materials manufacturer or contractor that his/her finished product will perform over a stated period of time. Most states have a minimum warranty period law for construction problems due to product failure or workmanship.

WET TRACK ABRASION TEST (ASTM D-3810)—The wet track abrasion test is a performance-related test conducted in a materials laboratory. The test measures the amount of material that is abraded or lost by a rotating rubber hose on a submerged cured sample of slurry seal or seal coat. The wet track abrasion test has been correlated to the wearing characteristics of field-applied slurry seal or seal coat material. After six minutes of exposure to the rotating rubber hose piece, the wearing characteristics can be determined under wet or saturated conditions. This test provides an expected wearing characteristic or longevity of the material being applied.

WORK—The entire completed construction or the various separately identifiable parts thereof required to be furnished under the contract documents and the scope of work. Work includes and is the result of performing or furnishing labor and furnishing and incorporating materials and equipment into the construction and performing or furnishing services and furnishing documents as required by the contract documents and the scope of work.

Resources

The following is a list of oranizations that provide professional training and certifications and is provided for reference. To obtain information about professional designations and education in an area of professional development, please contact the following organizations and agencies. This is only a partial listing. There are several other associations for restaurants, motels, hotels, mobile homes, medical and schools, to mention a few.

Property or Community Management

Institute of Real Estate Management (IREM)
Buidling Owners and Managers Association (BOMA)
Community Association Institute (CAI) for Homeowners Associations
Interanation Council of Shopping Centers (ICSC)
International Facilites Managers Association (IFMA)
National Apartment Association (NAA)
National Industrial Office Park Association (NIOPA)
Facility Plant Engineering

Associations available for Additional Information and Licensing

American Institute of Architecture (AIA)
American Society of Civil Engineers (ASCE)
American Public Works Association (APWA)
Transportation Research Board (TRB)
Association of Asphalt Paving Technologists (AAPT)
The Asphalt Institute (AI)
Association of General Contractors (AGC)

References

AMERICAN ASSOCIATION OF STATE HIGHWAY AND TRANSPORTATION OFFICIALS (AASHTO)
AASHTO standards and procedures are very similar to the ASTM procedures. AASHTO is established for use in highway and public works construction.
- Standard Specifications for Transportation and Methods of Sampling and Testing, 29th Eddition

AMERICAN SOCIETY FOR TESTING AND MATERIALS (ASTM)
There are several volumes of ASTM standards and procedures to test construction materials. When referring to the test methods in specifications, the test number designation determines where you can locate the test procedures. When referring to soils, concrete and asphalt pavement, you will need all three of these volumes. The following volumes pertain to materials used for soils, concrete and asphalt pavement.

- Volume 04.02 Concrete and Aggregates: test methods, specifications and practices
- Volume 04.03 Road and Paving Materials: test methods, specifications and practices
- Volume 04.08 Soil and Rock: geotechnical and geoenvironmental standards that cover soil testing

CALIFORNIA DEPARTMENT OF TRANSPORTATION
Tack Coat Guidelines April 2009 Division of Construction

FEDERAL HIGHWAY ADMINISTRATION (FHWA)

- Distresses For Pavements with Asphalt Concrete Surfaces
  Publication No. FHWA-RD-03-031, June 2003
- Slurry Seal Application Checklist
- Pavement Preservation Series #13, Publication No. FHWA-IF-06-014, November 2005

INDIANA DEPARTMENT OF TRANSPORTATION
Nuclear Guage Testing
All agencies and private geotechnical testing laboratories use nuclear density gauges to test moisture content and in-place compaction of soils, aggregate base courses, and hot-mixed asphalt pavement. This publication demonstrates proper procedures for using nuclear desnsity gauges. Manufacturers of gauges (Troxler, CPN, and Humbolt) have training courses available for certification and usage.

MINNESOTA DEPARTMENT OF TRANSPORATION
Asphalt Pavement Maintenance: Field Book, January 2002

NEBRASKA DEPARTMENT OF ROADS
NDOR Pavement Maintenance Manual, 2002
This publication was printed with assistance from Dean Testa, Kansas DOT; Larry Galehouse, Michigan DOT; Dick Clark, Montana DOT; and Tom Lorfeld, Wisconsin DOT

The Civil Engineering Handbook, Second Edition by W. F. Chen and J. Y. Richard Liew
The Civil Engineering Handbook has several sections pertaining to various phases of civil engineering; pavement design and pavement construction are a portion.

TRANSPORTATION RESEARCH BOARD (TRB)
- Transportation Research Circular E-C102, August 2006, Asphalt Emulsion Technologies
- Transportation Research Circular E-C105, September 2006, Factors Affecting Compaction of Asphalt Pavements

VIRGINA DEPARTMENT OF TRANSPORTATION
Unified Soil Classification System
The Federal Highway Administration, all state DOTs, municipalities, and counties use the Unified Soil Classification System as well as the AASHTO classification. This is the easiest and more visual setup of this classification system.

Index